Handbook on Soft Computing for Video Surveillance

CHAPMAN & HALL/CRC CRYPTOGRAPHY AND NETWORK SECURITY

Series Editor
Douglas R. Stinson

Published Titles

Jonathan Katz and Yehuda Lindell, Introduction to Modern Cryptography

Antoine Joux, Algorithmic Cryptanalysis

M. Jason Hinek, Cryptanalysis of RSA and Its Variants

Burton Rosenberg, Handbook of Financial Cryptography and Security

Shiu-Kai Chin and Susan Older, Access Control, Security, and Trust: A Logical Approach

Sankar K. Pal, Alfredo Petrosino, and Lucia Maddalena, Handbook on Soft Computing for Video Surveillance

Forthcoming Titles

Maria Isabel Vasco, Spyros Magliveras, and Rainer Steinwandt, Group Theoretic Cryptography

CHAPMAN & HALL/CRC
CRYPTOGRAPHY AND NETWORK SECURITY

Handbook on Soft Computing for Video Surveillance

Edited by

Sankar K. Pal
Indian Statistical Institute
Kolkata, India

Alfredo Petrosino
University of Naples Parthenope
Naples, Italy

Lucia Maddalena
National Research Council
Naples, Italy

CRC Press
Taylor & Francis Group
Boca Raton London New York

CRC Press is an imprint of the
Taylor & Francis Group an **informa** business
A CHAPMAN & HALL BOOK

CRC Press
Taylor & Francis Group
6000 Broken Sound Parkway NW, Suite 300
Boca Raton, FL 33487-2742

Printed in the United States of America on acid-free paper
Version Date: 20111117

International Standard Book Number: 978-1-4398-5684-0 (Hardback)

Visit the Taylor & Francis Web site at
http://www.taylorandfrancis.com

and the CRC Press Web site at
http://www.crcpress.com

Contents

Preface

Video surveillance is the area of computer science devoted to real-time acquisition, processing, and management of videos coming from cameras installed in public and private areas, in order to automatically understand events happening at the monitored sites, eventually sending up an alarm. Because of the rapidly increasing number of surveillance cameras, it has become a key technology for security and safety, with applications ranging from the fight against terrorism and crime, to private and public safety (e.g., in private buildings, transport networks, town centers, schools, and hospitals), and to the efficient management of transport networks and public facilities (e.g., traffic lights and railroad crossings). Video surveillance is an extremely interdisciplinary area, embracing the study of methods and algorithms for computer vision and pattern recognition, but also hardware for sensors and acquisition tools, computer architectures, wired and wireless communication infrastructures, and middleware. From an algorithmic standpoint, the general problem can be broken down into several steps, including motion detection, object classification, tracking, activity understanding, and semantic description, each of which poses its own challenges and hurdles for system designers. Moreover, the scope of video surveillance is being extended to offline multimedia analysis systems related to security and safety, thus entailing disciplines such as content-based video retrieval for visual data similarity retrieval and video mining for knowledge extraction; typical applications are in forensic video analysis and human behavior analysis.

Soft computing is a consortium of methodologies (working synergistically, not competitively) that, in one form or another, reflects its guiding principle: exploit the tolerance for imprecision, uncertainty, approximate reasoning, and partial truth to achieve tractability, robustness, low-cost solution, and close resemblance to human-like decision making. This provides flexible information processing capability for representation and evaluation of various real-life ambiguous and uncertain situations, and therefore results in the foundation for the conception and design of high MIQ (Machine IQ) systems. At this juncture, the principal constituents of soft computing are fuzzy sets, neurocomputing, genetic algorithms, probabilistic reasoning, and rough sets.

Recently, soft computing tools have shown enormous promise in many different tasks of video surveillance. For example, fuzzy logic has proved to be a powerful tool that allows one to handle the imprecision and uncertainty inherent in the background subtraction approach to moving object detection, as well as in the tracking process, and in visual sensor networks; approximate reasoning has proved beneficial for action recognition; fuzzy-rough interpretation of video data can help in dealing with the approximate, incomplete, and vague characteristics of surveillance videos for the analysis of usual and unusual events; neural networks allow self-organizing learning of a scene background model for moving object detection, tracking of image features in video image sequences, and learning of usual and unusual human behaviors. Several conference papers and journal articles on integrating soft computing techniques into video surveillance have been published in the past decade. Articles describing the challenging issues of soft computing, namely synergistically integrating the constituting components to achieve application specific merits, have also been reported. However, this scattering of information causes inconvenience for researchers, applied scientists, and practitioners.

With this volume we aim to bring together research work concerning the application of soft computing techniques to different tasks of video surveillance, investigating novel solutions, and discussing future trends of existing literature in this field. It includes both review and new material written by worldwide experts describing, in a unified way, the basic concepts, theories, algorithms, and applications that demonstrate why and how soft computing methodologies can be used in different video surveillance problems.

The book consists of eleven chapters. Chapter 1 presents an introduction to video surveillance systems, providing a nice reference for beginners in this area. It includes a fairly extensive and updated survey on such systems, tracing a brief history of their evolution and the actual state of the art. It also highlights the challenges that modern-day systems still face despite achieving significant advancements, and covers several sub-topics concerning video sensors, data fusion, and artificial intelligence techniques.

For the convenience of readers, a brief description of different soft computing tools is provided in Chapter 2, covering the basics of fuzzy sets, rough sets, neural networks, genetic algorithms, and probabilistic reasoning. Special focus is provided on the combination and the hybridization of rough and fuzzy sets, and their role for several image processing tasks, which represent the starting point of many algorithms employed in video surveillance.

Chapter 3 presents some examples of neural network–based approaches to the solution of video surveillance tasks provided in the literature, including moving object detection and tracking, crowd and traffic density estimation, anomaly detection, and behavior understanding. A specific neural-based approach is further described in order to give evidence of the advantages of its adoption for moving object detection, also showing possible uses in the context of other video surveillance tasks, such as stopped object detection, activity recognition, and anomaly detection.

Chapter 4 focuses on summarization, which helps generate movie trailers, sports and news video highlights, and keeps the records of interesting events for future inspection, playing an important role in the context of video surveillance, where processing huge chunks of video data for potential risk demands a huge amount of resources. It reviews existing summarization techniques in the context of their relation to significant content identification, and addresses the suitability of these techniques for surveillance video summarization along with the issue of personalization. Relevance of soft computing is also mentioned.

Chapters 5 through 11 cover a broad spectrum of video surveillance topics that adopt soft computing techniques. They are organized following the usual articulation of a video surveillance system, starting with the task of moving object detection, to tracking, to classification and the recognition of target objects.

Chapter 5 concerns the detection of moving objects in video streams, which is the first relevant step in information extraction in many computer vision applications, and specifically in video surveillance applications. It presents a fairly extensive and updated survey of research on background subtraction that exploits fuzzy techniques in order to handle imprecision and uncertainty inherent in all the steps for problem solution, ranging from background modeling, to foreground detection, to background maintenance. Some of the existing methods are thoroughly compared and future research directions are envisaged.

The problem of fusion of data acquired from multiple sensors, together with related methods and challenges, is considered in Chapter 6, with some applications to tracking objects in video sequences. Dempster–Shafer and Bayesian inference–based data fusion algorithms are considered, and the impact of the fusion of video sequences from different sensors for enhancing the tracking process in the presence of changeable illumination, shadows, and other ambiguous conditions, is specifically addressed. Open issues and other possible applications are also briefly mentioned.

Chapter 7 tackles the problem of human action modeling and recognition from video sequences of human silhouettes independent of the viewpoint. The modeling step relies on Kohonen self-organizing maps, trained from 2D motion templates recorded in different viewpoints and velocities, thus integrating spatial and temporal templates into a common framework and reducing their high dimensionality. Efficacy of the corresponding recognition system, adopting resampling to increase the number of training sequences, is evaluated both on virtual and real datasets.

The task of multi-perspective automated analysis of video data for learning usual event patterns, as well as detecting unusual events, is addressed in Chapter 8, based on clustering using both individual attributes and combination of several attributes, such as time, size, shape, and position of objects. A fuzzy–rough interpretation of the video data for automated video analysis is provided for the analysis of events from surveillance videos, which is helpful in dealing with the approximate, incomplete, and vague characteristics of the video data.

Chapter 9 concerns the task of detecting video shot boundaries to identify subsequences consisting of different video contexts, which is a prerequisite in several applications, including video surveillance, target tracking, and robotic maneuvering. The significance of a fuzzy entropy measure in determining the change in video context is highlighted, leading to an automatic unsupervised detection method.

Although many of the aforesaid chapters present specific applications of the reported methods, the last two chapters are mainly devoted to present comprehensive application settings, providing descriptions of advanced surveillance systems under development.

Chapter 10 presents a multi-camera platform to monitor the environment, integrated to a multi-robot platform, to enhance situation awareness. Two different vision methods are applied to monitor the environment in real-time, one based on a maximally stable detection and tracking algorithm, and the other using stereo depth. These lead to two different systems comprising the multi-camera platform, whose results are sent to the multi-robot system, which, on alarm, dispatches a robot to monitor the region of interest. The system has been tested in a number of real-world activity recognition tasks.

A surveillance system using multiple audio and video sensors is presented in Chapter 11, where the audio subsystem is used to cover large portions of the monitored area and exploited to guide the video subsystem toward the area of interest for further analysis. An approach for optimal coverage reconfiguration is also considered to reconfigure the sensors to maximize the coverage of the environment. Experimental results in real scenarios show the efficiency of the proposed framework.

A color insert gathers sixteen figures from different chapters, for the convenience of better understanding their color-related content.

The book is intended to be a reference for researchers and professionals, as well as graduate students, interested in the application of soft computing techniques to video surveillance and other related areas, including ambient intelligence, security and safety, civilian and military remote sensing, management of transport networks, and biometrics. It can also be considered as a suggested reading text to be adopted for teaching graduate courses in subjects like computer vision and pattern recognition, image processing, soft computing, and computational intelligence.

The editors of this volume extend their profound gratitude to the reviewers for their generosity and helpful comments concerning the chapters in this volume, which were extensively reviewed and revised before final acceptance. We also received useful suggestions from the reviewers of the original book proposal. In addition, we are very grateful for the help that we have received from Bob Stern and others at CRC Press during the preparation of this volume.

The editors have been supported by the FIRB Project *IntelliLogic* No. RBIP06MMBW of the Italian Research and Education Ministry (MIUR) and by the *AMIAV* Project of the Campania Regional Board. Sankar K. Pal acknowledges a MIUR-FIRB fellowship provided by the University of Naples Parthenope that allowed us to conceive the present book, in a wider collaborating agreement with the Indian Statistical Institute, and also acknowledges his J.C. Bose fellowship awarded by the Government of India.

Sankar K. Pal, Indian Statistical Institute
Alfredo Petrosino, University of Naples Parthenope
Lucia Maddalena, National Research Council of Italy

About the Editors

Sankar K. Pal (www.isical.ac.in/~sankar) is a distinguished scientist of the Indian Statistical Institute and a former director. He is also a J.C. Bose Fellow of the Government of India. He founded the Machine Intelligence Unit and the Center for Soft Computing Research: A National Facility in the Institute in Calcutta. He received a Ph.D. in radio physics and electronics from the University of Calcutta in 1979, and another Ph.D. in electrical engineering along with DIC from Imperial College, University of London in 1982. He joined his institute in 1975 as a CSIR senior research fellow where he later became a full professor in 1987, distinguished scientist in 1998, and the director for the term 2005–2010.

He worked at the University of California, Berkeley and the University of Maryland, College Park in 1986–1987; the NASA Johnson Space Center, Houston, Texas in 1990–1992 & 1994; and in the US Naval Research Laboratory, Washington DC in 2004. Since 1997 he has served as a distinguished visitor of the IEEE Computer Society (USA) for the Asia-Pacific Region, and held several visiting positions in Italy, Poland, Hong Kong, and Australian universities.

Prof. Pal is a fellow of the IEEE, USA, the Academy of Sciences for the Developing World (TWAS), Italy, the International Association for Pattern Recognition, USA, the International Association of Fuzzy Systems, USA, and all the four national academies for science/engineering in India. He is a co-author of 17 books and more than 400 research publications in the areas of pattern recognition and machine learning, image processing, data mining and web intelligence, soft computing, neural nets, genetic algorithms, fuzzy sets, rough sets and bioinformatics.

He received the 1990 S.S. Bhatnagar Prize (which is the most coveted award for a scientist in India), and many prestigious awards in India and abroad, including the 1999 G.D. Birla Award, 1998 Om Bhasin Award, 1993 Jawaharlal Nehru Fellowship, 2000 Khwarizmi International Award from the Islamic Republic of Iran, 2000–2001 FICCI Award, 1993 Vikram Sarabhai Research Award, 1993 NASA Tech Brief Award (USA), 1994 IEEE Trans. Neural Networks Outstanding Paper Award (USA), 1995 NASA Patent Application Award (USA), 1997 IETE-R.L. Wadhwa Gold Medal, the 2001 INSA-S.H. Zaheer Medal, 2005–2006 Indian Science Congress-P.C. Mahalanobis Birth Centenary Award (Gold Medal) for Lifetime Achievement, 2007 J.C. Bose Fellowship of the Government of India and the 2008 Vigyan Ratna Award from Science & Culture Organization, West Bengal.

Alfredo Petrosino (`cvprlab.uniparthenope.it/staff/alfpet`) has been associate professor of computer science at the University of Naples Parthenope since 2005. He received the Laurea degree cum laude in computer science from the University of Salerno, in 1989, supervisor E. R. Caianiello. During 1989–1994 he was a fellow researcher of the Italian National Research Council (CNR). In 1995 he was a contract researcher at International Institute of Advanced Scientific Studies (IIASS). He held positions as researcher of the National Institute for the Physics of Matter (INFM) (1996–2000), as researcher at the National Research Council (CNR) (2000–2002), and as senior researcher at CNR (2002–2004). He taught at the universities of Salerno (1991–2006), Siena (1997–1998), Naples Federico II (1999–2006), and Naples Parthenope (2001-today).

In 1994 he received the Academic Prize for Cybernetics from the Italian Academy of Science, Arts, and Literature. He is a senior member of the IEEE, USA, a member of the International Association for Pattern Recognition, USA, the International Neural Networks Society, USA. He co-edited 6 books and more than 100 research publications in the areas of computer vision, image and video analysis, pattern recognition, neural networks, fuzzy and rough sets, and data mining. He heads the research laboratory CVPRLab at University of Naples Parthenope. He is a permanent program committee member of the Workshop on Neural Networks (WIRN) and organized the biennal Workshop on Fuzzy Logic and Applications (WILF).

He is an associate editor of *Pattern Recognition* and *Pattern Recognition Letters* journals; member of the editorial board of *Internat. Journal of Knowledge Engineering and Soft Data Paradigms*; book editor of *WILF-Fuzzy Logic and Applications*, LNCS, Springer Verlag; guest editor of *Fuzzy Sets and Systems, Image and Vision Computing*, and *Parallel Computing* journals.

Lucia Maddalena (`www.na.icar.cnr.it/~maddalena.l`) has been a researcher at the Institute for High-Performance Computing and Networking of the Italian National Research Council since 1994. She received the Laurea degree cum laude in mathematics in 1990 and the Ph.D in applied mathematics and computer science in 1995, both from the University of Naples Federico II.

Initial research dealt with parallel computing algorithms, methodologies and techniques, and their application to computer graphics. Subsequent research is devoted to methods, algorithms and software for image processing and multimedia systems in high-performance computational environments, with application to real-world problems, mainly digital film restoration and video surveillance.

She taught at the University of Naples Federico II (1999–2006) and at the University of Naples Parthenope (2006–today). She is an associate editor of the *International Journal of Biomedical Data Mining*, serves as a reviewer for several international journals, and is a member of the IEEE and of the International Association for Pattern Recognition.

List of Contributors

Subhashis Banerjee, Indian Institute of Technology, Department of Computer Science & Engineering, Delhi, India

Siddhartha Bhattacharyya, RCC Institute of Information Technology, Burdwan, India

Thierry Bouwmans, University of La Rochelle, Laboratoire MIA, La Rochelle, France

Sergio Canazza, University of Padova, Department of Information Engineering, Padova, Italy

Biswanath Chakraborty, RCC Institute of Information Technology, Kolkata, India

Santanu Chaudhury, Indian Institute of Technology, Department of Computer Science & Engineering, Delhi, India

Ayesha Choudhary, Indian Institute of Technology, Department of Computer Science & Engineering, Delhi, India

Paramartha Dutta, Visva-Bharati University, Faculty of Computer & System Sciences, Santiniketan, India

Nour-Eddin El Faouzi, IFSTTAR – ENTPE, Traffic Engineering Laboratory - LICIT, Cedex, France

Alessio Ferone, University of Naples Parthenope, Department of Applied Science, Naples, Italy

Gian Luca Foresti, University of Udine, Department of Mathematics and Computer Science, Udine, Italy

Ashish Ghosh, Indian Statistical Institute, Center for Soft Computing Research, Kolkata, India

Raphael Grech, Kingston University, Digital Imaging Research Centre, London, United Kingdom

Elías Herrero, University of Zaragoza, Department of Electrical Engineering and Communications, Zaragoza, Spain

Luca Iocchi, University of Rome "La Sapienza," Department of Computer and System Sciences, Rome, Italy

Christopher King, University of Nevada, Department of Computer Science and Engineering, Reno, Nevada, USA

Lawrence A. Klein, Consultant, Santa Ana, California, USA

Lucia Maddalena, National Research Council, Institute for High-Performance Computing and Networking, Naples, Italy

Luca Marchetti, University of Rome "La Sapienza," Department of Computer Science and Systems, Rome, Italy

Francisco Martínez-Contreras, University of Zaragoza, Department of Electrical Engineering and Communications, Zaragoza, Spain

Christian Micheloni, University of Udine, Department of Mathematics and Computer Science, Udine, Italy

Lyudmila Mihaylova, Lancaster University, Department of Communication Systems, Lancaster, United Kingdom

Dorothy Monekosso, University of Ulster, Computer Science Research Institute, Newtownabbey, United Kingdom

Robert Mullen, Kingston University, Digital Imaging Research Centre, London, United Kingdom

Daniele Nardi, University of Rome "La Sapienza," Department of Computer and System Sciences, Rome, Italy

Mircea Nicolescu, University of Nevada, Department of Computer Science and Engineering, Reno, Nevada, USA

Carlos Orrite, University of Zaragoza, Department of Electrical Engineering and Communications, Zaragoza, Spain

Rajarshi Pal, Indian Statistical Institute, Center for Soft Computing Research, Kolkata, India

Sankar K. Pal, Indian Statistical Institute, Center for Soft Computing Research, Kolkata, India

Alfredo Petrosino, University of Naples Parthenope, Department of Applied Science, Naples, Italy

Claudio Piciarelli, University of Udine, Department of Mathematics and Computer Science, Udine, Italy

Hossein Ragheb, Kingston University, Digital Imaging Research Centre, London, United Kingdom

Tomi D. Räty, VTT Technical Research Centre of Finland, Software Architectures and Platforms, Oulu, Finland

Paolo Remagnino, Kingston University, Digital Imaging Research Centre, London, United Kingdom

Maria Valera, Kingston University, Digital Imaging Research Centre, London, United Kingdom

Sergio A. Velastin Kingston University, Digital Imaging Research Centre, London, United Kingdom

1

Introduction to Video Surveillance Systems

Tomi D. Raty
VTT Technical Research Centre of Finland, Oulu, Finland

A surveillance system can be defined as an implementation that serves as an extension or expansion to awareness. This service is provided to human users to facilitate and assist in the perception and detection of events in the purvey of interest. The capacity of human senses and minds has its limits in areas of cognizance and perspicacity. A wide range of diverse data will succumb rapidly to the most apt individuals. The rapidity and complexity of the received information will only further aggravate the predicament [28].

Currently well-known and modern surveillance systems include the elements of data collection and information analysis. Surveillance systems obtain data, such as video and audio, from their environment. Then the collected data is either processed or transmitted directly to a human or computer at a command center for decision execution. The instance performing the decisions can execute the correct consequential procedures, for example, raising alarms, initiating rescue operations, or informing other instances [27].

These functionalities demand high levels of scalability and usability to successfully and efficiently perform the required duties and responsibilities of the surveillance administrator. The correct information must be received by the correct people at the right time. There are multiple scientific fields and commercial enterprises that have incorporated these requirements to ascertain novel and ameliorate extant solutions. These fields include notable examples, such as signal processing, system engineering, and computer vision [36].

The established and recent solutions of surveillance systems have been applied to different scenarios and environments. These include endeavors to increase safety in [28]

- Transportation (railway stations, seaports, airports, motorways, and other forms of public commuting),
- Industrial applications (power plants and factories),
- Indoor and outdoor environments (banks, supermarkets, car parks, houses, and buildings), and
- Military applications (strategic infrastructures, and maritime and aerospace surveillance).

1.1 Generations of Surveillance Systems

Historically, in the dawn of early humans, soldiers and sentries would rely on the keenness of their five senses (sight, hearing, taste, smell, and touch) and their derivations to accumulate data from their environment, and they would utilize their minds to process the information and form outcomes of perceived events. Different areas progressed at different rates. For instance, weaponry and defense development preceded the development of surveillance. Catapults and shields were created earlier than observation balloons and telegraphs. Both of these latter contraptions preeminently heightened the capability to extend and expand the range of which was limited to an individual human being. The commencement of the twentieth century proved to provide considerable advancements to surveillance [30].

Military operations have exacerbated the development of surveillance to address difficulties and challenges in combat situations. Strategic movements of troops in relation to targets require both rapid and dynamic data obtainment and deductions to formulate correct decisions. Individual data elements must be collected from the field and delivered to the correct commanding officers to perform decisions. All the levels of information must be compact, precise, and comprehensive, and the information transmission must be executed with haste. The basic concept behind surveillance is the acquisition of information of an area of interest in the real world. Even as the technology has advanced with which surveillance is conducted, the fundamentals are still the same. Data is consolidated and transmitted to the appropriate individuals [25, 39, 40].

In general, the progression of surveillance systems can be decomposed into three distinct generations. There have been technological developments that have made each progressive step possible throughout history. Data storage, sensor processing, and communications are eminent enablers that have progressed in parallel with the development of surveillance systems. These include camera technologies that have progressed into eminent tools for situation comprehension and forensics. Increased capabilities in surveillance technologies have posed important ethical dilemmas pertaining to privacy rights. Another significant technological advancement has been the distribution of data and information with remote and widespread transmissions. A well-known application of both of these antecedent technologies is close circuit television systems (CCTV) [28].

1.1.1 First-Generation Video Surveillance Systems (1GSS)

In the 1960s, there was significant growth in the utilization of CCTV systems. This is considered the beginning of modern surveillance systems as remote surveillance became a tangible communal reality. The era of first-generation video surveillance systems (1GSS) prevailed from the 1960s to the 1980s. Video cameras were exploited and installed to procure visual signals from remote locations. These images were subsequently viewed from the control room. The human administration in the control rooms would view a large quantity of monitors and attempt to espy events of interest. 1GSS concentrated heavily on analog video signals and their transmission to a remote site for surveying. The main difficulties in 1GSS revolved around the natural human attention span, which resulted in a prominent amount of missed events. Technical difficulties consisted of bandwidth requirements of the sensors, the susceptibility to noise and degradation of contemporary playback mechanisms, and the storage of video surveillance tapes. In the 1970s, the growth in popularity of VCR (video cassette recording) alleviated the problems apposite to media storage. The amount of video cameras and bandwidth required were proportional to the vastness or complexity of the surveyed area. This typically has the consequence of having additional human administrators observing monitors in the control room. The interrelationships of 1GSS are presented in Figure 1.1 [28].

1.1.2 Second-Generation Video Surveillance Systems (2GSS)

The debut of second-generation surveillance systems (2GSS) occurred in the 1980s. This coincided with the development of enhanced resolution of video cameras and the emergence of inexpensive computers. These two improvements were important factors in video processing and event detection. Com-

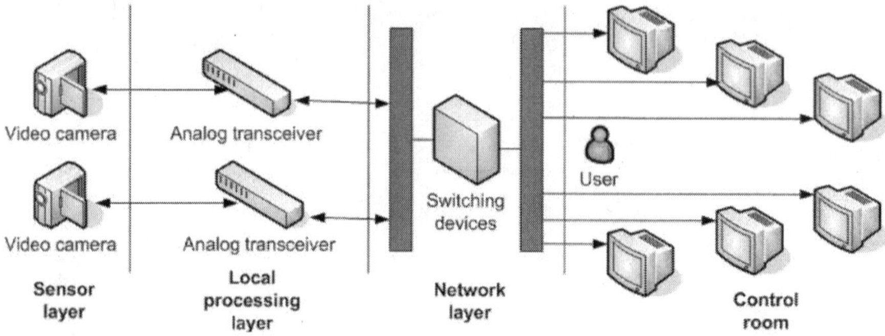

FIGURE 1.1 Structural example of 1GSS [28]. (From C.S. Regazzoni, V. Ramesh, and G.L. Foresti, Special Issue on Video Communications, Processing, and Understanding for Third Generation Surveillance Systems, *Proceedings of the IEEE*, Vol. 89, No. 10, October 2001 ©2001 IEEE. With permission.)

munication systems were able to provide higher quality at lower expense. The period of 2GSS belonged to the duration between 1980 and 2000. 2GSS can be seen as the maturing of technologies that were established in 1GSS. The utilization of digital technologies offered digital compression, robust transmission, reduced bandwidth, and processing methods. Notable results included intelligent video surveillance systems. In detail, these consisted of real-time analysis and tracking capabilities in two-dimensional (2D) images, human identification and behavioral comprehension, intelligent multi-sensor data fusion, wireless and wired networks, and high-performance algorithms. The majority of research focused on the development of automated real-time event detection. This would facilitate the duties and responsibilities of the human administrators of the control rooms through the refinement and presentation of essential information. Additionally, this would enable the amount of concurrent surveillance alarms automatically raised without necessarily any human intervention [28].

1.1.3 Third-Generation Video Surveillance Systems (3GSS)

Beginning in about 2000, third-generation video surveillance systems (3GSS) complete the digital conversion. There are still prevalent analog surveillance systems, but the majority of new surveillance systems are fundamentally digital solutions. 3GSS employs digital information through its entire range. From low-level sensors to high-level analysis and results projection, the amount of digital data is significant. The processing speeds of computer networks have continued to increase, and the expenses in the utilization of communications have decreased. The heterogeneity and mobility of exploited sensors has reached a new level, yet again. Supported communication media include local area networks (LANs), General Packet Radio System (GPRS) and 3rd generation (3G) mobile phones, and broadband media including optical fibers and coax cables. Figure 1.2 represents the connections of 2GSS [28].

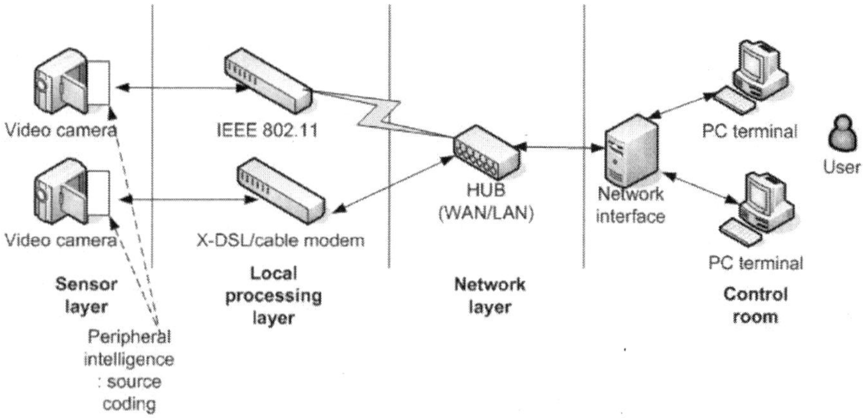

FIGURE 1.2 Structural example of 3GSS [28]. (From C.S. Regazzoni, V. Ramesh, and G.L. Foresti, Special Issue on Video Communications, Processing, and Understanding for Third Generation Surveillance Systems, *Proceedings of the IEEE*, Vol. 89, No. 10, October 2001 ©2001 IEEE. With permission.)

Another significant advancement is the dispersion of intelligence, viz. the transition from centralized intelligence to distributed intelligence. Numerous sensors and sensor types can perform intelligent and automatic data processing and refinement. They are able to handle and process data immediately upon obtainment from their environment. This decreases the amount of transmittable information over networks. Automated intelligent processing can be executed at control centers. The control center(s) or intermediate information processing station(s) that execute the fusion of multiple data inputs are relieved from addressing all the lower level data fusion and detection that the sensors can perform.

Research has advanced techniques on intelligent, open, and dedicated networks. Scientists have been able to exploit the augmentations of relatively inexpensive computational power, enhanced video processing and comprehension methods, and multi-sensor data fusion. This has produced constant and progressive developments in the form of novel tools, ideologies, and methods.

3GSS has facilitated the capability to handle profuse amounts of data communication and rendered the identification of interesting events more efficient. These amenities have benefited from automatic recognition functionalities and considerable quantities of digital communication information. The infrastructural architectures of 3GSS have provided robust, rapid digital information transmission across a wide array of different communications media.

These developments have considerably assisted human administrators in removing or curtailing their redundant and monotonous tasks and responsibilities.

The sensors have reached a degree of temporal and spatial scalability that addresses the deficit which was prevalent in 2GSS. Sensors, including video cameras, can alter their behavior according to the circumstances that are occurring. This promotes both a reduction in required devices and vicariously a decrease in the quantity of information that must be transmitted across a communicational medium. Despite reductions in transmittable information, the amount of transmitting sensors has grown and the demand for efficient network management has become paramount. Different modules and components that execute intelligence and information processing place high requirements on network performance and robustness.

As the types of sensors and their data have become extremely diverse, it enables meticulous automatic deductions on the sensor data. A combination of different sensors can fastidiously evaluate subtle alterations in the environment in comparison to a single sensor type, which typically results in the detection of similar events. Contemporary sensors consist of a vast range, including audio, visual, biometric, and radar units, which are utilized for disparate purposes, including fingerprint readings and aircraft trajectories [28].

3GSS presents complicated and advanced forms of interaction. Obtained sensor data can be utilized in different types of services and by miscellaneous users. Sensor data can be exploited individually or in various different collections. These requirements place heavy demands on the adaptability, flexibility, and modularity of developed sensors and systems. Novel functionality is progressively adapted to existing 3GSS systems in iterations to uphold management and maintainability. The primary functionalities that 3GSS presented were intelligent detection and tracking capabilities. Prototypical research solutions have evinced these functionalities. The complexity of the distinguished events is quite low. They usually are involved in the evaluation of trajectories of people or vehicles to identify potential anomalies. These are applied in light pedestrian traffic. Complicated event analysis requires the ability to address at least moderate flow conditions. This includes multiple object tracking, reasoning, and event interpretation [28].

Tseng et al. [32] present an illustration of their visual surveillance system in Figure 1.3. Miscellaneous end devices, such as handsets, PDAs (personal digital assistant), and Notebooks, are connected to servers and sensors through wireless connections, such as global system for mobile communication (GSM) and its GSM Short Message Service (SMS) server and Bluetooth. The GSM and Ethernet networks are connected to the Internet Protocol (IP) network through the Wireless Access Protocol (WAP) and short message peer-to-peer (SMPP) gateways [15].

A prominent predicament in 3GSS is the performance needed. 3GSS present multiple state-of-the-art solutions that have extremely high expectations from the end users and stakeholders. The false alarm rate must be kept low while attaining high probabilities of authentic alarms. A psychological requirement exists that human operators must be able to trust the surveillance

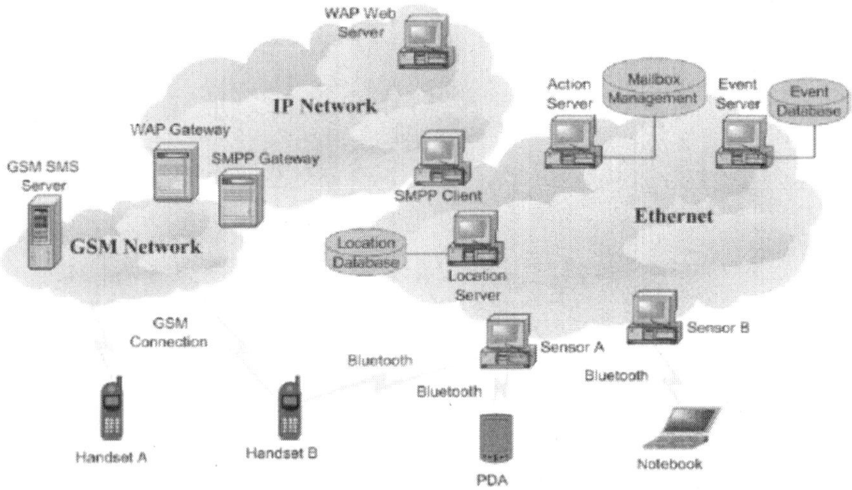

FIGURE 1.3 A contemporary example of the architecture of a realized 3GSS system [32]. (From Y.-C. Tseng, T.-Y. Lin, Y.-K. Liu, and B.-R. Lin, Event-driven messaging services over integrated cellular and wireless sensor networks: Prototyping experiences of a visitor system, *IEEE Journal on Selected Areas in Communications*, Vol. 23, No. 6, June 2005 ©2005 IEEE. With permission.)

system. If the surveillance system produces numerous false alarms, the tendency is for the system to be ignored and possibly disconnected. This problem is compounded when multiple event types occur. The false alarms may cause a cascade effect in which one false alarm triggers one or multiple other false alarms, etc. One must bear in mind that the ideal false alarm rate is next to zero, and the auspicious authentic alarm rate is near 100% in all weather conditions. The required detection time is instantaneous. There should not be virtually any reaction time between the occurrence and detection of an event [28]. Figure 1.4 presents an illustration of missed detections and false alarms. An incoming signal is obtained by an event detection system, which can categorize the signal as a false detection (FD), authentic detection (AD), indifferent missed detection (MD I), or a missed detection (MD). The MD adds to the global missed alarm rate. The event classification system can define the alarms to be an authentic alarm (AA), indifferent missed alarm (MA I), a missed alarm (MA), or a false alarm (FA). The missed alarm adds to the global missed alarm rate and the false alarm increases the global missed alarm rate [15].

Another preeminent dilemma related to surveillance systems is the inability to validate and verify the comprehensive surveillance systems according to all their details prior to deployment. The real world contains a plethora of cases, events, and circumstances that cannot be executed in advance in testing conditions. This naturally has the consequence of verified performance in

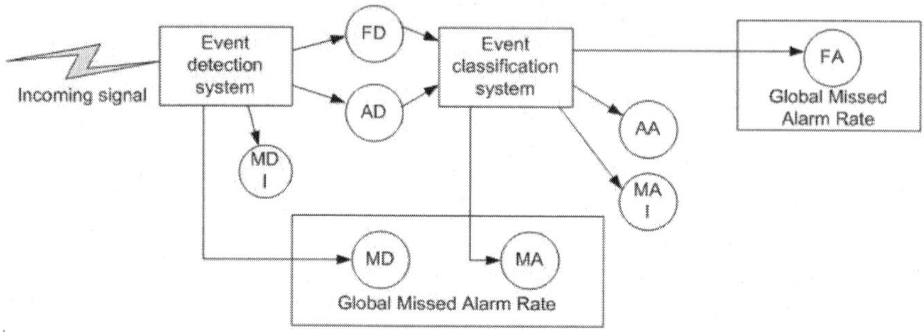

FIGURE 1.4 A depiction of global missed detection rate and global false alarm rate [15]. (From D. Istrate, E. Castelli, M. Vacher, L. Besacier, and J.F. Serignat, Information extraction from sound and medical telemonitoring, *IEEE Transactions on Information Technology in Biomedicine*, Vol. 10, No. 2, April 2006 ©2006 IEEE. With permission.)

untested conditions.

There are two types of self-awareness required from 3GSSs: 1) self-diagnosis and indication, and 2) self-degradation and indication. Sensors may become damaged, which results in faulty signals or a complete lack of signals. The surveillance system should have the capability of comprehending the functionality and operability of its sensors. A sensor that is transmitting aberrant signals or no signals at all should be excluded from a surveillance system as it could result in the production of false alarms or missed alarms. By raising an indication of an improperly functioning or completely dysfunctional sensor, human personnel can manually correct or replace the sensor. Graceful degradation is desirable when the performance of the surveillance system cannot be upheld. This can result from unmanageable complexity as the amount of data increases disproportionately for the surveillance system to handle. These types of behaviors have not attained the ideal level and require further research and examination [28].

The performance of surveillance systems is an open issue. There is a need to evaluate the performance of intelligent processing functions. These evaluations should be applicable to a wide range of different conditions to distinguish their performance in various scenarios, such as in the rain, fog, snow, and dry at nighttime, daytime, and in between their transitions. This will assist in the definition of the efficacy of the surveillance system in the satisfaction of its requirements. Testing and validation are arduous tasks that are normally executed manually. Through intelligent functionality, log information can be collected and analyzed in an automatic manner [28].

1.1.4 The Next Generation of Video Surveillance Systems

Technological progress will continue to have a significant impact on future generations of video surveillance systems. These are compounded with the transition of requirements into coveted systems, the efficiency of developed algorithms, the integration aspects of diverse systems, and fundamental and comprehensive validation and verification of target systems. This applies to all the different fields, ranging from hardware to software and the form of communication. The main area of 3GSS is public monitoring. These demands have resounded from public, legal, enforcement and political interest. It revolves around the delivery of correct personnel to an event in progress or recently incurred. Interoperability will be a fundamental requirement for future generations of surveillance systems. CCTV systems have been adopted and will still remain in use. Examples of their locations are subway systems, large public areas, and shopping centers. Their utilization has also permeated to small- and medium-scale establishments, including enterprises and households. These systems are still in worldwide use, even though the future will contain completely digital solutions. A challenge will reside in the incorporation of analog systems with digital ones. There is a clear discrepancy between the creation of novel surveillance systems and their adoption. There are strict demands and requirements for these systems. Their acceptance in the real world is relatively tardy. This has been mainly due to the prohibitive cost of these systems and to the end-user acceptance of these products. Regazzoni et al. present a collection of real-world application in Table 1.1. The table portrays the functional and cost/performance requirements [28].

To address such a widespread and multifaceted market, research innovations are demanded that enable end users and safety officials to exploit communications solutions, processing, and comprehension solutions. Different sensors include video and audio sensors. Sensor communication, processing, and comprehension can be perceived as fundamental requirements of surveillance applications [28].

Generically, sensor information can be collected, processed, and transmitted with different media. The plethora of information is crucial in surveillance systems. Data gathered from disparate sources is conveyed to a control center. When communication channels are utilized, a general axiom is that the bandwidth for the downlink (from the control center to the sensors) should be lower than the uplink (from the sensors to the control center.) This imposes stringent requirements on the media of transmission. In numerous cases, the information of surveillance systems must be transmitted across open networks for multiuser needs. This renders information protection as an important characteristic. As surveillance data can be used or heavily relied on by law enforcement, there are legal issues that unconditionally dictate requirements that must be followed. Exploited techniques and technologies include watermarking and data hiding to assure the identity, authenticity, and nonrepudiation of information. Traditional computer vision systems have grown in complicated video processing and comprehension. The amounts of irregular and variable monitored data have exploded in magnitude. Through the increase of information, the repercussions on processing capabilities have been considerable. Image processing algorithms have had to become ever more so-

TABLE 1.1 The real-world applications according to Regazzoni et al. [28].

Application domain	Fundamental benefits	Intelligent Functionality	Cost and performance requirements
Public area monitoring, Large facility monitoring	Safety, Security	Person/vehicle detection, tracking and event analysis	Low system cost, false alarm/ detection, rigid requirements
Construction of exterior and interior monitoring	Security, safety, access control, and building automation	Person/vehicle detection, car park monitoring, license plate recognition, face recognition	High-end market, high reliability of access control
Subway, highway, tunnel monitoring, transportation	Safety, security, resource management and ameliorated Quality of Service	People detection and tracking, vehicle detection and tracking, object classification, event recognition	Few extant high-end systems, very low false alarm rates, variations dependent on weather and illumination conditions
Indoor monitoring	Security and safety	Person detection, tracking, event analysis	Low-cost systems, minimal false alarms

Source: C.S. Regazzoni, V. Ramesh, and G.L. Foresti, Special Issue on Video Communications, Processing, and Understanding for Third Generation Surveillance Systems, *Proceedings of the IEEE*, Vol. 89, No. 10, October 2001 ©2001 IEEE. With permission.

phisticated by performing preprocessing and filtering. A common consequence of significantly variable scene conditions is the necessity to select between robust scene depiction and pattern recognition methods. A recurring desire from the end users is the automatic capability for surveillance systems to accommodate themselves to altering scene conditions and learn statistical models from normal event patterns. Typically, the learning capability uses a mechanism to indicate the ascertainment of potentially anomalous events through the evaluation of normal activity patterns. The most significant constraints that inhibit the adoption of the antecedent technologies into reality are real-time performance and low cost [28].

As 3GSS concentrated on intelligent alarm generation, future surveillance systems must address interoperability with legacy systems and proactive deductions. As contemporary systems are used as a forensic device, there is a desire to exceed the barrier of distinguishing the past, however recent, or present into detailed depictions of future events. This includes scrupulous behavioral activity analysis and ambient intelligence. There will be a stronger adoption of artificial intelligence, such as neural networks, applications of fuzzy logic, genetic algorithms, and gossip-type protocols. The future systems will be ameliorated to reach high rates of successful event detection and nonexistent false alarm rates.

As the systems grow in size and diversity, the complexity and inconsistency will increase, and the unreliability and nonresponsiveness will grow. The design and implementation of distributed real-time surveillance systems place essential challenges to their requirements to be fulfilled. In order to comprehend or create any complicated system, it must be divided into components and functions. Distributed systems can be perceived as independent concurrent activities that exchange data and interact with each other without undermining the predictability or performance of the system [35].

1.2 Video Sensors

The utilization of video sensors is popular and efficient in surveillance systems. Regazzoni et al. state that the expense of video sensors is low in comparison to other sensors that are used to address the coverage of a vast area and the event analysis functionality [28]. In authentic surveillance scenarios, a single video sensor has limited capability to cover a large area or track a moving object for a long period. Objects will be occluded by obstacles, such as trees and buildings, and the range of perception of video sensors has its own dimensional limits. The difficulty can be resolved through the utilization of a cooperative network of video sensors within an area and track individual objects seamlessly without a single line of view from an individual video sensor. This approach imposes specific challenges, including the ability to 1) actively control video sensors to function in collaboration; 2) fuse information coherently, consistently, and correctly; 3) observe the ongoing scene and instigate further processing or raise an alarm; and 4) offer the human administrative users a high-level interface for scene visualization [9].

Figure 1.5 illustrates the fundamentals of tracking. The system first captures an image and initializes the background estimates. Then the system begins to function in a loop in which tracks are predicted for every new image on capture. The pixels that contrast against the background are detected. The blobs that are related to actual targets are extracted with detection filters. The detection filters incorporate track predictions, edge detectors, and movement detectors. The background statistics for undetected pixels must be retained as they might initiate a detection event in the subsequent frame. The association process is used to consolidate one or multiple blobs with each target. Blobs that are unassociated initiate tracks if they coincide with requirements related to tracks. Each track is updated with its blobs while another function removes the tracks which have not been updated for a few previous captures [5].

Besada et al. [5] present a palpable realization of visual surveillance in

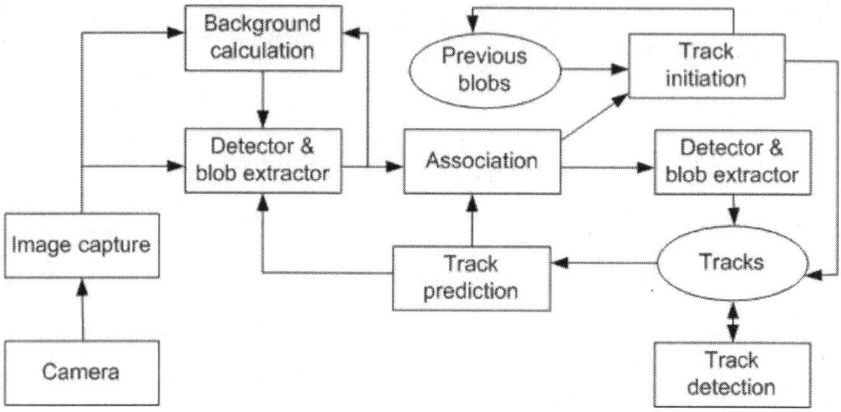

FIGURE 1.5 Depiction of a local tracker [5]. (From J.A. Besada, J. Garcia, J. Portillo, J.M. Molina, A. Varona, and G. Gonzalez, Airport surface surveillance based on video images, *IEEE Transactions on Aerospace and Electronic Systems*, Vol. 41, No. 3, July 2005 ©2005 IEEE. With permission.)

an airport traffic system. Airport surface movement must be surveyed to retain safety and throughput. Advanced surface movement and guidance systems must be capable of the identification of all aircraft and vehicles in the movement region of an airport. The moving targets typically consist of aircraft, fuel trucks, luggage convoys, buses, and cars. Their tracking requires precise and unequivocal information on their movement, including position, trajectory, and velocity, to avoid hazardous situations and to handle surface movement management. Examples of different sensors that can be utilized for these purposes are surface movement radar (SMR), multilateration systems (MS), differential GPS (DGPS), and television.

TABLE 1.2 Exemplary characteristics of different sensors [5].

Sensor	Cooperative	Identification	Mobile	Meteorological
SMR	No	No	All	All
MS	Yes	Yes	Equipped	All
DGPS	Yes	Yes	Equipped	All
TV	No	No	All	Clear

Source: J.A. Besada, J. Garcia, J. Portillo, J.M. Molina, A. Varona, and G. Gonzalez, Airport surface surveillance based on video images, *IEEE Transactions on Aerospace and Electronic Systems*, Vol. 41, No. 3, July 2005 ©2005 IEEE. With permission.

Table 1.2 illustrates the different capabilities of each sensor to provide cooperative behavior, identification, mobiles that can be tracked with the appropriate sensor, and meteorological conditions required for its usage. Clear meteorological conditions indicate that fog, rain, or snow cannot be dense. One must naturally contemplate costs against benefits. Normally, non-cooperative sensors are cheaper and easier to uphold [5].

Figure 1.6 illustrates the system of Besada et al. "It functions as a non-cooperative sensor with an integrated tracker suitable for tracking in dense regions, such as inner taxiways and apron areas. Occlusions and poor weather conditions are its greatest detriments. Other sensors would be required to accomplish all-weather requirements. In multi-target tracking systems, the local tracker must be able to perform detection and associations based on previous measurements." The intent of Besada et al. was "to conceive a tightly coupled detector/multi-target tracker which could address potential detection problems." Another important focus area of Besada et al. was "to devise a functional association process which could address previous measurements. This is appropriate for sensor behavior, particularly in extended detection splitting, overlapping and occlusion" [5].

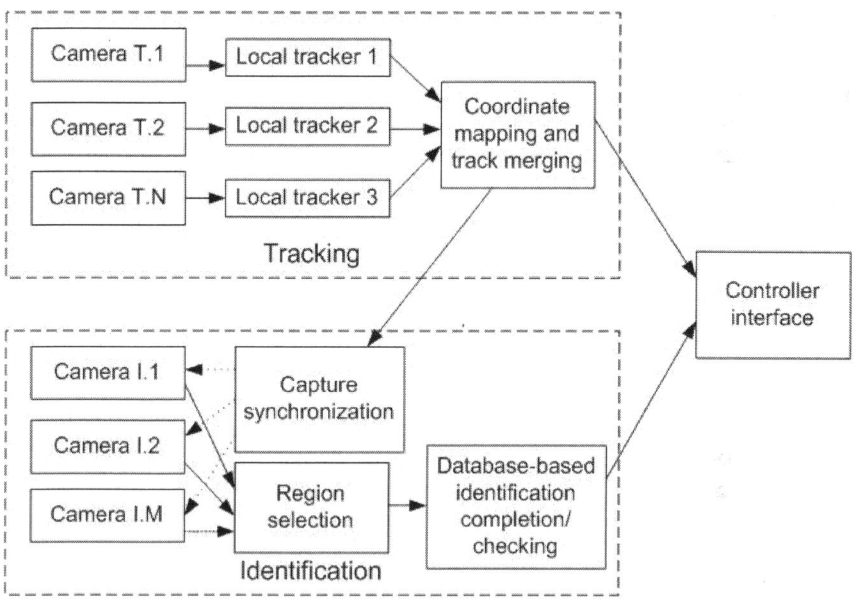

FIGURE 1.6 The structure of an image-based airport surveillance system [5]. (From J.A. Besada, J. Garcia, J. Portillo, J.M. Molina, A. Varona, and G. Gonzalez, Airport surface surveillance based on video images, *IEEE Transactions on Aerospace and Electronic Systems*, Vol. 41, No. 3, July 2005 ©2005 IEEE. With permission.)

The human administrative user cannot perceive all the events of a large area by viewing multiple screens of information as he will be completely overwhelmed with information that additionally requires often inadmissible amounts of network performance. Collins et al. [9] approached this dilemma with the development of graphical user interfaces (GUI). Figure 1.7 presents images that can be transmitted to GUIs for surveillance personnel [31].

GUIs can be used to provide a synthetic image of the area under surveillance. The system presents dynamic agents that represent people and vehicles on the display. This avoids the utilization of original resolution and viewpoints when perceiving events. The human administrative user has the liberty to select spatial relationships among multiple objects and scene characteristics. This approach presents an interesting benefit in the transmission rates through data compression. Only the symbolic information of the events are transmitted, not the raw video data. The symbolic information entails the object type, location, velocity, and measurements statistics, for example, timestamps and zoom, tilt, and pan information [9].

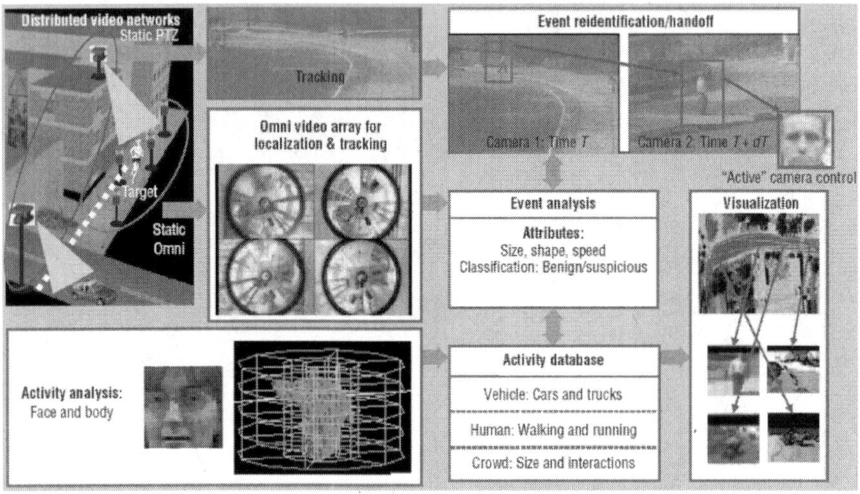

FIGURE 1.7 Active video capture and analysis for multilevel situation awareness [31]. (From M.M. Trivedi, T.L. Gandhi, and K.S. Huang, Distributed interactive video arrays for event capture and enhanced situational awareness, *IEEE Intelligent Systems*, Vol. 20, No. 5, September/October 2005 ©2005 IEEE. With permission.)

Normally, a large-scale distributed video surveillance system utilizes multiple video sources that are spread across a large area. These video sources transmit live video information to a control center for viewing and processing. Video sensors are usually networked digital video cameras that enable the deployment of large and complicated networks of surveillance cameras on the existent Internet Protocol (IP) network infrastructure. Multiple enterprises and facilities are capable of exploiting IP-based surveillance solutions. The implementation of fully functional intelligent, scalable, and widespread surveillance systems still is a considerable research problem [17].

The focus of research on surveillance systems is on autonomy and intelligence. Challenges include content comprehension, containing detecting, tracking, object classification, and the detection of unusual events. Scalability has not been a significant area of research. Systems typically consist of a centralized architecture and need huge amounts of computational power and network performance [17].

1.2.1 Situation Awareness in Video Surveillance

Situation awareness requires information that spreads across time and space. According to Hampapur et al. [13], security personnel must be aware of "who are the people and vehicles in a space" (identity tracking), "where are the people in a space" (location tracking), and "what are the people/vehicles/objects in a space doing" (activity tracking) [13]. Historical information may be applied to acquire the needed data. Intelligent video surveillance systems can exploit multiple scales of time and space in situation awareness. Unfortunately, these different technologies have been developed in isolation. For example, facial recognition is concentrated on having the subject facing the camera, and activity detection is being developed separately from tracking. Comprehensive, functional, and nonintrusive situation awareness requires the addressing of multiple sources across different scales of time and space.

Figure 1.8 presents the structure in which static cameras can be exploited to address an area of interest to depict a universal view. The PTZ (pan-tilt-zoom) cameras address detailed information of objects that are of interest in a scene. Video is used to detect and track multiple objects in either two or three dimensions. Fixed cameras can provide additional information on objects, for example, object classification or object attributes. Multiscale tracking systems offer information of objects in a surveyed region. The information is collected across multiple scales within a single framework. This is represented through the interactions and components of such a system [13].

Hampapur et al. itemize the difficulties of widespread intelligent surveillance as "1) the multiscale challenge, 2) the contextual event detection challenge, and 3) the large system deployment challenge" [13]. Once there are the technical capabilities to procure information from multiple scales, the challenges reside in amassing the information across multiple scales, and the interpretation of this information is crucial. Multiscale technologies include "camera control, processing information of mobile cameras, and task-based camera management" [13]. Contemporary systems are still in their infancy when considering their capabilities to perform automatic data analysis to distinguish events of interest and trends. Comprehensive, functional, and reliable automatic event detection will prove itself of significant assistance to human surveillance personnel. Large-system deployment challenges include "physical connections' costs, low-power hardware for battery-operated cameras, automatic camera calibration, automatic fault detection and sophisticated management tools" [13]. The challenges of context-based interpretation include video analysis, event discrimination through geometric paradigms and activity models, and learning techniques and technologies to ameliorate system performance and the detection of erratic events.

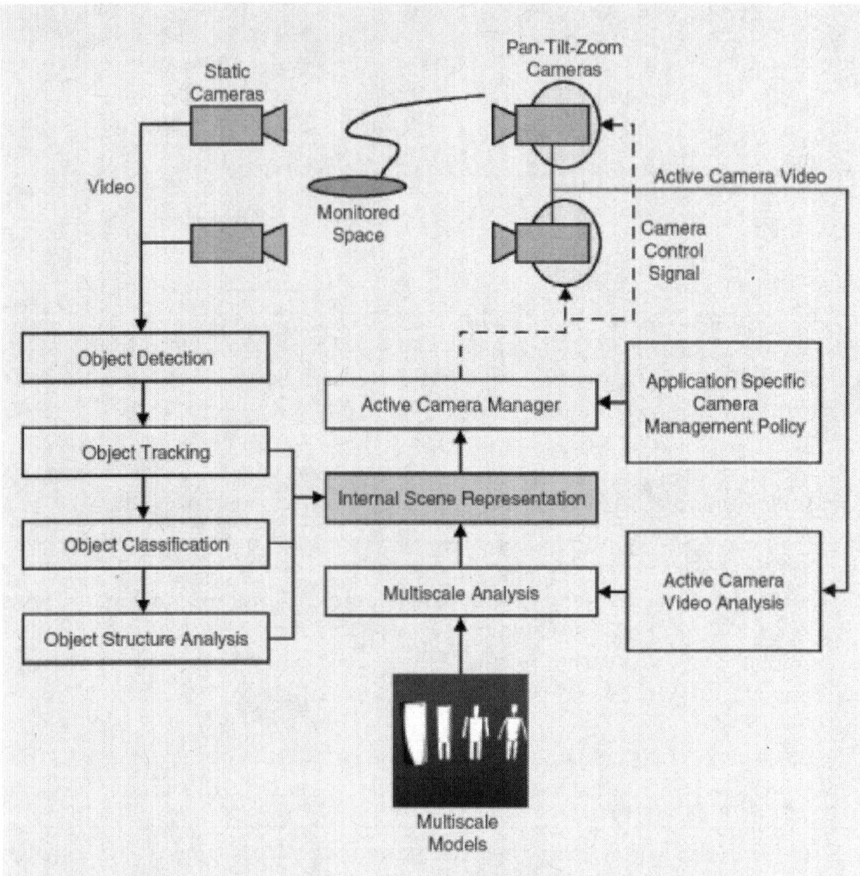

FIGURE 1.8 A general architecture of a multiscale tracking system [13]. (From A. Hampapur, L. Brown, J. Connell, A. Ekin, N. Haas, M. Lu, H. Merkl, S. Pankanti, A. Senior, C.-F. Shu, and Y.L. Tian, Smart video surveillance: Exploring the concept of multiscale spatiotemporal tracking, *IEEE Signal Processing Magazine*, Vol. 22, No. 2, March 2005 ©2005 IEEE. With permission.)

1.2.2 Real-Time Traffic in Video Surveillance

The observation of real-time traffic in video surveillance consists of the following different variables: flows, average speeds, and densities, in a designated timeframe or area. This provides real-time traffic measurement. Traffic can be perceived as any type of moving object, such as vehicles or people. Existing research has conceived traffic estimation algorithms that are based on traffic flow modeling and Kalman filtering. Early studies concentrated on short distance intervals and the models were relatively simple. Then, through the development of newer approaches, the models became more exhaustive, which enabled the utilization of larger distance intervals [38].

Surveillance tools endeavor to distribute comprehensive, real-time images of current and future network conditions. These tools can be applied to detect events in real-time freeway networks or applied to radars, video sensors, etc. The main intention of such tools is to provide surveillance tasks in a systematic and unified manner. A typical physical location of such a tool is the intermediate layer between the monitoring elements and the control system [38].

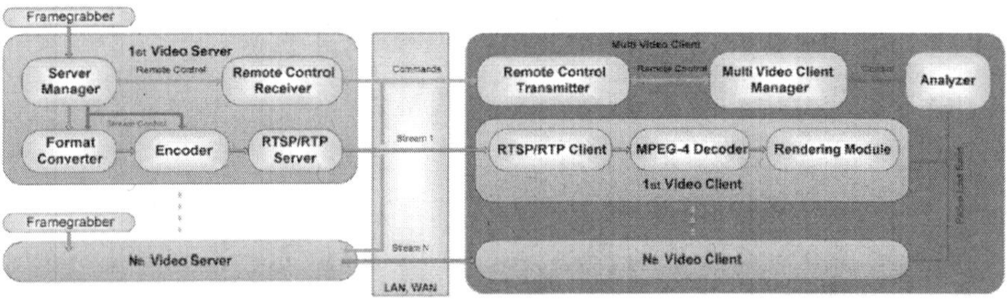

FIGURE 1.9 Video streaming transmission architecture [23]. (From K. Muller, A. Smolic, M. Drose, P. Voigt, and T. Wiegand, 3-D reconstruction of a dynamic environment with a fully calibrated background for traffic scenes, *IEEE Transactions on Circuits and Systems for Video Technology*, Vol. 15, No. 4, April 2005 ©2005 IEEE. With permission.)

Figure 1.9 illustrates a multiview video transmission system. It consists of two parts: 1) the autonomous server application for each surveillance camera and 2) a client application to consolidate multiple video streams. In this model, an exchangeable grabber module was implemented. Rapid motion estimation is performed within the video encoder. The bit streams are conveyed with RTP (Real-Time Transport Protocol). A remote control mechanism is used to configure the server and execute maintenance operations without resorting to expensive manual reparations. Control commands are conveyed over a designated Transmission Control Protocol (TCP) channel. The client application can receive an arbitrary amount of video streams. The limit is specified by the system performance, especially the processor performance and network bandwidth. The client contains one processing module for each stream. The processing module consists of a Real-Time Streaming Protocol (RTSP)/RTP client, a Moving Pictures Expert Group-4 (MPEG-4) decoder, and a rendering module. Packet loss rates are evaluated and new server configurations are issued to accommodate video streams to altering transmission conditions. The system can be applied to observe, for example, traffic, premises and parking lots. The transmitted information can be stored as MPEG-4 streams to record inherent and notable incidents [23].

1.3 Data Fusion

The utilization of data fusion techniques can enhance the estimation of performance and system robustness by exploiting the redundancy that is offered by multiple sensors surveying the same scene. In accordance with recent advancements in sensor and processing technologies, data fusion is of predominant interest. The most significant challenge of the required additional computational performance power has been removed through contemporary central processing units (CPUs). Sensors have become intelligent and are equipped with their own microprocessors and are able to perform distributed data processing and computation. This reduces the amount of computational activities that the central processing node must perform [29].

In a surveillance system that utilizes multiple sensors, there is the difficulty of choosing the most apposite sensor or collection of sensors to execute a particular task. The task could be target tracking, audio recording of a suspicious event, or triggering an alarm. It is desirable for the system to automatically select the correct sensor or collection of sensors. If required data from multiple sources is available and thus data fusion is possible, results could be influenced through malfunctioning sensors. There must be a method to evaluate the performance of existent sensors and to weigh their contribution in the fusion itself [29].

Modern surveillance systems utilize multiple asynchronous and miscellaneous sensors. The consolidation of information acquired from them to determine the events of the environment is a considerable research challenge. Information consolidation, or assimilation, is the process of assembling sensor and non-sensor information according to the context and history knowledge. Information is collected from multiple different sources and the assembly of information adds a new level of inferences in comparison with individual sources. There are multiple nested challenges that affect the successful creation and establishment of such systems. These include 1) the diversity and asynchrony of sensors, 2) the agreement or disagreement of media streams, and 3) the confidence of the media streams [2].

The diversity and asynchrony of sensors are caused by the different formats that are used, for instance, at different transmission rates. As an example, the video sensor could provide frames at different rates than at which audio is sampled. In addition, even two video sensors could offer frames at different rates. This is compounded by non-sensor information, for example, a record of past events, which is in a completely different format. These factors contribute to making the assimilation of information arduous along a timeline. The timeline is a measurable interval of time to which information can be assigned [2].

The agreement and disagreement of media streams are related to capturing the same environmental data that is in unison or discord with each other. The agreement and disagreement can both be exploited to corroborate the overall decision of the ongoing events in the environment. For instance, if two sensors have historical evidence of concurring events, greater emphasis can be placed on them in comparison with if they had contradictory historical evidence. This still places a considerable challenge on the utilization agreement and disagreement of historical information [2].

The confidence of media streams indicates the different levels of confidence that can be posed on the surveillance system according to disparate and distinct media streams in the detection of discrepant and discrete events. Confidence levels can influence the level of precision directly related to the media stream. For instance, if an event is detected correctly by a media stream 70% of the time, the particular media stream can be assigned a 70% level of confidence. The precision of a media stream includes both the precision of the sensor itself and the algorithms that are employed in the media streams. Levels of precision can be acquired through experimentation. Levels of confidence can also be fused. For instance, there are two separate media streams that, respectively, contain 70% and 80% levels of confidence. If the two media streams agree on an event, the consolidated outcome is more considerable than the two ones individually. This same approach is applicable in mutual disagreement. Contradictory disagreement or agreement would result in lower levels of confidence than what each media stream individually provides. These pose the difficulties of the essential challenges [2].

Fusion of data collected from different sources can be weighed. This is a similar approach to the determination of precision in confidence levels to fuse measurements in weighed manner. If there is no distinction between old and new information measurements, this could lead to precarious and erroneous estimates. This type of behavior is emphasized with malfunctioning sensors [22].

In data fusion, a surveillance task can be decomposed into four phases: 1) the detection of an event, 2) the representation of an event, 3) the recognition of an event, and 4) the query of an event. The detection phase addresses multi-source, spatio-temporal data fusion. The representation phase uses crude data to form hierarchical, invariant, and sufficient representations of events. Recognition incorporates event recognition and classification. Queries can be utilized to index and retrieve events according to matching query criteria [11].

In general, spatially distributed multi-sensor environments present promising possibilities and challenges for surveillance. There are studies revolving around data fusion techniques that address information sharing from different sensors. These challenges are compounded with communication aspects, such as bandwidth restrictions and the asymmetrical composition of communication. There are additional aspects that render the challenges difficult. Issues related to privacy and authentication are of importance. The desire is to advance automated learning capabilities to offer solutions of characterizing and recognizing events. A future challenge is to establish a wide-area, distributed multi-sensor surveillance system that is robust and entails real-time computational algorithms which require minimal reconfiguration for disparate applications. The systems should be applicable to different environments with discrepant weather, lighting, scene geometry and scene activity conditions, and support plug-and-play functionalities [36].

Generically, surveillance systems contain numerous sensors, for example, camera and radar, to gather data from the environment. There are two types of dilemmas: "1) the fusion of combinatory data from discrete sources in an optimal fashion, and 2) the management of multiple sensors which includes the global management of the entire system with all the individual operations of each sensor" [7]. Castanedo et al. [7] apply agent technologies to address these difficulties. "Agents are autonomous and are capable of monitoring their environment through a sensor" [7]. According to Castanedo et al., autonomous agents are able to cooperate to 1) attain higher performance for a specific surveillance task by collecting complementary information and combining data fusion techniques, and 2) achieve larger system coverage and task execution, which they are not capable of accomplishing individually [7].

1.3.1 The Architecture of Cooperative Sensor Agents

The data fusion process in multi-sensor networks constructs a coherent time-space description of interesting objects and events in a specific level. This is called the level of the fusion task. This provides an estimate of the reliability of the available sensors and processes to fused complementary data from different regions. Levels can be assigned to address different types of challenges, such as occlusions, contradictory information, and related data correlation. In addition to attaining higher spatial coverage and greater precision of events, robustness is used in cases of sensor malfunctions. To attain these goals, there are some universal requirements that must be addressed: 1) unification of environment and the association and synchronization of events, 2) dynamic alignment of events to assure unbiased information, 3) extraction of corrupted or erroneous data, 4) correct level of data association, and 5) the appropriate combination of information from different sensors [8].

"Cooperative sensor agents form a logical framework for autonomous agents which function in a network environment" [7]. Each agent must be able to cooperate within its neighborhood, which is a collection of agents that can establish a collaborative consolidation. Castanedo et al. [7] conceived an architecture of cooperative sensor agents. The architecture consists of two layers: 1) the sensor layer and 2) the coalition layer. On the sensor layer, "each sensor is controlled by an autonomous agent" [7]. Agents have different capabilities according to the sensor. This enables each autonomous agent to operate together with its neighbors to execute a task on the coalition layer.

The achievement of a distributed multi-agent approach presents its benefits: 1) intelligent cooperation enables wide-area surveillance with fewer sensors, 2) the improvement of robustness as agents that may fail to complete duty may be replaced with other agents, and 3) performance is resilient when tasks can be distributed. The likelihood of correct classification grows through the agreement and disagreement of multiple sensors [36].

The use of agent-based technologies for facial recognition has received attention. Active tracking of sequential images can detect human faces. Through the integration of multi-sourced information with real-time or near-real-time agents, the surveillance system is capable of tracking individuals in a room reliably and efficiently. The real-time agents acquire and transmit inherent information of captured images that are used to quickly detect and track faces in real-time. The near-real-time agents capture the facial region, determine the overall evaluation probability for facial objects, and detect the successive images. This system is straightforward, but has yet to be prove facial feature extraction and graph matching schemes [20].

1.4 Techniques of Artificial Intelligence

The recognition of human activities in different settings, such as airports and parking lots, is of interest in surveillance systems. In these areas, the form of the data varies greatly in quality and granularity. The identification and interpretation of human activities requires an activity model, that is 1) adequately rich to distinguish multi-agent interactions, 2) robust against uncertainties, and 3) capable of handling ambiguities. The design of algorithms that are capable of identifying human activities is still far from achieving systematic solutions. There is continuous interest in automatically detecting abnormalities, but the difficulties are emphasized by 1) the low-level detection of the unambiguous detection of primitive actions because of illumination, occlusion, or noise; 2) the high-level detection of complicated events through multi-agent interactions; and 3) the application of one semantic activity to different variations that do not conform to statistical or syntactic restrictions. One has prior knowledge on the type and semantic structures of the activities in a domain. There also might be the possibility to have a training set that is not exhaustive. Domain knowledge must be applicable to semantic activity models to uphold the robustness of statistical approaches. The fusion of different types of knowledge, such as statistical and structural, has yet to be successfully implemented [1].

Successful video surveillance is dependent on the capability to detect moving objects in the video stream. Every image is isolated and processed with an image analysis technique. This should be accomplished with a reliable and efficient approach to handle unconstrained environments, nonstationary background, and different object patterns. Numerous algorithms have been presented and adopted for object detection. They are all dependent on different assumptions, such as statistical models of the background, minimization of Gaussian differences, minimum and maximum values, or a consolidation of frame differences and statistical models. Unfortunately, there still is not enough information on the performance of all these approaches in the field [24].

Surveillance systems typically need real-time segmentation of all the moving objects in the video stream. Segmentation is required because it has a heavy influence on the performance of other modules, for example, object tracking, classification, and recognition. For example, precise detection is needed to produce the correct classification of the object. Background subtraction is a typical approach to determine the moving objects of a video sequence. "The fundamental concept is to remove the current frame from a background image in order to classify each pixel as either belonging to foreground or background" [24]. This is accomplished through comparison according to a certain threshold. Morphological operations are executed with component analyses to calculate the active regions within an image. In practice, multiple challenges can be identified. The background image can be compromised by noise resulting from camera motion or fluttering objects, such as waving trees, clouds, or shadows. Significant research endeavors have been undertaken to address shadows and nonstationary backgrounds. There are two types of changes that need to be solved: "1) slow alterations, which incur according to the time of day, and 2) rapid alterations, which incur suddenly, such as rain or abrupt changes in the static objects" [24]. The models must be adaptive and they must have thresholds that are capable of handling these alterations. Techniques can be utilized "to recursively update background parameters and thresholds to track the evolution of parameters in nonstationary operating conditions" [24]. Another significant challenge is the addressing of ghosts. Ghosts are false active regions that result from static objects belonging to the background. An example of this is a car, which is initially stationary, moves, and becomes active. This difficulty is normally handled with recent background subtraction based on frame differentiations or high-level functionalities [24].

In detail, some research has illustrated the utilization of a deterministic background model. This characterizes the permissible interval for each pixel of the background image and the maximum rate of changes in subsequent images or the median of the largest, absolute interframe differences. The majority of contemporary research does rely on statistical background models. This assumes that each pixel is a stochastic variable with a probability distribution estimated from the video stream. As an example, the Pfinder system (Person finder) [41] uses a Gaussian model to illustrate each pixel of the background image. Another, more generic approach can use a consolidation of Gaussians to present each pixel. This permits the use of multimodal distributions that exist in a natural scene, for example, in fluttering. Another collection of algorithms is based on spatio-temporal segmentation of the video stream. These methods attempt to identify the moving regions based on the temporal progression of the pixel intensities, colors, and specific characteristics. The segmentation is done in a 3D image-time space according to the temporal evolution of neighboring pixels. There are multiple approaches in which this can be conducted, for example, by using spatio-temporal entropy or morphological operations. Other approaches include 3D structural tensor, which is defined from the spatial and temporal derivatives of pixels in a time interval [24].

The application of comprehension and artificial intelligence began in the 1980s. Early research concentrated on the comprehension of vehicle tracking; 3D modeling expands modeling to address distorted information that could be parameterized and improve robustness. Recent years have witnessed pragmatic approaches to cognitive research. Activity monitoring incorporates low-level monitoring and tracking. The tracking of people is more challenging than the tracking of vehicles, because people have less rigidity and greater flexibility and capriciousness in motions [3].

The 1990s presented a dilemma in cognitive vision, that of automatic video interpretation. This addresses specifically the dynamic scenarios. A scenario can be a combination of states, events, or sub-scenarios. A conventional approach to address dynamic scenarios is the adoption of probabilistic/neural networks. A conventional way to recognize a scenario is to use a symbolic network, of which the nodes correspond to the Boolean logic scenario recognition. These techniques improve efficient scenario recognitions, but there are spatio-temporal restrictions that are difficult to address. The majority of these solutions require that the scenarios are confined in time or identical manner [6].

An area of definite interest is unsupervised behavioral learning and recognition. This comprises the capability of the surveillance system to learn and detect frequent scenarios without predefined behaviors. The majority of approaches in this field are applied to designated domains and utilize available domain knowledge to create proper models or select features. Markov models and Hidden-Markov models (HMMs) have been exploited to achieve unsupervised behavioral learning and recognition. Markov models provide good stochastic models but cannot address temporal relations. HMMs can be used to learn a topology through merging and splitting states, but the proven methodologies can handle only simple events and are unable to create concept hierarchies. Thus, there is no guarantee that the states of these models will attribute to inherent events [6].

Real-time tracking is mainly focused on appearance-based models. Haritaoglu et al. [14] used contour analysis to distinguish the limbs of people to perform tracking in image sequences. Oliver et al. [26] used Bayesian analysis to identify human interactions through trajectories. Video comprehension is dependent on camera positioning and the density of the camera network. Javed et al. [16] use multiple views to identify the trajectory of people across nonoverlapping cameras to link different perspectives together. Once a significant number of trajectories have been viewed, geometric and probabilistic models can be created for long-term predictions. Scene modeling improves the reliability of scene comprehension. For instance, people are more likely to be on the sidewalk and the vehicles on the road in video scene comprehension [3].

Vu et al. [37] present an automatic video interpretation system that recognizes predefined scenarios depicting human behaviors from video sequence. Vu et al. concentrate on the description of human behaviors, for example, scenario metaclasses according to scene context, human bodies, and actions. This enables the creation of generic models to depict specific scenarios, such as "meeting at the coffee machine" [37]. "Their automatic video interpretation system includes three principal modules: 1) detection of an individual, 2) tracking an individual, and 3) recognition of a scenario, i.e., behavior" [37]. Inputs are acquired from video and its outputs are recognized scenarios. This approach is challenging for developers as the scenarios must be carefully defined with experts.

The majority of different approaches use methods for detection and domain-specific events. For instance, these can be applied in the usage of gesture recognition or trajectory classification. A detriment of these approaches is the specific utilization of techniques that are only specific to a particular domain. This results in significant difficulties when attempting to apply these techniques to other regions. Some researchers have applied a two-step methodology to address this problem. The first step is to extract visual cues and primitive events. The second step is to use the extracted information to detect more complicated and abstract behavioral patterns [3].

1.4.1 Video Understanding

Video understanding encompasses the previous technologies to establish vision cooperation and behavioral recognition. The vision module addresses three separate tasks: 1) a motion detector and a frame-to-frame tracker establish a graph of objects for each camera, 2) the graphs calculated for each camera are consolidated into a global graph, and 3) the global graph is used for long-term tracking of individuals, vehicles, and groups of people as the scene evolves [6].

The behavior recognition module performs three levels of reasoning for each tracked object. They are 1) states, 2) events, and 3) scenarios. Bremond et al. used 3D scene models, that is, geometric models of empty scenes for each camera for contextual recognition of the scene under surveillance. The 3D positions and static dimensions were defined with static scene objects, such as benches, and the areas of interest, such as an entrance, for a scene model. The intention is to recognize specific behavior that happens in the scene. Great consternation is caused by the inability to define and reuse methods that recognize specific behaviors. The perception of the behaviors is highly dependent on the site, the point of view of the camera, and the individuals involved in the behaviors. There are approaches that require formalisms and extensive libraries from which functionalities can be drawn, but formalisms and libraries are rarely flexible or comprehensive enough for different types of behaviors [6].

1.4.2 Neurocomputing and Genetic Algorithms

The capability of learning from experience is an innate human quality for survival and evolution. Neuroscientists have proven that the human brain is capable of memorizing perceived information of objects and events, and there is interaction between entities and their environment. The design of cognitive systems, which are able to automatically adapt to "alterations, learning from experiences, and active interaction with external entities, is an active area of research. It ranges from computer vision to pure artificial intelligence" [12]. The approaches conceived are mimicked to the human brain and have offered promising applications [12].

A popular approach applied to classification tasks are neural networks that have similar structural design as the interconnections of neurons. "Numerous studies in the field of video surveillance have concentrated on learning the trajectories through statistical models to classify the activities of a monitored site or to detect abnormal behaviors" [12]. Additional work is required to analyze the learning procedures to avoid underestimating significant atypical events. Another dilemma is to accurately model the behavior of the object. For instance, a perpetrator will behave differently if a guard perceives his actions [12].

In the majority of cases, model predictions are influenced by uncertainties that result from the simplified presumptions of the model itself and the omission of complete knowledge of the model parameters. Acquisition of all the parameters is impossible. This typically results in the optimization or emphasis on information that is presumed to be of high importance. Both time and dynamism will alter the cases. In practice, it is more realistic to perceive the parametrical values as stochastic variables [21].

Modern numerical algorithms, such as genetic algorithms (GA), are numerical search tools that operate according to procedures which resemble natural selection and genetics. GAs have been applied successfully to multiple problems because of their flexibility and global disposition. The application of genetic and other evolutionary algorithms is a contemporary trend. While they have been applied to reliability and availability predicaments, they seldom address the uncertainties of the values of model parameters. In principle, there are multiple sources of estimations for these parameters, such as historical data. In practice, only few and sparse data is available. The parametric values typically lack the needed precision. This uncertainty has a direct influence on the output of the model. This forces resorting to "best estimates," as previously presented presumptions. Used values are acquired from estimations. Because it is an estimate, it is impacted by the degree of incertitude. This ultimately presents how usable or unusable the solution is. The higher degree of complexity renders this task more complicated to attain at some satisfactory level [21].

1.4.3 Probabilistic Reasoning

Greifenhagen et al. considered the use of quantitative statistical evaluation in a multi-sensor surveillance system. Their generic methodology decomposes a surveillance predicament into input submodules. Each submodule presents its discrete input and output relations. This shows how complex chains of modules can be predicted in a statistical sense based on the probabilistic knowledge of input data [28].

A challenge is tracking an individual target in a cluttered environment. Typically, the objective is to predict the state of the object based on noisy and equivocal measurements. The single target tracking problem can be specified and solved through Bayesian formulation by representing the target state probabilistically and using statistical paradigms to sense action and target state transition. This can be accomplished in practice with a ubiquitous Kalman filter, which is applicable and optimal when the measurements and state dynamics are Gaussian and linear. In settings in which nonlinear motions, non-Gaussian densities, or nonlinear measurement-to-target coupling is required, more sophisticated nonlinear filtering techniques are needed. Standard nonlinear filtering techniques include modifications to Kalman filters, such as the extended Kalman filter, the unscented Kalman filter, and the Gaussian sum approximations. All of these facilitate some of the linear assumptions of the Kalman filter [18].

The multi-target tracking problem has been addressed with different techniques, such as multiple hypothesis tracking (MHT) and joint probabilistic data association (JPDA). Both of these techniques interpret a measurement of the surveillance area into a collection of detections through thresholds. The detection is then associated with existent tracks, used to create new tracks, or considered false alarms. Kalman-filter algorithms may be used to update the existent tracks with new measurements once they have been associated with each other. The challenge is to identify the appropriate association between the measurements and the targets. Other approaches include the Bayesian perspective. A mathematical theory of multiple target tracking may be created from a Bayesian approach. As in single target tracking, fixed grid approaches for multi-target tracking are computationally expensive [18].

Tracking techniques can be further divided into two basic approaches: 1) 2D paradigms that do not have explicit shape models and 2) 3D paradigms. Numerous approaches can be utilized to classify new, detected objects. A conventional tracking method is to use a mechanism to predict the movement of the recognized object. The Kalman filter is the most typically used filter in surveillance systems. The placement of boxes or ellipses, which are normally known as "blobs," to form regions of high probability, is another tracking approach that is based on statistical paradigms. There is a tracking approach that uses connected-components to isolate changes in the scene into distinct objects without previous knowledge. This approach achieves satisfactory performance when the object is small, has a low-resolution approximation, and the location of the camera is carefully chosen [36].

1.5 Mobile Sensors and Robotics

Through the distribution of surveillance systems, the utilization of wide-area wireless sensors and their networks has grown. This also includes the ability of sensors to illustrate certain capabilities of self-awareness. This can be presented as tolerance against sensor failures and environmental conditions. Mobile sensor networks require self-configuration mechanisms, which assure adaptability, scalability, and optimal performance. The literature mainly contains sensor systems of completely autonomous mobile robots that function without human intervention. Human operators have resorted to semiautonomous robots to 1) examine detailed conditions, 2) uphold network communications, and 3) conduct maintenance activities, including reparations, replacements, or removals of sensors. These human-in-the-loop activities often have human errors that influence reliability. It has been stated that 40% of all the failures are attributable to the human operator [19].

In surveillance, robots equipped with cameras can identify obstacles and humans in the environment. The system can guide the robots around the obstacles. Normally, these systems draw their information from visual data. The handling of these quantities of data at high pace demands substantial computation or manpower. Wireless sensor networks (WSNs) can be used for object tracking. Objects emit signals that enable their tracking. Cameras do not necessarily always require mobility, but they offer different, dynamic options than static cameras, such as target following [33]. The recent progress in automated video technologies, which consists of both machine vision and robotics, should enable tangible alterations in industrial robots in the near future. These systems will introduce autonomous systems that can operate on sensor inputs and are capable of detection, recognition, and continuous tracking within the area of the robot [4].

Mobility and multi-purpose functionality normally reduce the amount of sensors needed to survey a given area. Additionally, mobile robots can be organized into teams that enable different types of strategic possibilities in surveillance covering larger areas. Traditionally, as different sensors are mounted on a robot, the tasks of navigation, exploration and surveillance present a challenging endeavor. In recent years, robots have proven to be dedicated to specific basic issues, such as operation in rough terrains, visual recognition, and sensor fusion [10].

A typical scenario in which autonomous mobile guard robots can be utilized is a building. An autonomous robot can perform the duties of a guard on a planned route. The robot can distinguish abnormal situations, such as leaking water or incipient fires. Because the robot has cameras, the images are transmitted to a monitoring station. After a routine patrol, the robot can be programmed to return to the docking station for recharging. Facial biometrics can be applied to the robot if the ability to automatically verify an intruder is required. Other needed activities could include raising alarms. Liu et al. [20] have developed the iBotGuard, which is an Internet-based intelligent robot system. It is capable of detecting intruders through face recognition. The two fundamental components of the system are 1) invariant face recognition and 2) intruder tracking. When an intruder transgresses a prohibited area, iBotGuard will intermittently capture images of the intruder and through face recognition identify that the object is a human. Figure 1.10 presents the basic functionality of the iBotGuard system. The robot can be controlled remotely in response to real-time images that it captures and transmits [20].

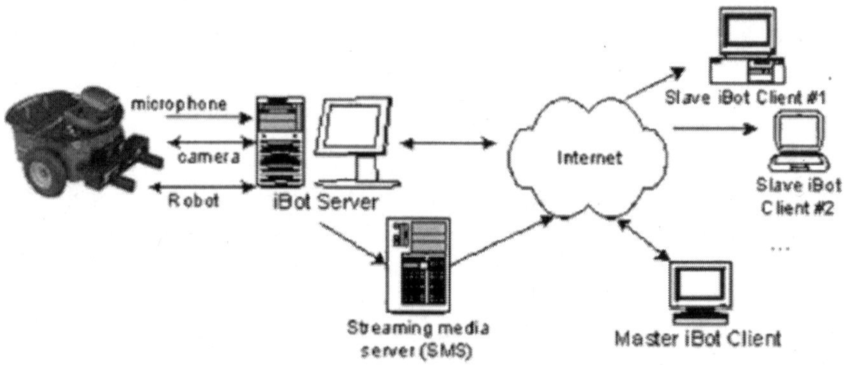

FIGURE 1.10 The iBotGuard system [20]. (From J.H.K. Liu, M. Wang, and B. Feng, iBotGuard: An Internet-based intelligent robot security system using invariant face recognition against intruder, *IEEE Transactions on Systems, Man, and Cybernetics, Part C: Applications and Reviews*, Vol. 35, No. 1, February 2005 ©2005 IEEE. With permission.)

The adoption of artificial intelligence in robotics has further enhanced the capability to detect a wider range of objects, to cover a wider area, and respond more individually to the situation at hand. Control in automated surveillance tends to remain centralized. Data is transmitted to a central node at which it is processed and decisions are issued. This type of architecture is simple, but it still promotes the difficulties of scalability, performance, and robustness. Distributed architectures enable the system nodes to present a certain amount of autonomy to conduct their own local and independent decisions. This inhibits profuse data communication of the overall system if nodes can illustrate successful actions in minor and independent cases. This improves both the performance and efficiency [34].

1.6 Conclusion

The main components of surveillance include data collection from multiple various sensors, such as video and audio, and the processing of individual data and information elements into high-level and broad informative knowledge. Throughout the years of research and development in the sphere of surveillance, the attained steps of progression have been enormous. The plethora of different sensor types and their information has multiplied and it will continue to grow. The transmission speeds and rates have been ameliorated, but naturally are always coveted to be increased.

The main advancements in the area of surveillance have been the growth in quantity and quality of the data procured by sensors. There have been huge advancements in all the fields of different sensors. Data collection, levels of intelligence, and transmission media have all led to considerable progress. These augmentations have enabled further and further detailed decisions based on collected data and information. As the number of sensors and sensor information increase and the processing speeds continue to improve, the basic challenge still remains the same: appropriately addressing all the information and being capable of drawing the correct conclusions for auspicious actions. Prioritization, levels of certainty, and information conflict situations will remain arduous dilemmas. Ambiguous situations and events are extremely difficult to categorize into correct deductions, as the line between a harmless and threatening situation is not always clear, nor does it always follow the same or even similar course. The rate of false alarms must be retained as low as possible and the rate of authentic alarms must attain high probability. Otherwise, the surveillance systems will be neglected.

The key item has been and will continue to be the collection of individual data elements from the field under surveillance, its consolidation and refinement into succinct, comprehendible, and relevant information according to which the humans in command can issue the appropriate, prudent, and above all, correct decision to protect, uphold, and enhance safety.

References

1. M. Albanese, R. Chellappa, V. Moscato, A. Picariello, V.S. Subrahmanian, P. Turaga, and O. Udrea. A constrained probabilistic Petri net framework for human activity detection in video. *IEEE Transactions on Multimedia*, 10(6):982–996, Oct. 2008.
2. P.K. Atrey, M.S. Kankanhalli, and R. Jain. Timeline-based information assimilation in multimedia surveillance and monitoring systems. In *Proceedings of the Third ACM International Workshop on Video Surveillance & Sensor Networks*, VSSN '05, pages 103–112, New York, NY, USA, 2005. ACM.
3. A. Avanzi, F. Brémond, C. Tornieri, and M. Thonnat. Design and assessment of an intelligent activity monitoring platform. *EURASIP J. Appl. Signal Process.*, 2005:2359–2374, January 2005.
4. A. Bakhtari, M.D. Naish, M. Eskandari, E.A. Croft, and B. Benhabib. Active-vision-based multisensor surveillance - An implementation. *IEEE Transactions on Systems, Man, and Cybernetics, Part C: Applications and Reviews*, 36(5):668–680, Sept. 2006.
5. J.A. Besada, J. Garcia, J. Portillo, J.M. Molina, A. Varona, and G. Gonzalez. Airport surface surveillance based on video images. *IEEE Transactions on Aerospace and Electronic Systems*, 41(3):1075–1082, July 2005.
6. F. Bremond, M. Thonnat, and M. Zuniga. Video Understanding Framework for

Automatic Behavior Recognition. *Behavior Research Methods*, 3:416–426, 2006.

7. F. Castanedo, M. A. Patricio, J. García, and J. M. Molina. Extending surveillance systems capabilities using bdi cooperative sensor agents. In *Proceedings of the 4th ACM International Workshop on Video Surveillance and Sensor Networks*, VSSN '06, pages 131–138, New York, NY, USA, 2006. ACM.
8. F. Castanedo, M.A. Patricio, J. Garcia, and J.M. Molina. Robust data fusion in a visual sensor multi-agent architecture. In *Proceedings of the 10th International Conference on Information Fusion*, pages 1–7, July 2007.
9. R.T. Collins, A.J. Lipton, H. Fujiyoshi, and T. Kanade. Algorithms for cooperative multisensor surveillance. *Proceedings of the IEEE*, 89(10):1456–1477, Oct. 2001.
10. D. Di Paola, D. Naso, A. Milella, G. Cicirelli, and A. Distante. Multi-sensor surveillance of indoor environments by an autonomous mobile robot. In *Proceedings of the 15th International Conference on Mechatronics and Machine Vision in Practice*, M2VIP 2008, pages 23–28, Dec. 2008.
11. Z. Dimitrijevic, G. Wu, and E.Y. Chang. SFINX: A multisensor fusion and mining system. In *Proceedings of the 2003 Joint Conference of the Fourth International Conference on Information, Communications and Signal Processing and the Fourth Pacific Rim Conference on Multimedia*, 2:1128–11322, Dec. 2003.
12. A. Dore, M. Pinasco, and C.S. Regazzoni. A bio-inspired learning approach for the classification of risk zones in a smart space. In *Proceedings of the IEEE Conference on Computer Vision and Pattern Recognition*, CVPR '07, pages 1–8, June 2007.
13. A. Hampapur, L. Brown, J. Connell, A. Ekin, N. Haas, M. Lu, H. Merkl, and S. Pankanti. Smart video surveillance: Exploring the concept of multiscale spatiotemporal tracking. *IEEE Signal Processing Magazine*, 22(2):38–51, March 2005.
14. I. Haritaoglu, D. Harwood, and L.S. Davis. W4: Real-time surveillance of people and their activities. *IEEE Transactions on Pattern Analysis and Machine Intelligence*, 22(8):809–830, August 2000.
15. D. Istrate, E. Castelli, M. Vacher, L. Besacier, and J.-F. Serignat. Information extraction from sound for medical telemonitoring. *IEEE Transactions on Information Technology in Biomedicine*, 10(2):264–274, April 2006.
16. O. Javed, Z. Rasheed, K. Shafique, and M. Shah. Tracking across multiple cameras with disjoint views. In *Proceedings of the Ninth IEEE International Conference on Computer Vision*, 2:952–957, Oct. 2003.
17. P. Korshunov and W.T. Ooi. Critical video quality for distributed automated video surveillance. In *Proceedings of the 13th Annual ACM International Conference on Multimedia*, MULTIMEDIA '05, pages 151–160, New York, NY, USA, 2005. ACM.
18. C. Kreucher, K. Kastella, and A.O. Hero. Multitarget tracking using the joint multitarget probability density. *IEEE Transactions on Aerospace and*

Electronic Systems, 41(4):1396–1414, Oct. 2005.

19. J.-S. Lee. A Petri net design of command filters for semiautonomous mobile sensor networks. *IEEE Transactions on Industrial Electronics*, 55(4):1835–1841, April 2008.
20. J.N.K. Liu, M. Wang, and B. Feng. iBotGuard: An Internet-based intelligent robot security system using invariant face recognition against intruder. *IEEE Transactions on Systems, Man, and Cybernetics, Part C: Applications and Reviews*, 35(1):97–105, Feb. 2005.
21. M. Marseguerra, E. Zio, and L. Podofillini. Optimal reliability/availability of uncertain systems via multi-objective genetic algorithms. *IEEE Transactions on Reliability*, 53(3):424–434, Sept. 2004.
22. C. Micheloni, G.L. Foresti, and L. Snidaro. A network of co-operative cameras for visual surveillance. *IEE Proceedings - Vision, Image and Signal Processing*, 152(2):205–212, April 2005.
23. K. Muller, A. Smolic, M. Drose, P. Voigt, and T. Wiegand. 3-D reconstruction of a dynamic environment with a fully calibrated background for traffic scenes. *IEEE Transactions on Circuits and Systems for Video Technology*, 15(4):538–549, April 2005.
24. J.C. Nascimento and J.S. Marques. Performance evaluation of object detection algorithms for video surveillance. *IEEE Transactions on Multimedia*, 8(4):761–774, Aug. 2006.
25. H. A. Nye. The problem of combat surveillance. *IRE Trans. Mil. Electron.*, MIL-4(4):551–555, Oct. 1960.
26. N.M. Oliver, B. Rosario, and A.P. Pentland. A Bayesian computer vision system for modeling human interactions. *IEEE Transactions on Pattern Analysis and Machine Intelligence*, 22(8):831–843, Aug. 2000.
27. R. Ott, M. Gutiérrez, D. Thalmann, and F. Vexo. Advanced virtual reality technologies for surveillance and security applications. In *Proceedings of the 2006 ACM International Conference on Virtual Reality Continuum and Its Applications*, VRCIA '06, pages 163–170, New York, NY, USA, 2006. ACM.
28. C.S. Regazzoni, V. Ramesh, and G.L. Foresti. Special issue on video communications, processing, and understanding for third generation surveillance systems. *Proceedings of the IEEE*, 89(10):1355–1539, October 2001.
29. L. Snidaro, R. Niu, G.L. Foresti, and P.K. Varshney. Quality-based fusion of multiple video sensors for video surveillance. *IEEE Transactions on Systems, Man, and Cybernetics, Part B: Cybernetics*, 37(4):1044–1051, Aug. 2007.
30. W. M. Thames. From eye to electron — Management problems of the combat surveillance research and development field. *IRE Trans. Mil. Electron.*, MIL-4(4):548–551, Oct. 1960.
31. M.M. Trivedi, T.L. Gandhi, and K.S. Huang. Distributed interactive video arrays for event capture and enhanced situational awareness. *IEEE Intelligent Systems*, 20(5):58–66, Sept.-Oct. 2005.
32. Y.-C. Tseng, T.-Y. Lin, Y.-K. Liu, and B.-R. Lin. Event-driven messaging

services over integrated cellular and wireless sensor networks: Prototyping experiences of a visitor system. *IEEE Journal on Selected Areas in Communications*, 23(6):1133–1145, June 2005.

33. Y.-C. Tseng, Y.-C. Wang, K.-Y. Cheng, and Y.-Y. Hsieh. iMouse: An integrated mobile surveillance and wireless sensor system. *Computer*, 40(6):60–66, June 2007.
34. J.J. Valencia-Jimenez and A. Fernandez-Caballero. Holonic multi-agent systems to integrate independent multi-sensor platforms in complex surveillance. In *IEEE International Conference on Video and Signal Based Surveillance*, AVSS '06, page 49, Nov. 2006.
35. M. Valera and S.A. Velastin. Real-time architecture for a large distributed surveillance system. In *IEE Intelligent Distributed Surveilliance Systems*, pages 41–45, Feb. 2004.
36. M. Valera and S.A. Velastin. Intelligent distributed surveillance systems: A review. In *IEE Proceedings – Vision, Image and Signal Processing*, 152(2):192–204, April 2005.
37. V.-T. Vu, F. Brémond, and M. Thonnat. Human behaviour visualisation and simulation for automatic video understanding. In *WSCG'02*, pages 485–492, 2002.
38. Y. Wang, M. Papageorgiou, and A. Messmer. A real-time freeway network traffic surveillance tool. *IEEE Transactions on Control Systems Technology*, 14(1):18–32, Jan. 2006.
39. A. S. White. Application of signal corps radar to combat surveillance. *IRE Trans. Mil. Electron.*, MIL-4(4):561–565, Oct. 1960.
40. C.E. Wolfe. Information system displays for aerospace surveillance applications. *IEEE Transactions on Aerospace*, 2(2):204–210, April 1964.
41. C.R. Wren, A. Azarbayejani, Darrell T., and Pentland A.P. Pfinder: Real-time tracking of the human body. *IEEE Transactions on Pattern Analysis and Machine Intelligence*, 19(8):51–56, July 1997.

2

The Role of Soft Computing in Image Analysis: Rough-Fuzzy Approach

Alessio Ferone
University of Naples Parthenope, Naples, Italy

Sankar K. Pal
Indian Statistical Institute, Kolkata, India

Alfredo Petrosino
University of Naples Parthenope, Naples, Italy

Soft computing is a group of methodologies that works synergistically to provide flexible information processing capability for handling real-life ambiguous situations. The aim is to exploit the tolerance for imprecision, uncertainty, approximate reasoning, and partial truth in order to achieve tractability, robustness, low-cost solutions, and close resemblance with human-like decision-making. Soft computing methodologies (involving Fuzzy Sets, neural networks, genetic algorithms, Rough Sets, and probabilistic reasoning) have been successfully employed in various image processing tasks, including image segmentation, enhancement, and classification, both individually and in combination with other soft computing techniques. The reason for such success has its motivation in the fact that they provide powerful tools to describe uncertainty, naturally embedded in images, which can be exploited in various image processing tasks. The chapter is focused on theories of Rough and Fuzzy Sets, their synergic operations, and their applications in the field of image processing.

2.1 Introduction

The digital revolution guided by the development of new hardware and software solutions has made available a huge amount of data. As a result, traditional statistical data summarization and database management techniques are no longer adequate for handling such data volumes and for extracting information and knowledge useful in the support of the decision-making processes. This massive amount of data is generally heterogeneous, being characterized by the presence of numeric, textual, symbolic, pictorial, and other types of data, and may contain redundancy, errors, imprecision, and so on. In this scenario, soft computing provides a group of methodologies that work synergistically for handling real-life ambiguous situations. Its aim is to exploit imprecision, uncertainty, approximate reasoning, and partial truth in order to achieve tractability, robustness, and low-cost solutions. The guiding principle is to devise methods of computation that lead to an acceptable solution at low cost, by seeking an approximate solution to both precise and imprecise formulated problems. Soft computing methodologies, involving Fuzzy Sets (FSs), Artificial Neural Networks (ANNs), Genetic Algorithms (GAs), Rough Sets (RSs), and Probabilistic Reasoning (PR), are most widely applied in the data mining process [33], where Fuzzy Sets [75] provide a natural framework to deal with uncertainty, Artificial Neural Networks [44] and Rough Sets [46] are widely used for classification and rules generation, Genetic Algorithms are involved in various optimization and search processes [8], while Probabilistic Reasoning deals with uncertainty and propagation of belief.

Each soft computing methodology has its own peculiarities and offers different advantages. For instance, FSs are often used to model human reasoning and provide a natural mechanism for dealing with uncertainty. ANNs are robust to noise and have a good ability to model highly nonlinear relationships. GAs are particularly useful for optimal search. Rough Sets are very efficient in attribute reduction and rule extraction. PR provides a rigorous framework for representation of a probabilistic knowledge, for modeling random phenomena, and for analyzing them. On the other hand, these techniques also present shortcomings that do not allow their individual application in some cases: FSs are dependent on expert knowledge; the training time of ANNs can be long when the input data is large, and most neural network systems lack explanation facilities; the theoretical basis of GAs is weak, especially on algorithm convergence; RSs are sensitive to noise and lead to NP problems in the choice of optimal attributes reduct and optimal rules; the PR approach cannot discriminate between ambiguity and uncertainty, leading to lack of complete knowledge. In order to cope with the drawbacks of individual approaches and leverage the performance of data mining systems, it seems natural to develop hybrid systems by integrating two or more soft computing technologies, each contributing a distinct methodology for addressing problems in its domain, in a cooperative, rather than a competitive, manner. The results are more intelligent and robust systems providing a human-interpretable, low-cost, approximate solution, as compared to traditional techniques.

In this chapter we present a brief introduction to the mentioned soft computing techniques. In particular, we focus on rough and fuzzy hybridization and its role in the field of image processing, which represents the starting point of many algorithms employed in video surveillance. The rest of the chapter is organized as follows. Section 2.2 briefly introduces soft computing techniques in terms of constituent parts: FSs, RSs, ANNs, GAs, and PR. Section 2.3 concentrates on the synergic use of Fuzzy and Rough theories, namely Fuzzy-Rough Sets and Rough-Fuzzy Sets. Section 2.4 addresses the topic of soft computing techniques in image analysis, while Sections 2.5 and 2.6 present an overview of applications of Rough and Fuzzy theories by differentiating between their combination and their hybridization, respectively. Finally, Section 2.7 concludes the chapter.

2.2 Soft Computing Components

2.2.1 Fuzzy Set Theory

The development of fuzzy logic has led to the rise of soft computing, becoming the earliest and most widely reported constituent in this field. The aim is the modeling of imprecise and qualitative knowledge, as well uncertainty management at various stages [29]. Fuzzy logic is able to encode, to a certain extent, human reasoning in natural form. Despite a growing versatility of knowledge discovery systems, there is an important component of human interaction that is inherent to any process of knowledge representation, manipulation, and processing. Fuzzy logic has been extensively used in many application fields to exploit its characteristic features, for instance, knowledge discovery in databases [12], clustering [49, 59, 67], web mining [38], and image retrieval [13, 42]. The basic concept is represented by Fuzzy Sets that are inherently inclined to cope with linguistic domain knowledge and produce more interpretable solutions.

Let us assume U is the universe of discourse. A *Fuzzy Set* F of U is defined as a mapping

$$\mu_F : U \to [0, 1]$$

that, for each $x \in U$, associates the membership degree of x in F. If we consider U as a set of objects of concern and a crisp subset of U as a non-vague concept imposed on objects in U, then a Fuzzy Set F of U can be thought of as a mathematical representation of a vague concept described in linguistic terms. The *support set* of Fuzzy Set F is the crisp set that contains all the elements of U that have a nonzero membership value in F, while the *core* of Fuzzy Set F is the crisp set that contains all the elements of X whose membership value in F is 1 (one). The mapping function $\mu_F(\cdot)$, called the *membership function*, represents the degree of similarity to an ideal element. A large value of the membership function represents a high degree of membership, and hence high similarity to an ideal element. Membership functions can also represent the uncertainty using some particular functions, allowing the handling of linguistic variables by means of numerical computations.

2.2.2 Rough Set Theory

Rough Set theory [47] has emerged as a major mathematical tool for handling uncertainty that arises from granularity in the domain of discourse; this is done by managing the indiscernibility between objects in a set. It offers powerful tools to extract hidden patterns from data without any *a priori* knowledge. Recently, Rough Set theory has been extensively employed in various application fields, although its use generally proceeds along two main directions:

1. Decision rule induction based on generation of discernibility matrices and reducts [36, 62]
2. Data filtration by extracting elementary blocks from data based on equivalence relation [57]

The theory of Rough Sets is based upon the notion of *approximation space*, which is a pair $< U, R >$, where U is a nonempty set (the universe of discourse), and R is an equivalence relation on U, that is, R is reflexive, symmetric, and transitive. The purpose of relation R is to decompose the set U into disjoint classes so that two elements x and y are in the same class if and only if $(x, y) \in R$ or equivalently xRy. Given the approximation space $< U, R >$, U/R is defined as the quotient set of U by the relation R, that is,

$$U/R = \{T_1, \ldots, T_i, \ldots, T_p\},$$

where T_i is an equivalence class of R, $i = 1, \ldots, p$, i.e., the class of elements $x, y \in U$ such that xRy. If two elements x and y in U belong to the same equivalence class $T_i \in U/R$, they are said to be *indistinguishable*.

Given an arbitrary set $T \in 2^U$, in general, it may not be possible to describe T precisely in $< U, R >$. One can characterize T by a pair of approximation sets defined as

$$\underline{T} = \{[x]_{\mathcal{R}} | \; [x]_{\mathcal{R}} \cap T \neq \emptyset\}$$

$$\overline{T} = \{[x]_{\mathcal{R}} | \; [x]_{\mathcal{R}} \subseteq T\},$$

where $[x]_{\mathcal{R}}$ denotes the class of elements $y \in U$ such that $x\mathcal{R}y$, with $x \in U$. $\overline{T}$ and $\underline{T}$ are the *upper* and *lower approximation* of T by $\mathcal{R}$, respectively, that is,

$$\underline{T} \subseteq T \subseteq \overline{T}$$

Hence, the interval $[\overline{T}, \underline{T}]$ represents a set T in the approximation space $< U, R >$ and is called the *Rough Set* of T.

2.2.3 Neural Networks

Neural networks have proved to be a powerful tool for mining data, although they were earlier considered unsuitable for this task because of the lack of information for verification or interpretation by humans [18]. This has not prevented neural networks from being used, and sometimes abused, for nearly every classification and regression task, both in supervised and unsupervised versions. Recently a growing interest, aimed at filling this hole of knowledge, has arisen by extracting it from the trained networks in the form of symbolic rules [66]. In this way it is possible to identify the attributes that are the most significant determinants of the decision or classification. The main contribution of neural nets in the field of data mining is for rule extraction, classification, and clustering. In general, the first step to extract knowledge from a connectionist model is to provide a representation of the trained neural network, in terms of its nodes and links. One or more hidden and output units are automatically selected to derive the rules, which can be combined to gain a more comprehensible rule set. Neural nets then provide high parallelism and optimization capability in the data domain. First a network is trained to achieve the required accuracy and then redundant connections are pruned. Classification rules are generated by analyzing the network in terms of link weights and activation values of the hidden units [25].

2.2.4 Genetic Algorithms

Genetic Algorithms are adaptive, robust, and efficient optimization methodologies based on the principles of nature [19]. They can also be viewed as searching algorithms, suitable in situations where the search space is large, because they explore a space using heuristics inspired by nature, such as natural selection, crossover, and mutation [15].

A Genetic Algorithm is executed iteratively on a population of coded solutions by means of three basic operators: selection, crossover, and mutation. Starting from a set of solutions, at each iteration the algorithm evolves by producing a new set of solutions, that is, each iteration corresponds to a new generation of candidate solutions. Hence, evolution takes place on the encoded possible solutions of the problem. At each iteration, an objective function is used to evaluate the new generation of solutions, and, based on these values, some of the solutions are selected for reproduction. The idea is that solutions with high fitness values will, on average, reproduce more often than those with low fitness values. The solutions allowed to reproduce will procreate a new generation of solutions depending on their fitness values. In this way the good individuals are selected, while the bad ones are eliminated.

GAs do not require or use derivative information and hence the most appropriate applications are in problems where gradient information is unavailable or costly to obtain. Reinforcement learning is an example of such a domain. In GAs, the only feedback used by the algorithm is information about the relative performance of different individuals.

2.2.5 Probabilistic Reasoning

Probabilistic Reasoning [48] is based on two major paradigms: Bayesian belief networks and Dempster-Shafer theory (also known as theory of belief).

A probabilistic network is a graphical representation of dependence and independence relations between random variables, able to represent and process probabilistic knowledge. Bayesian networks are probabilistic networks based on Bayes' rule suited to represent knowledge in many situations involving reasoning under uncertainty. They provide model-based domain descriptions, where the model reflects properties of the problem domain and probability calculus is used to compute with uncertainty. The components of a probabilistic network are qualitative and quantitative. The qualitative component encodes a set of conditional dependent and independent statements among a set of random variables. The quantitative component, on the other hand, specifies the strengths of dependence relations using probability theory.

The Dempster-Shafer theory is a generalization of Bayesian theory. While Bayesian theory requires probabilities for each topic of interest, belief functions allow one to assign degrees of belief for one topic based on probabilities for a related question. The Dempster–Shafer theory is based on the idea of obtaining degrees of belief for one topic from subjective probabilities for a related topic, and on the Dempster rule to combine such degrees of belief. Implementation of Dempster–Shafer theory involves two problems. First, uncertainties in the problem must be sorted into *a priori* independent items of evidence and then Dempster's rule must be computed.

2.3 Rough and Fuzzy Hybridization

The set-oriented view of Rough Sets is defined over a classical set algebra and associates a Fuzzy Set to each subset of the universe. Vagueness arises in concept representation from the lack of information when defining a precise concept. In this view, Rough membership functions can be thought of as a special type of Fuzzy membership function, defined as probabilities derived by cardinalities of sets. In fact, one can use a probability function on the universe to define rough membership functions [80]. From a fuzzy logic point of view, lower and upper approximations can be defined with respect to a Fuzzy Set

A as

$$\underline{A} = \{x|\mu_A(x) = 1\} \tag{2.1}$$

$$\overline{A} = \{x|\mu_A(x) > 0\}, \tag{2.2}$$

that is, $\underline{A}$ and $\overline{A}$ are the core and the support of the Fuzzy Set A, respectively. Although, in the theory of Fuzzy Sets, the membership value of an element does not depend on other elements, in the theory of Rough Sets, with respect to an equivalence relation, the membership value of an element depends on other elements [6]. In the study of Fuzzy Sets, many types of fuzzy membership functions have been proposed to implicitly specify the membership value of one element with respect to other elements [23].

It is clear that both theories provide means to handle vague concepts, even if from different points of view, and hence it is not surprising that many efforts have been made to combine Rough and Fuzzy approaches to obtain more general and powerful tools.

The two theories seem to complement each other, and hence researchers have explored a variety of different ways in which the theories can interact with each other. The origins of both theories were essentially logical, and, hence much of the hybridization between Fuzzy and Rough Set theory is logically based.

Nevertheless, these kinds of approaches only address the vagueness or uncertainty present in an image. In recent years a new trend emerged to try to exploit both theories at the same time. These new approaches evolved along two distinct research lines. Techniques belonging to the first one try to *combine* the two theories in different steps of the algorithm, thus exploiting fuzzyness and roughness separately. The second one, which can be considered a more general approach, aims to *hybridize* both theories, thus exploiting fuzzyness and roughness at the same time. As will be made clear, Rough- and Fuzzy-based techniques have proved effective in the field of image analysis, as well as in other fields of pattern recognition.

Two types of combinations of Rough Set theory and Fuzzy Set theory lead to distinct generalizations of classical sets theory. Using an equivalence relation on the universe of discourse, one can introduce lower and upper approximations in Fuzzy Set theory to obtain an extended notion, called *Rough-Fuzzy Sets* [11]. Alternatively, a Fuzzy similarity relation can be used to replace an equivalence relation, which results in another notion called *Fuzzy-Rough Sets* [11].

The expressions for the lower and upper approximations of a set X depend on the type of relation R and whether X is a crisp or a Fuzzy Set. When X is a crisp or a Fuzzy Set and the relation R is a crisp or a fuzzy equivalence relation, the expressions for the lower and upper approximations of the set X are given by [61]

$$\underline{R}X = \{(u, \underline{M}(u))|u \in U\} \tag{2.3}$$

$$\overline{R}X = \{(u, \overline{M}(u))|u \in U\}, \tag{2.4}$$

where

$$\underline{M}(u) = \sum_{Y \in U/R} m_Y(u) \times \inf_{\varphi \in U} \max(1 - m_Y(\varphi), \mu_X(\varphi)) \quad (2.5)$$

$$\overline{M}(u) = \sum_{Y \in U/R} m_Y(u) \times \sup_{\varphi \in U} \min(m_Y(\varphi), \mu_X(\varphi)). \quad (2.6)$$

The membership function m_Y is the membership degree of each element $u \in U$ to a granule $Y \in U/R$ and takes values in $[0, 1]$, and μ_X is the membership function associated with X and takes values in $[0, 1]$. When X is a crisp set, μ_X would take values only from the set $\{0, 1\}$. Similarly, when R is a crisp equivalence relation, m_Y would take values only from the set $\{0, 1\}$. Fuzzy union and intersection are chosen based on their suitability with respect to the underlying application of measuring ambiguity. The pair of sets $\langle \underline{R}X, \overline{R}X \rangle$ and the approximation space U/R are referred to differently, depending on whether X is a crisp or a Fuzzy Set and the relation R is a crisp or a Fuzzy equivalence relation. The different combinations are listed in Table 2.1 [61].

TABLE 2.1 Different combinations of Rough and Fuzzy Sets.

X	R	$\langle \underline{R}X, \overline{R}X \rangle$	U/R
crisp	crisp	Rough Set of X	crisp
fuzzy	crisp	Rough-Fuzzy Set of X	crisp
crisp	fuzzy	Fuzzy-Rough Set of X	fuzzy
fuzzy	fuzzy	Fuzzy-Rough-Fuzzy Set of X	fuzzy

Hence, the approximation of a crisp set in a fuzzy approximation space is called a *Fuzzy-Rough Set*, and the approximation of a Fuzzy Set in a crisp approximation space is called a *Rough-Fuzzy Set*, making the two models complementary [73]. In this framework, the approximation of a Fuzzy Set in a fuzzy approximation space is considered a more general model, unifying the two theories. In [58] a broad family of Fuzzy-Rough Sets is constructed, substituting min and max operators by different implicators and t-norms, and the properties of three well-known classes of implicators (S-, R-, and QL-implicators) are investigated. Further research in the area of Rough and Fuzzy hybridization from different perspectives can be found in [9, 65, 71, 74]. In [54] the authors present a model of hybridization of Rough and Fuzzy Sets that has been observed to possess a viable and effective solution to some of the most difficult problems in image analysis. The model exhibits a certain advantage of having a new operator to compose Rough-Fuzzy Sets, called the $\mathcal{RF}$-product, able to produce a sequence of compositions of Rough-Fuzzy Sets in a hierarchical manner. Theoretical foundations and properties, together with an example of application for image compression are also described. In [70] the properties of generalized Fuzzy-Rough Sets are investigated, defining a pair of dual generalized Fuzzy approximation operators based on arbitrary Fuzzy relations, while in [30] a new approach introduces definitions for generalized Fuzzy lower and upper approximation operators determined by a residual implication. Assumptions are found that allow a given Fuzzy Set theoretic operator to represent a lower or upper approximation from a Fuzzy relation. Different types of fuzzy relations produce different classes of Fuzzy-Rough Set algebras.

Other generalizations are possible in addition to the previous hybridization approaches. One of the first attempts at hybridizing the two theories is reported in [71], where the negative, boundary, and positive regions of a Rough Set are expressed by means of a fuzzy membership function. All objects in the positive region have a membership of one, those belonging to the boundary region have a membership of 0.5, while those contained in the negative region have zero membership (i.e., they do not belong to the Rough Set). This construction leads one to express a Rough Set as a Fuzzy Set, with suitable modifications to the Rough union and intersection operators. Another approach that exploits the similarities between Rough and Fuzzy Sets has been proposed in [50], where the author introduces the concept of shadowed set. The main idea comes from the consideration that a numeric Fuzzy Set representation may be too precise, because a concept can be described only once its membership function has been defined. It is like requiring excessive precision in order to describe imprecise concepts. Shadowed sets do not use exact membership values, but adopt basic truth values and a zone of uncertainty (the unit interval), where elements may belong with certainty (membership of 1), possibility (unit interval) or not at all (membership of 0). This can be seen as analogous to the Rough Sets definitions for the positive, boundary, and negative regions.

2.4 Soft Computing in Image Analysis

Soft computing offers a novel approach to manage uncertainty in discovering data dependencies, relevance of features, mining of patterns, feature space dimensionality reduction, and classification of objects. Consequently, these techniques have been successfully employed for various image processing tasks, including image segmentation, enhancement, and classification, both individually or in combination with other soft computing techniques. Over the years, the combination of two or more techniques has been proved effective in image processing, yielding algorithms that have overcome classical approaches. The reason for such success has its motivation in the fact that soft computing techniques provide powerful tools to describe uncertainty, naturally embedded in images, which can be exploited in various image processing tasks.

Concepts represented in an image, for example, a region, are not always crisply defined; hence, uncertainty can arise within any processing phase, and any decision made at a particular level will have an impact on all higher level activities. A recognition or vision system should have sufficient provision for representing and manipulating the uncertainties involved at every processing stage, so that the system can retain as much information content of the data as possible. The output of the system will then possess minimal uncertainty and, unlike conventional systems, will not be biased/affected much by lower level decisions [40]. For instance, a gray tone image possesses ambiguity within pixels because of the possible multi-valued levels of brightness in the image. This indeterminacy, both in grayness and spatially, is due to inherent vagueness rather than randomness and, hence, many basic concepts of image analysis (e.g., edges, corners, boundary regions) do not lend themselves well to precise definition.

Over the years, many algorithms have been proposed to cope with intrinsic uncertainty in image analysis by exploiting either Fuzzy or Rough Set theory. Here we give just two examples.

Example 1: Fuzzy Segmentation

Fuzzy C–Means (FCM) [3] is a clustering algorithm based on the minimization of the following functional

$$J = \sum_{i=1}^{N} \sum_{i=j}^{C} u_{ij}^{m} ||x_i - c_j||^2,$$

where u_{ij} is the membership degree of the pattern x_i to the cluster j, m is a real number greater than 1, c_j is the centroid of the j–th cluster, and $|| \cdot ||$ is the Eucledian norm. When applied to an image, minimization of the functional J yields a segmentation of the image (see Figure 2.1).

(a) (b) (c)

FIGURE 2.1 a) Test image; b) segmentation performed employing Fuzzy C–Means [3]; c) segmentation performed employing Otsu's method [39].

Example 2: Rough Image Segmentation

Foreground object and background can be viewed as two sets with their Rough representation [45]:

- $\underline{O}_T$ is the lower approximation of the object
- $\overline{O}_T$ is the upper approximation of the object
- $\underline{B}_T$ is the lower approximation of the background
- $\overline{B}_T$ is the upper approximation of the background

where T is a threshold to separate the foreground from the background. Segmentation is performed by means of a definition of Rough Entropy,

$$RE_T = -\frac{e}{2}[R_{O_T}\log_e(R_{O_T}) + R_{B_T}\log_e(R_{B_T})],$$

where R_{O_T} and R_{B_T}, defined as

$$R_{O_T} = 1 - \left|\frac{\underline{O}_T}{\overline{O}_T}\right| \quad \text{and} \qquad R_{B_T} = 1 - \left|\frac{\underline{B}_T}{\overline{B}_T}\right|,$$

are the roughness of the object and of the background, respectively. Maximization of homogeneity in both object and background is achieved through maximization of Rough Entropy, thus yielding an effective segmentation (see Figure 2.2).

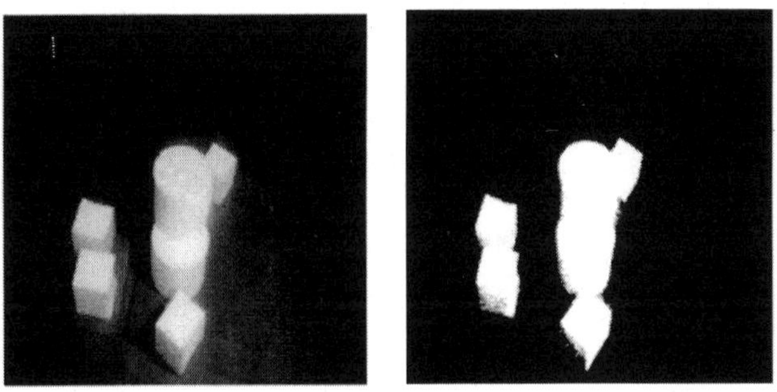

(a) (b)

FIGURE 2.2 a) Test image; b) segmentation performed employing Rough Entropy [45].

2.5 Combined Use of Rough and Fuzzy Sets

In this section an overview of methods that combine the theories of Rough and Fuzzy Sets is presented with regard to different tasks in image processing. As stated in Section 2.3, these techniques try to exploit both theories separately, but in a coordinated way.

2.5.1 Image Segmentation

In [34, 35], the fusion of Rough Set theory and FCM is used for color image segmentation. The technique aims to segment natural images characterized by regions with gradual color variations. Core centers are evaluated through approximations obtained by Rough Set theoretics, to reduce the computational complexity required by standard FCM. FCM-based segmentation strategies require *a priori* information about the number of clusters and their means as initialization points. The proposed technique extracts color information from the image employing Rough Set approximations on the segments and presents it as input to FCM for the soft evaluation of the segments. The advantage of the proposed technique is to analyze colors employing the three-dimensional vector (RGB) as one entity, where many algorithms work on single bands. However, employing FCM requires the definition of a distance between colors that, due to nonperceptive uniformity of the RGB color space, can lead to inconsistencies.

The problem of segmentation of multispectral satellite images is addressed in [41, 43]. The proposed method aims to integrate Rough Set theory for faster convergence and for avoiding the local minima problem; the Expectation Maximization (EM) is used to extract a statistical model of the data and handles the associated measurement and representation uncertainties; a Minimal Spanning Tree (MST) allows one to determine non-convex clusters on Gaussians determined by granules, rather than on the original data points. The proposed technique exploits Rough Set theoretic logical rules to obtain an initial approximation of Gaussian mixture model parameters. The mixture model is refined through EM to obtain accurate clusters. Here, Rough Set theory offers a fast and robust solution to the initialization, hence reducing the local minima problem of iterative refinement clustering. Also, the problem of choosing the number of mixtures is circumvented, as the number of Gaussian components to be used is automatically decided by Rough Set theory. The problem of modeling non-convex clusters is addressed by constructing an MST with each Gaussian component as nodes and Mahalanobis distance between them as edge weights. Because MST clustering is performed on the Gaussian models rather than the individual data points and because the number of models is much less than the data points, the computational time requirement is significantly less.

In [4], the authors propose a Fuzzy decision algorithm to face segmentation in a color image. The main characteristic of the proposed algorithm is the use of Fuzzy decision-making to segment images without user interaction, while Rough Sets are adopted to merge segments and choose the face region in each image. Use of different color quantizations in YCbCr color space and the Fuzzy decision algorithm allows Rough Sets to correctly merge face skin regions, but only partially addressing the problem of detection in bigger images.

2.5.2 Feature Selection

A method to select an optimal group of bands in hyperspectral images based on Rough Sets and FCM clustering is proposed in [64]. First, the FCM clustering algorithm is used to classify the original bands into equivalent band groups, as adjacent bands in hyperspectral images always show strong correlation. The concept of attribute dependency in Rough Sets is used to define the distance between a group and the cluster center. Then the data is reduced by selecting only the band with a maximum grade of Fuzzy membership from each of the groups. In this way, the number of bands is decreased, while preserving the most useful information. The final step consists of either selecting only one from each of the groups, or linearly composing the images in each group. The proposed technique exploits one of the most useful characteristics of Rough Sets, that is, attribute dependency, but employs a clustering algorithm that needs the number of clusters to be known a priori.

Hassanien [17] introduced a hybrid scheme that combines the advantages of Fuzzy Sets and Rough Sets in conjunction with statistical feature extraction techniques. The first step consists of a Fuzzy image preprocessing technique to enhance the contrast of the whole image, to extract the region of interest, and then to enhance the edges surrounding the region of interest. Next, features from the segmented regions of the interested regions are extracted using the grey-level co-occurrence matrix. Rough Set is used for the generation of all reducts, which contain a minimal number of features, and hence rules. Although Rough Set rules generation allows one to identify significant attributes very accurately, the major drawback of this technique is the need to choose the number of clusters to segment the image, which can vary depending on the image.

2.5.3 Image Quality Evaluation

Due to the complexity of fused image quality evaluation, in [72] a hybrid model of knowledge reduction is constructed by means of Rough Set theory and Fuzzy Support Vector Machine (FSVM). The proposed model combines the reduction ability of Rough Sets with the classification ability of the FSVM. A reduced information table is obtained by reducing the number of evaluation criteria, without information loss, through the Rough Set method. The reduced information is used to develop classification rules and train the FSVM.

2.5.4 Detection

Gao et al. [14] present a feature reduction method based on Rough Set theory and FCM to extract rules for shot boundary detection. Based on the classification capability of various features for different shot transitions, the correlation between features can be defined using the classification ability of attributes (or dependence between attributes) in Rough Set theory. Then, by employing the FCM algorithm, the optimal feature reduction can be obtained. The first step consists of extracting conditional attributes from video sequences. Then, by calculating their correlation, the importance of conditional attributes can be computed. Selected features are obtained by clustering features with FCM. For each class, the Fuzzy if-then rule is generated for decision with Fuzzy inference. This technique presents the same limits of the other approaches based on FCM, that is the need to predetermine the number of clusters.

In [24], two combined classifiers were discussed in the field of landmines detection. In the first classifier, Hebb Net learning is used in combination with Rough Set theory, while in the second one Fuzzy filter neural network is used. Rough Sets have been applied to classify the landmine data because in this theory no prior knowledge of rules is needed, hence these rules are automatically discovered from the database. The rough logic classifier uses lower and upper approximations for determining the class of the objects. The neural network is for training, and has been used, in particular, to avoid the boundary rules given by the Rough Sets that do not classify the data with high probability. Although the combined use of Rough Set and Fuzzy filter classifier gives good results, it can only partially reduce the problem of ambiguous patterns belonging to the boundary regions.

In [2], a method for object labeling, based on the uncertainty measurement of a Fuzzy similarity, is presented. The labeling is performed on objects detected in a scene, based on information provided by a set of different sensors. First, the Fuzzy similarity is computed between the detected object and a Rough Set of possible prototypes, followed by a measurement of the uncertainty induced by the observation. For all results obtained from each sensor, the global uncertainty, corresponding to the most likely label, is computed. The proposed technique aims to improve the labeling process by suppressing the inconsistent observations and making new labeling determinations. Different prototypes are used in this process, corresponding to different observation distances and positions. Also, from these observations, uncertainty variation can be analyzed, as determined by the switch from one prototype to another. Problems may arise in complex scenes where inconsistencies can be faced at different resolutions, resulting in erroneous label assignments.

2.6 Hybridization of Rough and Fuzzy Sets

In this section techniques that exploit a different approach to the combination of Rough and Fuzzy theories are summarized. These methods mainly employ the concept of Rough-Fuzzy Sets and Fuzzy-Rough Sets as a generalization of their constituent theories. The aim of Rough and Fuzzy hybridization is to exploit, at the same time, uncertainty and vagueness as a whole, thus leading to better results.

2.6.1 Image Segmentation

Mitra et al. [31] introduced a hybrid clustering architecture, in which several subsets of patterns can be processed together to find a common structure. A detailed clustering algorithm is developed by integrating the advantages of both Fuzzy Sets and Rough Sets, and a measure for quantitative analysis of the experimental results is provided for synthetic and real-world data. Rough Sets are used to model clusters in terms of upper and lower approximations, which are weighted by a pair of parameters while computing cluster prototypes. The use of Rough Sets helps to control uncertainty among patterns in the boundary regions, during collaboration between the modules. Memberships are used to enhance the robustness of clustering as well as collaboration. The main limitation of the proposed Rough FCM relies on the optimal selection of the parameters, which can vary among different datasets.

In [26], the development of a generalized methodology, which integrates c-means algorithm, Rough Sets, and probabilistic and possibilistic memberships of Fuzzy Sets, is presented by Maji and Pal. This formulation is geared toward maximizing the utility of both Rough and Fuzzy Sets with respect to knowledge-discovery tasks. The concept of crisp lower and Fuzzy upper regions is used, thereby significantly reducing the computational time, as required by the technique proposed in [31] and FCM. Several measures are defined based on Rough Sets to evaluate the performance of Rough-Fuzzy clustering algorithms. The effectiveness of the proposed algorithm is demonstrated by comparing it with other related algorithms for the task of image segmentation. Also in this case, as in [31], the main drawback is optimal selection of the parameters.

In [61], Rough and Fuzzy Sets theories are used to measure the ambiguities in images. Rough Set theory is used to capture the indiscernibility among nearby gray values and nearby pixels, whereas Fuzzy Set theory is used to capture the vagueness in the boundaries of the various regions. Different Rough-Fuzzy entropy measures using various generalizations of Rough Sets are proposed to quantify image ambiguity. By using them, a characteristic measure of an image, called the *average image ambiguity*, is presented. The Rough-Fuzzy entropy measure is used to perform various image processing tasks, such as object/background separation, multiple region segmentation, and edge extraction. The performance is compared to those obtained using existing Fuzzy and Rough Set theory-based image ambiguity measures. It is seen that generalization improves image segmentation. Although the results are very promising, finding the optimal values of the input parameters can be a tricky task.

In [32], an application of Rough-Fuzzy clustering is presented for synthetic as well as CT scan images of the brain. The algorithm generates good prototypes even in the presence of outliers. The Rough-Fuzzy clustering simultaneously handles overlap of clusters and uncertainty involved in class boundary, hence yielding the best approximation of a given structure in unlabeled data.

The number of clusters is automatically optimized in terms of various validity indices. Comparison with other partitive algorithms is also presented. Experimental results demonstrate the effectiveness of the proposed method in CT scan images, and is validated by medical experts. The hybrid approach proposed in this paper aims to maximize the utility of both Fuzzy and Rough Sets so to improve the performance of FCM and Rough C–Means. Nevertheless, the computational complexity is increased due to the simultaneous use of the two models.

In [27], a comprehensive investigation into Rough Set entropy-based thresholding image segmentation techniques was performed. Simultaneous combination of entropy-based thresholding with Rough Sets results in the Rough Entropy thresholding algorithm. Standard RECA (Rough Entropy Clustering Algorithm) and Fuzzy RECA, combined with Rough entropy-based partitioning routines, have been proposed. Rough entropy clustering incorporates the notion of Rough entropy into a clustering model, taking advantage of uncertainty in the analyzed data. Based on the test reported in the article, Standard and Fuzzy RECA seem to have similar performances. This result could be more thoroughly investigated because the Fuzzy version, at least in principle, should better capture uncertainty in data and hence lead to better segmentation.

In [77], an improved hybrid algorithm, called the *Rough-Enhanced Fuzzy C-Means* (REnFCM) algorithm, is presented for segmentation of brain MR images. The enhanced FCM algorithm can speed up the segmentation process for gray-level images, especially for MR image segmentation. Experimental results indicate that the proposed algorithm is more robust to noise and faster than many other segmentation algorithms.

Jiangping et al. [21] propose a Fuzzy-Rough approximation method for image segmentation. Based on graph theory combined with the shortest path algorithm of watershed transformation, the paper presents a shortest path segmentation algorithm based on the Rough-Fuzzy grid, where to each Fuzzy-Rough grid of the digital image is assigned a shortest path. The proposed method was applied in a Traditional Chinese Medicine (TCM) tongue image segmentation experiment, where the algorithm has proved to avoid over-segmentation of the image.

In [20], a method to segment tongue images based on the theory of Fuzzy-Rough Sets is presented. The proposed method, called *Fuzzy-Rough Clustering Based on Grid*, extracts condensation points by means of Fuzzy-Rough Sets, and quarters the data space layer by layer. The algorithm has been used in tongue image segmentation of TCM. Results indicate that the algorithm avoids over-segmentation of the image.

A multi-thresholding algorithm for color image segmentation is presented [37] using the concept of *A-IFS histon* obtained from Atanassov's Intuitionistic Fuzzy Set (A-IFS) representation of the image. A-IFS histon, an encrustation of the histogram, consists of the pixels that belong to the set of similar color pixels. In a Rough Set theoretic sense, A-IFS histon and the histogram can be correlated to upper and lower approximations, respectively. A multi-thresholding algorithm, using a roughness index, is then employed to get optimum threshold values for color image segmentation. The qualitative and quantitative comparisons of the proposed method against the histogram-based and the conventional histon-based segmentations prove its superiority. Performance of the proposed algorithm also demonstrates that exploiting uncertainty and vagueness can lead to good results when dealing with difficult concepts like colors.

2.6.2 Feature Selection

In [51], a new coding/decoding scheme based on the properties and operations of Rough-Fuzzy Sets is presented. By normalizing pixel values of an image, each pixel value can be interpreted as the degree of belonging of that pixel to the image foreground. The image is then subdivided into blocks, which are partitioned and characterized by a pair of approximation sets. Feature extraction is based upon Rough-Fuzzy Sets and performed by partitioning each block into multiple Rough-Fuzzy Sets that are characterized by two approximation sets, containing inf and sup values over small portions within the block. The method is shown to efficiently encode images in terms of high peak signal-to-noise ratio values, while alleviating the blocking problem.

2.6.3 Edge Detection

Petrosino et al. [55] presented a multi-scale method based on the hybrid notion of Rough-Fuzzy Sets, coming from the combination of Rough Sets and Fuzzy Sets. Marrying both notions leads one to consider, for instance, approximation of sets by means of similarity relations or Fuzzy partitions. The most important features are extracted from the scale spaces by unsupervised cluster analysis, to successfully tackle image processing tasks. [53] describes a feed-forward layered ANN whose operations are based on those of C-calculus [11], able to operate on a single image at a time. Within this framework, C-calculus arises as a method of representing Fuzzy image subsets [5]. Its applications to shrinking, expanding, and filtering [1] naturally lead to using it as a mathematical framework for designing a hierarchical neural network to deal with image analysis. The employed ANN is provided with a feedback mechanism and is structured in a hierarchical architecture. Application of the proposed network to edge detection is reported. The advantage of the proposed framework relies on the possibility of building a hierarchy of Rough-Fuzzy Sets, that is, the possibility of exploiting uncertainty and vagueness at different resolutions.

Authors in [60] present a histogram thresholding technique based on the beam theory to minimize ambiguity in information. This beam theory-based process considers a distance measure in order to modify the shape of the histogram. The ambiguity in the overall information, given by the modified histogram, is minimized to obtain a threshold value, employing the theories of Fuzzy and Rough Sets. The proposed scheme is applied to object and edge extraction in images and compared with those of a few existing classical and ambiguity minimization-based schemes for thresholding. In the case of Fuzzy Set theory-based ambiguity minimization, the Fuzzy membership value of each element corresponding to the modified histogram is computed. Rough Set theory-based ambiguity minimization is carried out by employing upper and lower approximations of the two ambiguous classes regarding bi-level thresholding of the modified histogram. Once roughness in the information corresponding to the two classes is calculated, the Rough Entropy of the information given by the modified histogram is measured across the modified histogram. The element value for which the value of Rough Entropy is minimum is considered the Rough-optimal threshold value.

In [68], image processing based on Rough Set theory is discussed in detail. The paper presents a binary Fuzzy-Rough Set model based on a triangle modulus, which describes a binary relationship by upper approximation and lower approximation. Given an image described by the binary relationship, the upper approximation and lower approximation can be used to represent the image. An edge detection algorithm by the upper approximation and the lower approximation of the image is presented, and image denoising is also discussed. The proposed model is well fit for processing images that have gentle gray changes.

2.6.4 Texture Segmentation

Traditional hybridization of Fuzzy and Rough Sets can be found in [52], where a texture segmentation algorithm is proposed to solve the problem of unsupervised boundary localization in textured images using Rough-Fuzzy Sets and hierarchical clustering.

In [78], Rough Set theory is applied to multiple-scale texture-shape recognition. The multiple-scale texture-shape recognition approach tries to cluster textures and shapes. In a multiresolution approach, texture and shape should be analyzed at different levels. It becomes evident that an exact representation is not a feasible option (both practically and conceptually); therefore one needs to look at some viable approximation. This is obtained by means of Rough Sets to construct the generalized approximate space by employing its Fuzzy function and Rough inclusion function. In this paper, according to the dataset extracted from images, the Fuzzy function and Rough inclusion function of generalized approximate space are defined. Also, statistical measures to denote the threshold of the Fuzzy function and the importance degree of each extracted feature are used. The image texture recognition algorithm is also compared with many other methods.

In [10], the authors propose a Rough content-based image quality measure. The image is partitioned into three parts: edges, textures, and flat regions, according to their gradient. In each part, the Rough-Fuzzy integral is applied as the Fuzzy measure of the similarity. The overall image quality metric is calculated based on the different importance of each part.

Based on the Fuzzy Rough Model (FRM), in [76] a rough neural network suitable for decision system modeling is proposed. It can implement a smooth Fuzzy partition of universe space by the adaptive G-K (Gaustafason-Kessel) clustering algorithm, which overcomes the defects of traditional reduct calculation based on Rough data analysis method. By taking advantage of the characteristics of Fuzzy clusters obtained by the adaptive G-K clustering algorithm, significant generalization ability enhancement is achieved. By making full use of the learning ability of neural networks, FRM_RNN_M improves the adaptability and achieves a comprehensive soft decision-making ability. Classification of Brodatz texture images indicates that FRM_RNN_M is superior to traditional Bayesian and learning vector quantization methods.

2.6.5 Image Classification

Mao et al. [28] proposed a fuzzy Hopfield net model based on Rough Set reasoning for the classification of multispectral images. The main purpose was to embed a Rough Set learning scheme into the fuzzy Hopfield network to construct a classification system, called *Rough-Fuzzy Hopfield Net* (RFHN). The classification system is a paradigm for the implementation of fuzzy logic and rough systems in neural network architecture. Instead of all the information in the image being fed into the neural network, the upper- and lower-bound gray levels, captured from a training vector in a multispectral image, are fed into a Rough-Fuzzy neuron in the RFHN. Therefore, only 2/N pixels are selected as the training samples if an N-dimensional multispectral image is used.

Wang et al. [69] proposed a nearest-neighbor classification algorithm based on Fuzzy-Rough Set theory. First, they make every training sample Fuzzy-Rough and use the nearest-neighbor algorithm to remove training sample points in class boundary or overlapping regions. Then the mountain clustering method is used to select representative cluster center points, and finally the Fuzzy-Rough Nearest-Neighbor algorithm is applied to classify the test data. The proposed method is applied to hand gesture recognition, and the results show that it is more effective and performs better than other nearest-neighbor methods.

In [63], a combined approach of neural network classification systems with a Fuzzy-Rough Sets-based feature reduction method is presented. Unlike transformation-based dimensionality reduction techniques, this approach retains the underlying semantics of the selected feature subset. This is very important to ensure that classification results are understandable by the user. Following this approach, the conventional multi-layer feed-forward networks, which are sensitive to the dimensionality of feature patterns, can be expected to become effective in the classification of images whose pattern representation may otherwise involve a large number of features. The proposed scheme has been applied to the real problem of normal and abnormal blood vessel image classification involving different cell types.

Authors in [7] present a study of the classification of a large-scale Mars McMurdo panorama image. Three-dimensional reduction techniques, based on Fuzzy-Rough Sets, information gain ranking, and principal component analysis, respectively, are each applied to this challenging image dataset to support learning of effective classifiers. The work allows the induction of low-dimensional feature subsets from feature patterns of a much higher dimensionality. To facilitate comparative investigations, two types of image classifiers are employed, namely multi-layer perceptrons and K-nearest neighbors. Experimental results demonstrate that feature selection helps to increase the classification efficiency by requiring considerably less features, while improving the classification accuracy by minimizing redundant and noisy features.

The focus of the study in [22] is on analysis of the effect of granularity on the indiscernibility relation of objects. In this study, the authors have applied the Rough Set theory to handle the imprecision due to granularity of the structure of satellite images. Rough Set and Rough-Fuzzy theory offer a better and more transparent choice to have faster, comparable, and effective results.

2.6.6 Detection in Real-Life Applications

Han et al. [16] present a feature reduction method based on Rough-Fuzzy Sets, by which the dissimilarity function for shot boundary detection in news video is obtained. By calculating the correlation between conditional attributes, the importance of conditional attributes in the Rough Set can be obtained. Due to the ambiguity of the set of features, the class precision of Rough-Fuzzy Set is given. Then, the importance of conditional attributes is defined as the Rough-Fuzzy operator by the product of the importance of conditional attributes in the Rough Set and the class precision of the Rough-Fuzzy Set. According to the proportion of each feature, the top k features can be obtained. The dissimilarity function is generated by weighting these important features.

Human face detection plays an important role in applications such as video surveillance, human-computer interface, face recognition, and face image database management. In [79], an attribute reduction method based on Fuzzy-Rough Sets is applied for face recognition. This paper mainly quotes attribute reduction of Fuzzy-Rough Sets to deal with the face data, while the recognition process uses neural network ensemble. The method avoids losing information caused by Rough Set attribute reduction. A Fuzzy similarity relation is employed to replace an equivalence relation so that the dispersion of data is avoided. As a result, the recognition accuracy improves.

In [56], the authors present a scheme for human faces detection in color images under unconstrained scene conditions, such as the presence of a complex background and uncontrolled illumination. The proposed method adopts a specialized unsupervised neural network to extract skin color regions in the Lab color space, obtained from the integration of the Rough-Fuzzy Sets-based scale space transform and neural clustering. A correlation-based method is then applied for the detection of ellipse regions. Experiments on three benchmark face databases demonstrate the ability of the proposed algorithm in detecting faces also in difficult conditions.

2.7 Conclusions

In this chapter we reported how soft computing techniques can lead to effective solutions of most problems in image analysis. Soft computing methodologies have been successfully employed in various image processing tasks, including image segmentation, enhancement, and classification, both individually and in combination with other soft computing techniques to exploit their respective peculiarities. The chapter mainly concentrated on the role of combined and hybridized use of Rough Set and Fuzzy Set theories in those tasks of image analysis that represent the starting point of algorithms employed in many application fields, for instance, video surveillance. The large number of applications and the obtained results, supported by solid theories, make them a privileged tool to exploit, and at the same time, properties like coarseness and vagueness typical of concepts represented in images.

References

1. A. Apostolico, E.R. Caiainiello, E. Fischetti, and S. Vitulano. C-calculus: An elementary approach to some problems in pattern recognition. *Pattern Recognition*, 10:375–387, 1973.
2. C. Barna. Object labeling method using uncertainty measurement. In *Proc. Intl. Workshop on Soft Computing Applications*, SOFA 09, pages 225–228, 2009.
3. J.C. Bezdek. *Pattern Recognition with Fuzzy Objective Function Algorithms.* Plenum, NewYork, 1981.
4. O. Byun, I. Park, D. Baek, and S. Moon. Video object segmentation using color fuzzy determination algorithm. In *Proc. 12th IEEE International Conference on Fuzzy Systems*, FUZZ 03, 2:1305–1310, 2003.
5. E.R. Caianiello. A calculus of hierarchical systems. In *Proc. 1st Inf. Congress on Pattern Recognition*, 1973.
6. S. Chanas and D. Kuchta. Further remarks on the relation between rough and fuzzy sets. *Fuzzy Sets and Systems*, 47(3):391–394, 1992.
7. S. Changjing, D. Barnes, and S. Qiang. Effective feature selection for mars mcmurdo terrain image classification. In *Proc. Intl. Conf. on Intelligent Systems Design and Applications*, ISDA 09, pages 1419–1424, 2009.
8. L. Davis, Editor. *Handbook of Genetic Algorithm.* Van Nostrand Reinhold, New York, 1991.
9. M. De Cock, C. Cornelis, and E.E. Kerre. Fuzzy rough sets: Beyond the

obvious. In *Proc. IEEE International Conference on Fuzzy Systems*, 1:103–108, 2004.

10. W. Dong, Q. Yu, C.N. Zhang, and H. Li. Image quality assessment using rough fuzzy integrals. In *Proc. 27th International Conference on Distributed Computing Systems Workshops*, ICDCSW 2007, pages 1–5, 2007.
11. D. Dubois and H. Prade. Rough fuzzy sets and fuzzy rough sets. *International Journal of General Systems*, 17:191–209, 1990.
12. U.M. Fayyad, G. Piatetsky-Shapiro, P. Smyth, and R. Uthurusamy, Editors. *Advances in Knowledge Discovery and Data Mining*. Menlo Park, CA: AAAI/MIT Press, 1996.
13. H. Frigui. Adaptive image retrieval using the fuzzy integral. In *Proc. NAFIPS 1999*, pages 575–579, 1999.
14. X. Gao, B. Han, and H. Ji. A shot boundary detection method for news video based on rough sets and fuzzy clustering. In *Image Analysis and Recognition, LNCS*, volume 3656, pages 231–238. 2005.
15. D.E. Goldberg. *Genetic Algorithms in Search, Optimization and Machine Learning*. Addison-Wesley, Reading, MA, 1989.
16. B. Han, X. Gao, and H. Ji. A shot boundary detection method for news video based on rough-fuzzy sets. *International Journal of Information Technology*, 11(7):101–111, 2005.
17. A.E. Hassanien. Fuzzy-rough hybrid scheme for breast cancer detection. *Image and Computer Vision Journal*, Elsevier, 25(2):172–183, 2007.
18. S. Haykin. *Neural Networks: A Comprehensive Foundation*. Macmillan, New York, 1994.
19. J.H. Holland. *Adaption in Natural and Artificial Systems*. University of Michigan Press, Ann Arbor, MI, 1975.
20. L. Jiangping, P. Baochang, and W. Yuke. Tongue image segmentation based on fuzzy rough sets. In *Proc. Intl. Conf. on Environmental Science and Information Application Technology*, ESIAT 09, 3:367–369, 2009.
21. L. Jiangping and W. Yuke. A shortest path algorithm of image segmentation based on fuzzy-rough grid. In *Proc. Intl. Conf. on Computational Intelligence and Software Engineering*, CiSE 09, pages 1–4, 2009.
22. M. Juneja, E. Walia, P.S. Sandhu, and R. Mohana. Implementation and comparative analysis of rough set, artificial neural network (ann) and fuzzy-rough classifiers for satellite image classification. In *Proc. Intl. Conf. on Intelligent Agent & Multi-Agent Systems*, IAMA 09, pages 1–6, 2009.
23. G.J. Klir and B. Juan. *Fuzzy Sets and Fuzzy logic, Theory and Applications*. Prentice-Hall, New Jersey, 1995.
24. S. Kumar, S. Atri, and L. Mandoria. A combined classifier to detect landmines using rough set theory and Hebb net learning & fuzzy filter as neural networks. In *Proc. Intl. Conf. on Signal Processing Systems*, pages 423–427, 2009.
25. H.J. Lu, R. Setiono, and H. Liu. Effective data mining using neural networks. *IEEE Trans. Knowledge Data Eng.*, 8:957–961, 1996.
26. P. Maji and S.K. Pal. Rough set based generalized fuzzy c-means algorithm and

quantitative indices. *IEEE Trans. on Systems Man and Cybernetics*, 37(6):1529–1540, 2007.

27. D. Malyszko and J. Stepaniuk. Standard and fuzzy rough entropy clustering algorithms in image segmentation. In C.-C. Chan, J. Grzymala-Busse, and W. Ziarko, Editors, *Rough Sets and Current Trends in Computing, LNCS*, volume 5306, pages 409–418. Springer Berlin / Heidelberg, 2008.
28. C.-W. Mao, S.-H. Liu, and J.-S. Lin. Classification of multispectral images through a rough-fuzzy neural network. *Optical Engineering*, 43:103–112, 2004.
29. J.M. Mendel. *Uncertain Rule-Based Fuzzy Logic Systems: Introduction and New Directions*. Prentice Hall PTR, 2001.
30. J.S. Mi and W.X. Zhang. An axiomatic characterization of a fuzzy generalization of rough sets. *Information Sciences*, 160:235–249, 2004.
31. S. Mitra, H. Banka, and W. Pedrycz. Rough-fuzzy collaborative clustering. *IEEE Trans. on Systems Man and Cybernetics*, 36(4):795–805, 2006.
32. S. Mitra and B. Barman. Rough-fuzzy clustering: An application to medical imagery. In *Rough Sets and Knowledge Technology, LNCS*, 5009:300–307, 2008.
33. S. Mitra, S.K. Pal, and P. Mitra. Data mining in soft computing framework: A survey. *IEEE Trans. on Neural Networks*, 13(1):3–14, 2002.
34. A. Mohabey and A.K. Ray. Fusion of rough set theoretic approximations and FCM for color image segmentation. In *Proc. IEEE Intl. Conf. on Systems Man and Cybernetics*, 2:1529–1534, 2000.
35. A. Mohabey and A.K. Ray. Rough set theory based segmentation of color images. In *Proc. 19th International Conference of the North American Fuzzy Information Processing Society*, NAFIPS 2000, pages 338–342, 2000.
36. T. Mollestad and A. Skowron. A rough set framework for data mining of propositional default rules. In *Lecture Notes Computer Science*, 1079:448–457, 1996.
37. M.M. Mushrif and A.K. Ray. A-ifs histon based multithresholding algorithm for color image segmentation. *IEEE Signal Processing Letters*, 16(3):168–171, 2009.
38. O. Nasraoui, R. Krishnapuram, and A. Joshi. Relational clustering based on a new robust estimator with application to web mining. In *Proc. 18th International Conference of the North American Fuzzy Information Processing Society*, NAFIPS 1999, pages 705–709, 1999.
39. N. Otsu. A threshold selection method from gray-level histograms. *IEEE Trans. Sys., Man., Cyber.*, 9(1):62–66, 1979.
40. S.K. Pal. Fuzzy image processing and recognition: Uncertainties handling and applications. *Int. J. Image Graphics*, 1(2):169–195, 2001. (Invited Paper).
41. S.K. Pal. Rough-fuzzy granulation, rough entropy and image segmentation. In *Proc. First Asia International Conference on Modelling & Simulation*, AMS 07, pages 3–6, 2007.

42. S.K. Pal, A. Ghosh, and M.K. Kundu, Editors. *Soft Computing for Image Processing*. Physica-Verlag, Heidelberg, Germany, 2000.
43. S.K. Pal and P. Mitra. Multispectral image segmentation using the rough-set-initialized EM algorithm. *IEEE Trans. on Geoscience and Sensing*, 40(11):2495–2501, 2002.
44. S.K. Pal and S. Mitra. *Neuro-Fuzzy Pattern Recognition: Methods in Soft Computing*. Wiley, New York, 1999.
45. S.K. Pal, B.U. Shankar, and P. Mitra. Granular computing, rough entropy and object extraction. *Pattern Recognition Letters*, 26(16):2509–2517, 2005.
46. Z. Pawlak. Rough sets. *Int. J. of Information and Computer Sciences*, 11(5):341–356, 1982.
47. Z. Pawlak. *Rough Sets, Theoretical Aspects of Reasoning about Data*. Kluwer, Dordrecht, 1991.
48. J. Pearl. *Probabilistic Reasoning in Intelligent Systems: Networks of Plausible Inference*. Morgan Kaufmann Publishers Inc., San Francisco, CA, 1988.
49. W. Pedrycz. Conditional fuzzy c-means. *Pattern Recognition Lett.*, 17:625–632, 1996.
50. W. Pedrycz. Shadowed sets: bridging fuzzy and rough sets. In *Rough-Fuzzy Hybridization: A New Trend in Decision Making*, pages 179–199. Springer Verlag, Singapore, 1999.
51. A. Petrosino. Rough fuzzy set-based image compression. *Fuzzy Sets and Systems*, 160(10):1485–1506, 2009.
52. A. Petrosino and M. Ceccarelli. Unsupervised texture discrimination based on rough fuzzy sets and parallel hierarchical clustering. In *Proc. IEEE International Conference on Pattern Recognition*, pages 1100–1103, 2000.
53. A. Petrosino and P. Feng. A fuzzy hierarchical neural network for image analysis. In *Proc. Intl. Conf. Systems Engineering in the Service of Humans*, 2:657–662, 1993.
54. A. Petrosino and A. Ferone. Feature discovery through hierarchies of rough fuzzy sets. In W. Pedrycz and S.-M. Chen, Editors, *Granular Computing and Intelligent Systems: Design with Information Granules of Higher Order and Higher Type*, 13:57–73. Springer-Verlag, 2011.
55. A. Petrosino and G. Salvi. Rough fuzzy set based scale space transforms and their use in image analysis. *Int. J. of Approximate Reasoning*, 41(2):212–228, 2006.
56. A. Petrosino and G. Salvi. A rough fuzzy neural based approach to face detection. In *Proc. Intl. Conf. on Image Processing, Computer Vision and Pattern Recognition*, IPCV'10, 2010.
57. L. Polkowski and A. Skowron. *Rough Sets in Knowledge Discovery 1 and 2*. Physica-Verlag, Heidelberg, Germany, 1998.
58. A.M. Radzikowska and E.E. Kerre. A comparative study of fuzzy rough sets. *Fuzzy Sets and Systems*, 126(2):137–155, 2002.
59. S. Russell and W. Lodwick. Fuzzy clustering in data mining for telco database marketing campaigns. In *Proc. 18th International Conference of the*

North American Fuzzy Information Processing Society, NAFIPS 1999, pages 720–726, 1999.

60. D. Sen and S.K. Pal. Histogram thresholding using beam theory and ambiguity measures. *Fundamenta Informaticae*, 75:483–504, 2007.
61. D. Sen and S.K. Pal. Generalized rough sets, entropy, and image ambiguity measures. *IEEE Trans. on Systems Man and Cybernetics*, 39(1):117–128, 2009.
62. N. Shan and W. Ziarko. Data-based acquisition and incremental modification of classification rules. *Comput. Intell.*, 11:357–370, 1995.
63. C. Shang and Q. Shen. Aiding neural network based image classification with fuzzy-rough feature selection. In *Proc. IEEE Intl. Conf. on Fuzzy Systems FUZZ-IEEE 2008 - IEEE World Congress on Computational Intelligence*, pages 976–982, 2008.
64. H. Shi, Y. Shen, and Z. Liu. Hyperspectral bands reduction based on rough sets and fuzzy c-means clustering. In *Proc. 20th IEEE IMTC 03*, 2:1053–1056, 2003.
65. H. Thiele. Fuzzy rough sets versus rough fuzzy sets – An interpretation and a comparative study using concepts of modal logics. Technical Report CI-30/98, University of Dortmund, 1998.
66. A.B. Tickle, R. Andrews, M. Golea, and J. Diederich. The truth will come to light: Directions and challenges in extracting the knowledge embedded within trained artificial neural networks. *IEEE Trans. Neural Networks*, 9:1057–1068, 1998.
67. I.B. Turksen. Fuzzy data mining and expert system development. In *Proc. IEEE Int. Conf. Syst., Man, Cybern.*, pages 2057–2061, 1998.
68. D. Wang and M. Wu. Binary fuzzy rough set model based on triangle modulus and its application to image processing. In *Proc. IEEE Intl. Conf. on Cognitive Informatics*, ICCI 09, pages 249–255, 2009.
69. X. Wang, J. Yang, X. Teng, and N. Peng. Fuzzy-rough set based nearest neighbor clustering classification algorithm. In *Lecture Notes in Computer Science*, 3613:370–373, 2005.
70. W.Z. Wu, J.S. Mi, and W.X. Zhang. Generalized fuzzy rough sets. *Information Sciences*, 151:263–282, 2003.
71. M. Wygralak. Rough sets and fuzzy sets – Some remarks on interrelations. *Fuzzy Sets and Systems*, 29(2):241–243, 1989.
72. G. Xian and B. Zeng. A novel hybrid model for information processing basing on rough sets and fuzzy SVM. In *Proc. Intl. Conf. on Multimedia and Ubiquitous Engineering*, MUE'08, pages 320–323, 2008.
73. Y.Y. Yao. Combination of rough and fuzzy sets based on a-level sets. In T.Y. Lin and N. Cereone, Editors, *Rough Sets and Data Mining: Analysis of Imprecise Data*, pages 301–321. Kluwer Academic Publishers, Norwell, MA, 1997.
74. S.D. Yeung, D. Chen, E.C.C. Tsang, J.W.T. Lee, and W. Xizhao. On the generalization of fuzzy rough sets. *IEEE Trans. on Fuzzy Systems*, 13(3):343–361, 2005.

75. L.A. Zadeh. Fuzzy sets. *Information and Control*, 8:338–353, 1965.
76. D. Zhang, Y. Wang, and H. Huang. Rough neural network modeling based on fuzzy rough model and its application to texture classification. *Neurocomputing*, 72:2433–2443, 2009.
77. W. Zhang, C. Li, and Y. Zhang. A new hybrid algorithm for image segmentation based on rough sets and enhanced fuzzy c-means clustering. In *Proc. IEEE Intl. Conf. on Automation and Logistics*, ICAL'09, pages 1212–1216, 2009.
78. Z. Zheng, H. Hu, and Z. Shi. Rough set based image texture recognition algorithm. In *Knowledge-Based Intelligent Information and Engineering Systems, LNCS*, 3213:772–778, 2004.
79. L. Zhou, W. Li, and Y. Wu. Face recognition based on fuzzy rough set reduction. In *Proc. International Conference on Hybrid Information Technology*, ICHIT'06, 1:642–646, 2006.
80. W. Ziarko. Probabilistic rough sets, rough sets, fuzzy sets, data mining, and granular computing. In *LNCS*, 3641:283–293, 2005.

3

Neural Networks in Video Surveillance: A Perspective View

Lucia Maddalena
National Research Council, Naples, Italy

Alfredo Petrosino
University of Naples Parthenope, Naples, Italy

3.1 Introduction

Artificial neural networks (ANNs) are among the soft computing tools most frequently adopted for several video surveillance tasks due to their well-known advantages, such as adaptivity, parallelism, and learning [43]. Indeed, an ANN can modify its connection weights using some training algorithms or learning rules; by updating the weights, the ANN can optimize its connections to adapt to changes in the environment. Moreover, when the information is fed to an ANN, it is distributed to different neurons for processing, and neurons can work in parallel and synergistically, if they are activated by the inputs. The capability of neural networks in emulating many unknown functional links by learning offline a limited set of representative examples allows one to infer a function from observations. This allows one to learn representations of the input that capture the salient input distribution features. In the general context of data classification, the neural network approach is particularly attractive, as it overcomes the difficulty of defining a single statistical model for different data types, which is the main problem associated with most conventional methods based on multivariate models. Also, neural approaches based on pyramidal structures have been pursued in the field of image processing and recognition, due to the compactness of multi-scale representations, which produce good textural features for landscape characterization, also providing support of efficient coarse-to-fine search [5]; efforts toward developing pyramid-based neural network techniques for recognition purposes arise as a way to handle problems of scaling with input dimensionality.

The aim of this chapter is to give some examples of neural network-based approaches to the solution of video surveillance tasks provided in the literature. A summary of our recent research activities related to video surveillance is also provided in order to show some of the advantages that can be gained by the adoption of such tools for moving object detection and other video surveillance tasks.

Section 3.2 reports examples of neural network-inspired solutions that have been proposed for video surveillance in past years, subdividing them according to the tackled video surveillance task. In Section 3.3 we describe the neural nonparametric model that we proposed in [32] for moving object detection. Section 3.4 describes how such a model can also be adopted in the context of other video surveillance tasks, including stopped object detection, activity recognition, and anomaly detection. Section 3.5 draws some conclusions.

3.2 ANN Approaches to Video Surveillance

3.2.1 Moving Object Detection

Moving object detection, as a specific case of object segmentation, can be viewed as a clustering problem in a feature space derived from the color and motion information, and therefore is well suited for unsupervised modeling approaches. Unsupervised learning is in general preferred over supervised learning because the latter requires a set of training samples, which may not be available, especially when the image features are unknown or when a certain degree of automation is desired. Many approaches have been explored; among them, neural network-based solutions have received considerable attention due to the fact that these methods are usually more effective and efficient than traditional ones.

One of the first research efforts proposing the use of neural networks for foreground detection is the one by Schofield et al. [47], who adopt a RAM-based neural networkin order to identify background elements in the current image and thus to isolate changed regions containing moving persons. Their objective of people counting is then achieved by searching the peaks of the produced gray-scale *score images*, whose high values correspond to foreground regions.

Pajares [42] proposes a Hopfield neural network-based algorithm for change detection that does not employ background modeling to achieve segmentation. A difference image is obtained by subtracting, pixel by pixel, two images and the network topology is built so that each pixel in the difference image is a node in the network. Each node is characterized by its state, which determines if a pixel has changed. An energy function is derived, so that the network converges to stable states, integrating spatial-contextual and self-data information, thus allowing for each pixel a trade-off between the influence of its neighborhood and its own criterion.

Culibrk et al. [13] present a background modeling and subtraction approach for video object segmentation based on a feed-forward neural network called BNN (Background Neural Network. The new structure represents a combination of a probabilistic neural network (PNN)and a winner-take-all neural network. Rules for temporal adaptation of the weights of the network are set based on a Bayesian formulation of the segmentation problem, reflecting the observed statistics of the background. Such a network is able to serve both as an adaptive model of the background in a video sequence and a Bayesian classifier of pixels as background or foreground. A modification of BNN has been proposed by Zhiming et al. [52], called ABPNN (Adaptive Background PNN). Every pixel in a video frame is classified as foreground or background by its conditional probability of being a background pixel, where background probability is estimated by a Parzen estimator in HSV feature space. Foreground is further classified into motion region and shadows by shadow detection.

Later in this chapter we describe the approach proposed by Maddalena and Petrosino in [32], called SOBS (Self-Organizing Background Subtraction), based on a self-organizing network for background and foreground modeling from video sequences. For each image pixel, a neuron map is built based on its color components, and the information stored in each pixel is updated only if its best matching weight vector is close enough to the background model based on a predefined distance; otherwise the pixel is considered as belonging to a moving object. The approach is at the basis of further neural network-based research aimed at simplifying SOBS (see Chacon et al. [6, 7]) and implementing it on GPUs (see Fauske [17]).

Luque et al. [31] propose a pixel-based technique for classifying each pixel of a specific frame as belonging to the foreground or background. In order to perform this task, a neural network architecture based on Adaptive Resonance Theory (ART) is used, where the inputs are the pixel color components. Each neuron is associated with one of the two classes and each class can be composed of several neurons, thus allowing the handling of multimodal backgrounds.

Lopez et al. [30] propose a new kind of probabilistic background model that is based on probabilistic self-organising maps, so that background pixels are modeled with more flexibility. Moreover, a statistical correlation measure is used to test the similarity among nearby pixels, so as to enhance the detection performance by providing a feedback to the process.

Other neural network-based approaches for the more general problem of object segmentation from color images and image sequences include [12, 15, 25, 38]. Further research relying on neural networks for different detection problems includes, for example, infrared target detection [8] and face detection [51].

3.2.2 Moving Object Tracking

After the identification of the object of interest, the aim of tracking is to generate the trajectory of the object in time by locating its position in every frame of the video. Several neural network solutions have been proposed for tracking objects in image sequences.

In [16], Drumea and Frezza-Buet apply a variation of Growing Neural Gas (GNG) to track rigid objects in video sequences in a way that supports smooth updating of tracked moving objects and fast adaptation to deal with changes in the number of tracked objects.

Luque et al. [31] track rigid objects with a Growing Competitive Neural Network (GCNN). Each object in a scene is assigned to a neuron, and that neuron represents and identifies that particular object. Neurons are added or deleted when new objects enter or exit the scene in order to get a one-to-one association between objects currently in the scene and neurons. This association is kept in each frame, what constitutes the foundations of this tracking system. In general, Competitive Neural Networks (CNNs) are suitable for data clustering, because each neuron in a CNN is specifically designed to represent a single cluster. In the field of object tracking in video sequences, such clusters correspond to moving objects. Thus, it seems reasonable to use CNNs as trackers. However, due to the dynamic nature of a video sequence, objects are constantly appearing and disappearing from the scene, and the method used to track objects should take care of this situation. Consequently, the use of GCNNs becomes a good approach for tracking, as this kind of network is able to generate new process units (neurons) when needed, in order to get a better representation of the input space.

García-Rodríguez et al. [18] use a kind of self-organizing network, the GNG, to represent non-rigid objects as a result of an adaptive process by a topology-preserving graph that constitutes an induced Delaunay triangulation of their shapes. The neural network is used to build a system able to track image features in video image sequences. The system automatically keeps correspondence of features among frames in the sequence, following the dynamic of the net and using the neurons to predict and readjust the representation among frames.

Other specific tracking problems, such as lips tracking [37, 48] or deformable contour tracking [12], have also been tackled based on neural networks.

3.2.3 Crowd and Traffic Density Estimation

People counting and crowd density estimation are crucial and challenging problems in visual surveillance. An accurate and real-time estimation of people in a shopping mall can provide valuable information for managers. Automatic monitoring of the number of people in public areas is also important for safety control and urban planning [22]. The behavioral analysis of crowded scenes can be used to develop crowd management strategies, to provide guidelines for the design of public spaces, and to validate or increase the performance of the mathematical models used in crowd simulations [23]. Density estimation is of interest not only for crowd analysis, but also to obtain useful information for traffic management, such as real-time traffic density and number of vehicle types moving along the roads [41]. Also, this video surveillance task can benefit from the adoption of neural networks, as shown by related literature.

Cho et al. [9] present a method based on neural networks to estimate the crowd density in subway stations. Estimation is carried out by extracting a set of significant features from sequences of images. Those feature indexes are modeled by a single hidden layer neural network to estimate the crowd density. The approach is implemented using a hybrid method for global learning that combines the least-squares method with three types of global optimization approaches, which are capable of providing the global search characteristic and fast convergence speed.

Kong et al. [29] present a method based on learning to estimate the number of people in crowds. Edge orientation and the histogram of the object areas (extracted from foreground objects through a background subtraction algorithm) are used as image features. A normalization procedure is performed to account for camera perspective, and the training model used to relate the detected features with the number of people is based on a feed-forward neural network. Because features are normalized, eventual changes in the camera setup do not require a new training phase.

Hou and Pang [22] present a method for estimating the number of people, based on a neural network, and locate each individual, based on the EM algorithm, in complicated scenes. The relationship between foreground pixels and the number of people is determined via a neural network adopting three different methods, based on manually annotated training images from a similar scene.

Ozkurt and Camci [41] use feed-forward neural networks for the identification of vehicle types (i.e., big, medium, and small vehicles, or not a vehicle) for the purpose of traffic density estimation.

3.2.4 Anomaly Detection and Behavior Understanding

In automated visual surveillance applications, detection of anomalous and suspicious human behaviors is of great practical importance in order to alert relevant authorities for attention. This task involves modeling and classification of human behaviors with certain rules. However, modeling human behavior is not a trivial task, as the observed input space of human movements can be very large due to the apparent randomness and complexity in human behavior. ANN classifiers can perform well, because these stateless data-modeling methods can be trained heuristically to nonlinearly partition the input space into some discrete states of the human movements, and can then be used to classify them appropriately [24].

Examples of effective use of neural network-based classifiers for video-based behavior understanding applications include research in [44–46] by Sacchi et al., where multi-layer perceptron neural networks with back-propagation learning rule are employed for recognizing abandoned objects in unmanned railway stations, counting the number of persons walking through a tourist passage point, and detecting vandal behaviors in metro stations, respectively. Neural networks are thus confirmed as useful tools in the implementation of advanced video-surveillance systems able in assisting human operators even in the recognition of complex dangerous situations.

Owens and Hunter [39] use a self-organizing feature map to learn normal trajectory patterns. While classifying trajectories, if the distance of the trajectory to its allocated class exceeds a threshold value, the trajectory is identified as anomalous. Therefore the approach is largely model-free: there is no explicit modeling of normal or abnormal behavior, which is instead learned by the neural network. However, this also entails a high sensitivity of the approach to training data.

A hierarchical self-organizing neural network is described by Owens et al. [40] for the detection of unusual pedestrian behavior in video-based surveillance systems. The system is trained on a normal dataset, with no prior information about the scene under surveillance, thereby requiring minimal user input. Nodes use a trace activation rule and feed-forward connections, modified so that higher layer nodes are sensitive to trajectory segments traced across the previous layer. Top layer nodes have binary lateral connections and corresponding *novelty accumulator* nodes. Lateral connections are set between co-occurring nodes, generating a signal to prevent the accumulation of the novelty measure along normal sequences. In abnormal sequences, the novelty accumulator nodes are allowed to increase their activity, generating an alarm state.

Jan et al. [24] introduce a data-based modeling neural network, the Modified Probabilistic Neural Network, which nonlinearly partitions the decision space, consisting of trajectory-related information regarding the velocity of people head, in order to achieve reliable classification, however still with acceptable computational requirements.

Khalid [26] presents a classification technique for motion activity and anomaly detection using object motion trajectories. A novel mechanism is proposed for modeling various patterns that are present in motion dataset, where a pattern is modeled by a set of cluster centers of mutually disjunctive sub-classes, referred to as *mediods*, within the pattern. The algorithm for identification of mediods is based on the adaptation of the neural gas-based learning rule. The resulting models of identified patterns can then be used to classify new unseen trajectory data to one of the modeled classes.

3.3 Neural-Based Moving Object Detection: SOBS

In this and the next section we resume some of the activities related to video surveillance that we carried out in the past few years, whose aim is the analysis and design of algorithms for detection, tracking, classification, and recognition of moving objects into digital image sequences, and their applications. The basic idea consists of exploiting the available knowledge concerning the self-organized learning behavior of the brain, which is the foundation of human visual perception, traducing it into models and algorithms that can accurately solve the afforded problems. The main staple, common to the reported research, is a self-organizing neural network that has proven to accurately model image sequences and their variations in time. This provides a background model particularly suitable for moving object detection, allowing us to robustly deal with typical problems of background subtraction, whose applications also span stopped object detection, activity recognition, and anomaly detection.

3.3.1 Rationale

Human visual perception and the brain make up the most complex cognition system and the most complex of all biological organs. Visual inputs contribute to most parts of the total information from all kinds of sensors (e.g., visual-, auditory-, motor-, or somato-sensory) entering the brain. Visual information is processed in both the retina and brain, but it has been widely verified that most processing is done in the retina, such as extracting lines, angles, curves, contrasts, colors, and motion. The retina is a complex neural network, consisting of more than 100 million photosensitive cells that process in parallel the raw images, encode the information, and send it to the brain cortex.

In such neural networks there exist both short-range excitatory interactions between nearby cells, due to horizontal connections among them, as well as long-range inhibitory interactions between distant neighbors, due to lateral inhibition.

External stimuli received by various sensors are coded by the living neural networks and projected through axons onto the cerebral cortex, often to distinct parts of the cortex. Therefore, the different areas of the cortex (cortical maps) often correspond to different sensory inputs, though some brain functions require collective responses. Topographically ordered maps are widely observed in the cortex. The main structures (primary sensory areas) of the cortical maps are established before birth in a predetermined topographically ordered fashion. Other more detailed areas (associative areas), however, are developed through self-organization gradually during life and in a topographically meaningful order.

Many are the approaches concerning the study of self-organized learning behavior of brains, including Hebb's learning law [21]; Marr's theory of the cerebellar cortex [36]; Willshaw, Buneman, and Longnet-Higgins's nonholographic associative memory [49]; Gaze's studies on nerve connections [19]; von der Malsburg and Willshaw's self-organizing model of retina-cortex mapping [50]; Amari's mathematical analysis of self-organization in the cortex [1]; Kohonen's self-organizing map [27]; and Cottrell and Fort's self-organizing model of retinotopy [11]. Of special interest is the basic idea developed by Von der Malsburg and Willshaw [50]:

> ...the geometrical proximity of presynaptic cells is coded in the form of correlations in their electrical activity. These correlations can be used in the postsynaptic sheet to recognize axons of neighboring presynaptic cells and to connect them to neighboring postsynaptic cells, hence producing a continuous mapping ...

Based on these considerations, we proposed to adopt a biologically inspired problem-solving method based on visual attention mechanisms [32]. The aim is to obtain the objects that keep the user attention in accordance with a set of predefined features, including gray level, motion, and shape features. Our approach defines a method for the generation of an active attention focus to monitor dynamic scenes for surveillance purposes. The idea is to build the background model by learning in a self-organizing manner many background variations, that is, background motion cycles, seen as trajectories of pixels in time. Based on the learned background model through a map of motion and stationary patterns, our algorithm can detect motion and selectively update the background model. Specifically, a novel neural network mapping method is proposed in which one uses a whole trajectory incrementally in time fed as an input to the network. This makes the network structure much simpler and the learning process much more efficient. The achieved background model is particularly suitable for moving object detection, allowing us to robustly deal with typical problems of background subtraction.

3.3.2 SOBS Model

In [32], moving object detection is achieved through background subtraction, where the background model is built and updated learning in a self-organizing manner background variations, seen as trajectories of pixels in time. Variants of the basic 2D neural model [32] allow us to handle uncertainty deriving from the need to choose model parameter values [34] and to achieve an even more compact and significant representation of an image sequence [33].

The self-organizing neural network is arranged as a 2-D flat grid of neurons. Each neuron computes a function of the weighted linear combination of incoming inputs, with weights resembling the neural network learning, and can be therefore represented by a weight vector obtained by collecting the weights related to incoming links. An incoming pattern is mapped to the neuron whose set of weight vectors is most similar to the pattern, and weight vectors in a neighborhood of this node are updated; such learning of the neuronal map allows us to adapt the neural model to scene modifications.

Specifically, for each pixel $\mathbf{x}$ we build a neuronal map consisting of $n \times n$ weight vectors $m_0^i(\mathbf{x}), i = 1, \ldots, n^2$, where each weight vector is a 3D vector initialized to the color components of the corresponding pixel of the first sequence frame I_0. An example of such a neuronal map is given in Figure 3.1, where for pixel $\mathbf{x}$, identified by the square on the left, we have the nine weight vectors $m_i = m_0^i(\mathbf{x}), i = 1, \ldots, 9$ (choosing $n = 3$), arranged as the 2D neural model shown on the right.

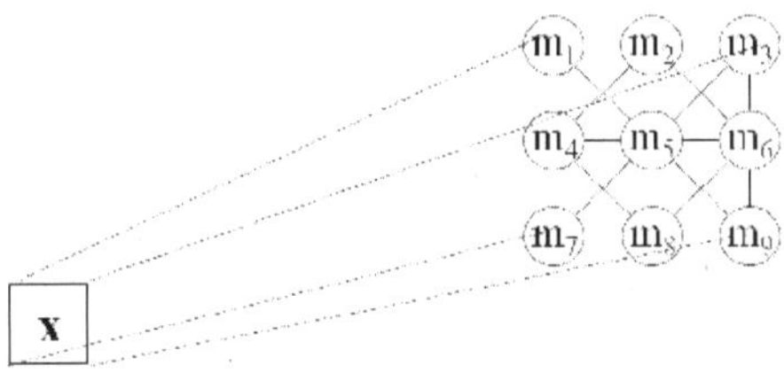

FIGURE 3.1 The 2D neuronal map for pixel $\mathbf{x}$.

The complete set of weight vectors for all pixels of an image I_0 with N rows and M columns is represented as a 2D neuronal map B_0 with $n \times N$ rows and $n \times M$ columns, where adjacent blocks of $n \times n$ weight vectors correspond to adjacent pixels in image I_0.

By subtracting the current image I_t from the background model B_{t-1} at each subsequent time instant t, every pixel $\mathbf{x}$ of I_t is compared to the current pixel weight vectors $(m_{t-1}^1(\mathbf{x}), \ldots, m_{t-1}^{n^2}(\mathbf{x}))$ to determine the weight vector $m_{t-1}^{BM}(\mathbf{x})$ that best matches it according to a metric $d(\cdot)$:

$$d(m_{t-1}^{BM}(\mathbf{x}), I_t(\mathbf{x})) = \min_{i=1,\ldots,n^2} d(m_{t-1}^i(\mathbf{x}), I_t(\mathbf{x})). \tag{3.1}$$

The best matching weight vector is used as the pixel's encoding approximation, and therefore $\mathbf{x}$ is detected as foreground if the distance in Equation (3.1) exceeds a background/foreground segmentation threshold ϵ; otherwise it is classified as background.

Learning of the neuronal map, allowing to adapt the background model B_{t-1} to scene modifications, is achieved by updating for background pixels

$\mathbf{x}$ the best matching weight vector $m_{t-1}^{BM}(\mathbf{x})$, supposed to be found in position $\mathbf{z}$ of background model B_{t-1}, together with other weight vectors in its neighborhood $N_{\mathbf{z}}$ of B_{t-1} according to the weighted running average:

$$B_t(\mathbf{y}) = (1 - \alpha_t(\mathbf{y}, \mathbf{z}))B_{t-1}(\mathbf{y}) + \alpha_t(\mathbf{y}, \mathbf{z})I_t(\mathbf{x}), \quad \forall \mathbf{y} \in N_{\mathbf{z}}. \tag{3.2}$$

Here, $\alpha(\mathbf{y}, \mathbf{z}) = \gamma_t \cdot G(\mathbf{y}\text{-}\mathbf{z})$, where γ_t represents the learning rate, chosen as a monotonically decreasing function during model training to guarantee the convergence of the neural network to the background model, and as a constant value depending on scene variability during the adaptation. Gaussian weights given by the Gaussian function $G(\cdot) = G(\cdot; 0, \Sigma^2)$ allow us to reinforce the contribution to the background model of the best matching weight vector, at the same time smoothly taking into account spatial relationships between the incoming pixel $\mathbf{x}$ and weight vectors of pixels in a neighborhood of $\mathbf{x}$.

(a) (b)

FIGURE 3.2 Moving object detection on sequence *Msa*: (a) original frame and (b) result of 2D-SOBS algorithm.

In Figure 3.2 we report achieved results on sequence *Msa* (publicly available in the Download section of `http://cvprlab.uniparthenope.it`), where we can observe that foreground objects (the man and the bag) are almost perfectly detected.

Model uncertainty deriving from the need for choosing model parameter values has been handled in [34] by relating the learning factor in Equation (3.2) to the relative distance of the best matching weight vector from the current pixel value, so that the more the current pixel is well modeled by the background model, the more it contributes to the model update. Moreover, the learning factor has also been related to a spatial coherency factor [14], which gives an indication of the percentage of image pixels adjacent to current pixel under examination that are well modeled by the neural network. The resulting 2D-SOBS_CF algorithm allows us to enhance detection results in terms of less false negatives and false positives than the 2D-SOBS algorithm. This can be exemplified in Figure 3.3, where the strong unsteadiness in sequence *Seq00* (available at `http://www.cs.cmu.edu/~yaser/Data/`) leads to some false positives due to an excessively moving background, and the camouflage of the gray car with the gray street leads to some false negatives in 2D-SOBS results (see Figure 3.3-(center)). Spatial coherency and uncertainty handling, instead, allow us to obtain more accurate detection results (see Figure 3.3-(right)).

FIGURE 3.3 Moving object detection on sequence *Seq00*: original frame (left), and results of 2D-SOBS algorithm (center) and of 2D-SOBS_CF algorithm (right).

A natural evolution of the 2D neural model is a 3D self-organizing neural model [33]. This can be visualized as a set of images contained in n layers that gives an even more compact representation of an image sequence, still preserving topological properties of the input patterns.

Specifically, given an image sequence $I_0, \ldots, I_{T-1}$, for each pixel $\mathbf{x}$ we build a neuronal map consisting of n weight vectors $m_t^i(\mathbf{x}), i = 1, \ldots, n$. The complete set of weight vectors $\{m_t^1(\mathbf{x}), \ldots, m_t^n(\mathbf{x})\}$ for all pixels $\mathbf{x}$ of the t-th sequence frame is organized as a 3D neuronal map $\mathcal{M}_t$ with n layers.

Initialization and learning of the 3D neural model are carried out in a manner similar to that of the 2D neural model. However, the model update here is done not only in a 2D intra-layer neighborhood of the best matching weight vector — smoothly preserving spatial relationships of input pixels — but also in a 1D neighborhood of weight vectors for the current pixel in adjacent model layers — smoothly reinforcing their contribution to the model.

Comparisons with several other existing methods on different real image sequences have been carried out not only in [32–34], but also by other authors (e.g., [4, 6, 7, 30, 32]). One of the most recent comparisons has been given by Brutzer et al. [4], who evaluated the performance of nine methods for background subtraction with respect to the challenges of video surveillance. Even though each of the tested background subtraction approaches did exhibit drawbacks in some experiments, 2D-SOBS algorithm was found to be one of the most promising methods for different background subtraction issues, also due to its regional diffusion of background information in the update step.

3.4 SOBS for Other Video Surveillance Tasks

3.4.1 Stopped Object Detection

The neural model described in the previous section has also been adopted for the detection of stopped objects, that is, objects that enter the scene as foreground objects and then stop into it [33]. The proposed *Stopped Foreground Subtraction* approach is based on the ideá of adopting, besides the usual background model $\mathcal{B}_t$, also a foreground model $\mathcal{F}_t$, to be used for detecting a prolonged permanence in the scene of stationary foreground objects.

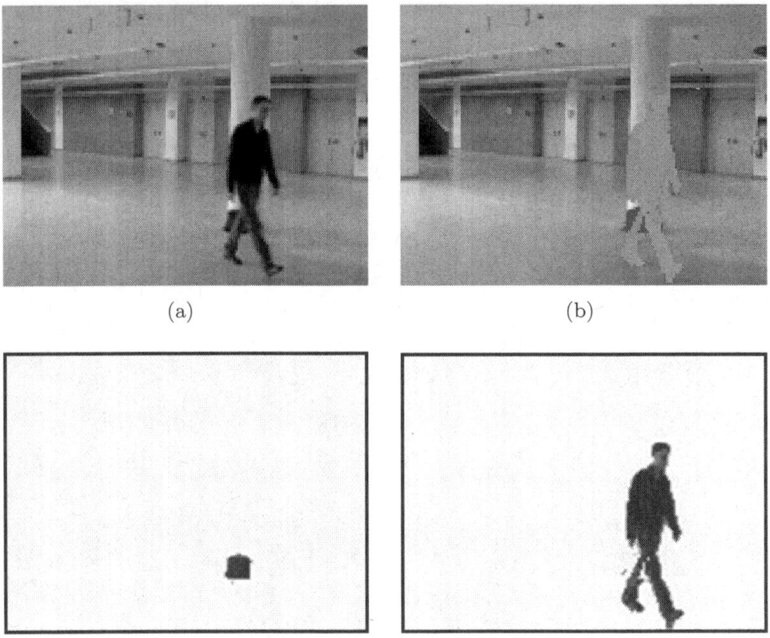

FIGURE 3.4 Stopped Foreground Subtraction for sequence *Msa*: (a) original frame, (b) original frame with moving (green) and stopped (red) foreground objects, (c) moving foreground model, and (d) stopped foreground model. (See color insert.)

A suitable layering mechanism for the obtained models allows us to handle cases where several stopped or moving objects appear as overlapped, as in Figure 3.4 the moving person who passes in front of the stationary bag of sequence *Msa*. The usual moving object detection mask would have not allowed the distinction of the two objects. In Figure 3.4-(c) and (d), we report the built moving foreground and the stopped foreground 3D neural models adopted for the segmentation.

The model-based approach is independent of the chosen model and allows us to obtain an initial segmentation of moving and stopped objects as an inexpensive by-product of moving object detection, as it requires only a small overhead compared to background subtraction alone. The approach allows us to obtain high correct detection rates for stationary objects, also as compared to higher level approaches based on tracking.

Moreover, in [20] we adopted a dual background approach to stopped object detection in digital image sequences based on our 2D neural self-organizing model, proposing a data parallel algorithm, specifically suitable for SIMD architectures. Parallel performance of our CUDA GPU implementation can be considered quite satisfactory, because we achieved significant speedup as compared to our serial implementations, allowing for real-time stopped object detection. Selected results are reported in Table 3.1.

TABLE 3.1 Results of the data parallel algorithm for stopped object detection: sequential (T_{SEQ}) and parallel (T_{PAR}) times (in ms), speedup, and number of frames per second (fps), varying the image size and the optimal block size.

Image Size	Block Size	T_{SEQ}	T_{PAR}	Speedup	fps
720 × 480	16 × 8	431.68	20.55	21.0x	48.64
960 × 720	8 × 8	862.94	40.96	21.0x	24.90
1200 × 960	16 × 8	1430.30	65.41	21.8x	15.00

3.4.2 Activity Recognition

Concerning human activity recognition, a recently proposed texture descriptor [28], used for building visual words according to a Bag-of-Words approach to the identification of spatio-temporal interest points, is adopted in order to achieve high accuracy in activity classification.

Different from other part-based approaches, points of interest are searched only in foreground areas, where foreground is achieved by 2D-SOBS, as shown in Figure 3.5(a) for sequence *run* of the Weizmann dataset [3].

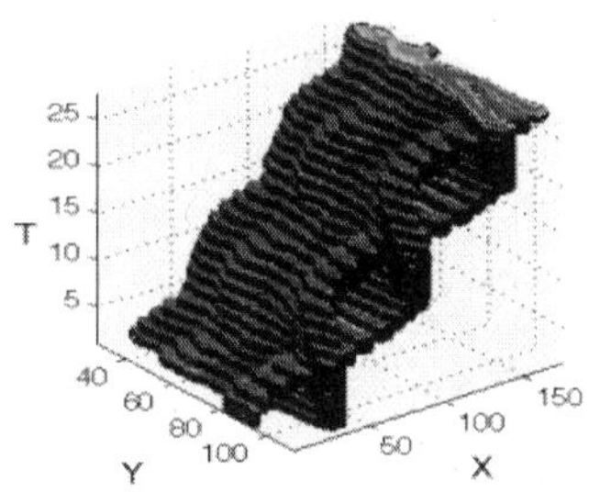

(a)

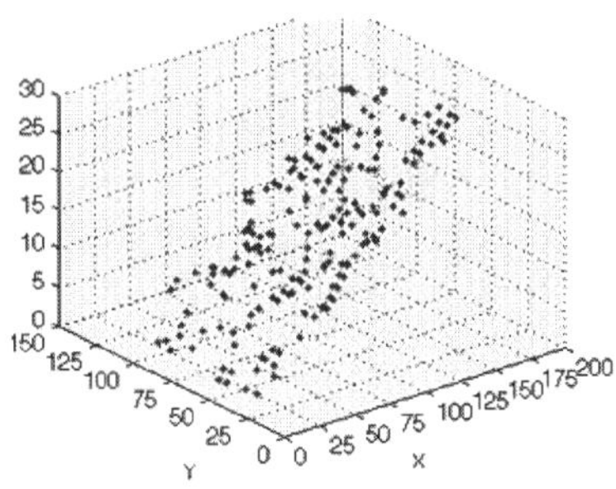

(b)

FIGURE 3.5 Results for sequence *run*: volumetric view of (a) foreground search areas and (b) detected interest points.

The feature extraction from interest points consists of a *detector*, where the spatio-temporal interest points are detected by applying separately spatial linear filters along both the spatial dimensions and the temporal dimension as in [2], and of *descriptors* to describe identified interest points. Besides the texture descriptor in [28], we considered also the 3D image gradient proposed in [2], as well as the combination of the two.

A reduction method based on the k-means algorithm allows us to cluster all interest points belonging to a particular type of activity into k clusters. The number of points of interest is thus reduced to k for each activity, and a video sequence is then represented by a histogram.

The classification is finally performed using an SVM classifier with linear kernel, extended to a multiclass classifier. Given the training set of video sequences, an SVM is constructed for each class of activity. The complete dataset of 93 video sequences has been divided into two parts: the training set for the SVM multiclass, and the test set for testing the generalization abilities. Table 3.2 reports the achieved results, using individually the descriptor based on 3D gradient and the textural descriptor, and using the concatenation of both methods. Such results show that the adoption of the textural descriptor alone leads to the highest performance.

TABLE 3.2 Performance results of the proposed method for activity recognition, varying the adopted descriptor and the relative dimension of the training set on the complete Weizmann dataset.

Descriptor	40%	50%	60%
3D gradient	92.05%	95.56%	97.25%
Texture	94.49%	98.93%	99.02%
3D gradient + texture	92.48%	97.75%	98.86%

3.4.3 Anomaly Detection

The 2D-SOBS neural model is being adopted for the problem of anomaly detection based on trajectory descriptors [35]. Given a large number of moving object trajectories computed in digital image sequences, the goal of trajectory-based motion learning is to learn a model that is capable of detecting *normal* motion patterns, while identifying instances representing *anomalous* behaviors. In this context, *anomalous* behavior stands for atypical behavior patterns that are not represented by sufficient samples in training data and are infrequently occurring or unusual [26].

Specifically, trajectory extraction is achieved by moving object detection based on 2D-SOBS, blob detection, and mean shift tracking [10]. Example results are provided in Figure 3.6.

For the trajectories of each moving object, we adopt the representation scheme described in [26], based on time series representation and modified DFT.

(a) (b) (c)

FIGURE 3.6 Trajectory extraction: (a) sequence frame no. 202, (b) moving object mask for sequence frame no. 202 computed by 2D-SOBS, and (c) trajectory obtained by Mean Shift tracking.

Modeling of trajectory-based motion patterns is achieved by *m-mediods modeling* [26], which models a class containing n members with m mediods (set of cluster centers of mutually disjunctive sub-classes) known *a priori*.

The classification of a new trajectory is performed by checking the closeness of the trajectory to the models of different classes, and assigning it to the class containing most of closest mediods. If the new trajectory is reasonably close to the closest activity pattern, then it is classified as *normal trajectory*; otherwise, it is considered an *anomalous trajectory*. The criterion is based on the merged anomaly detection procedure described in [26], where a significance parameter τ — ranging from 1 to m — determines the sensitivity of the algorithm to anomalies.

For the experiments, we prepared a dataset consisting of 120 labeled video sequences for indoor surveillance, where 100 sequences have normal trajectories (60 for training and 40 for testing) and 20 sequences present anomalous trajectories. Because the conference room under surveillance should not be entered through the main door, but only through lateral doors, trajectories that are considered anomalous are those entering the main door. Examples of normal and anomalous trajectories are reported in Figure 3.7.

(a) (b)

FIGURE 3.7 Anomaly detection: (a) normal trajectory and (b) anomalous trajectory.

Fixing $m = 20$ and varying the significance parameter τ, we obtained the ROC curve of Figure 3.8(a) and the Precision-Recall curve of Figure 3.8(b). Optimal values for the threshold τ are greater than 6. Choosing $\tau = 8$, we

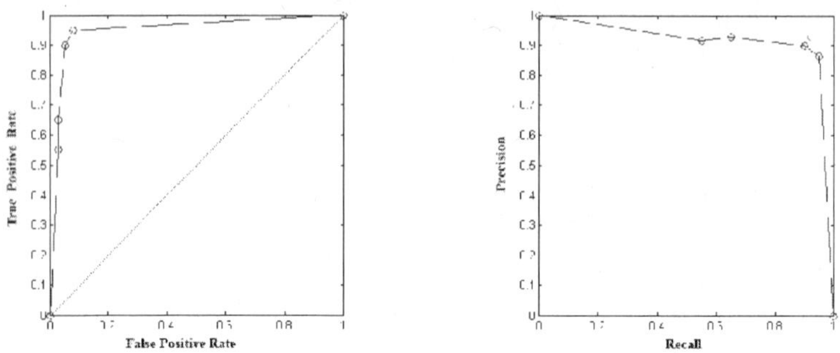

(a) (b)

FIGURE 3.8 Anomaly detection accuracy: (a) ROC curve and (b) Precision-Recall curve.

achieved the classification results reported in the confusion matrix of Table 3.3, from which we deduce an accuracy of 0.95%.

TABLE 3.3 Confusion matrix for anomaly detection.

	Ground Truth	
	Normal	Anomalous
Normal	39	1
Anomalous	2	18

3.5 Conclusions

In this chapter we have presented some examples of neural network-based approaches to the solution of video surveillance tasks provided in the literature, including moving object detection and tracking, crowd and traffic density estimation, anomaly detection, and behavior understanding. A specific neural-based approach has been further described in order to give evidence of advantages of its adoption for moving object detection, showing also possible uses in the context of other video surveillance tasks, such as stopped object detection, activity recognition, and anomaly detection.

Acknowledgments

This work has been partially supported by the FIRB Project IntelliLogic No. RBIP06MMBW of the Italian Research and Education Ministry and by the

AMIAV Project of the Campania Regional Board.

References

1. S.I. Amari. Topographic organisation of nerve fields. *Bulletin Mathematical Biology*, 42:339–364, 1980.
2. L. Ballan, M. Bertini, A. Del Bimbo, L. Seidenari, and G. Serra. Recognizing human actions by fusing spatio-temporal appearance and motion descriptors. In *Proc. of IEEE International Conference on Image Processing*, ICIP'09, pages 3569–3572, Cairo, Egypt, November 2009.
3. M. Blank, L. Gorelick, E. Shechtman, M. Irani, and R. Basri. Actions as space-time shapes. In *Proc. Tenth IEEE International Conference on Computer Vision*, ICCV'05, 2:1395–1402, Oct. 2005.
4. S. Brutzer, B. Hoferlin, and G. Heidemann. Evaluation of background subtraction techniques for video surveillance. In *Proc. of Computer Vision and Pattern Recognition*, CVPR'11, pages 1937–1944, 2011.
5. V. Cantoni and A. Petrosino. 2-D object recognition by structured neural networks in a pyramidal architecture. In *Proceedings of the Fifth IEEE International Workshop on Computer Architectures for Machine Perception*, pages 81–86, 2000.
6. M. Chacon and S. Gonzalez. An adaptive neural-fuzzy approach for object detection in dynamic backgrounds for surveillance systems. *IEEE Transactions on Industrial Electronics*, PP(99):1, 2011.
7. M.I.M. Chacon, G.D. Sergio, and V.P. Javier. Simplified SOM-neural model for video segmentation of moving objects. In *Proceedings of International Joint Conference on Neural Networks*, IJCNN'09, pages 474–480, June 2009.
8. B. Chen, W. Wang, and Q. Qin. Infrared target detection based on fuzzy art neural network. In *Proceedings of the Second International Conference on Computational Intelligence and Natural Computing*, CINC'10, 2:240–243, Sept. 2010.
9. S.-Y. Cho, T.W.S. Chow, and C.-T. Leung. A neural-based crowd estimation by hybrid global learning algorithm. *IEEE Transactions on Systems, Man, and Cybernetics, Part B: Cybernetics*, 29(4):535–541, Aug. 1999.
10. D. Comaniciu and P. Meer. Mean shift: A robust approach toward feature space analysis. *IEEE Transactions on Pattern Analysis and Machine Intelligence*, 24(5):603–619, May 2002.
11. M. Cottrell and J.C. Fort. A stochastic model of retinotopy: A self organizing process. *Biol. Cybern.*, 53:405–411, April 1986.
12. A.M. Cretu, E.M. Petriu, P. Payeur, and F.F. Khalil. Deformable object segmentation and contour tracking in image sequences using unsupervised networks. In *Proc. of Canadian Conference Computer and Robot Vision*, pages 277–284. IEEE, 2010.
13. D. Culibrk, O. Marques, D. Socek, H. Kalva, and B. Furht. Neural network

approach to background modeling for video object segmentation. *IEEE Transactions on Neural Networks*, 18(6):1614–1627, Nov. 2007.

14. J. Ding, R. Ma, and S. Chen. A scale-based connected coherence tree algorithm for image segmentation. *IEEE Transactions on Image Processing*, 17(2):204–216, Feb. 2008.
15. G. Dong and M. Xie. Color clustering and learning for image segmentation based on neural networks. *IEEE Transactions on Neural Networks*, 16(4):925–936, July 2005.
16. G. A. Drumea and H. Frezza-Buet. Tracking fast changing non-stationary distributions with a topologically adaptive neural network: Application to video tracking. In *Proceedings of the 15th European Symposium on Artificial Neural Networks*, ESANN 2007, pages 43–48, Bruges, Belgium, April 25–27, 2007.
17. E. Fauske, L.M. Eliassen, and R.H. Bakken. A comparison of learning based background subtraction techniques implemented in cuda. In *Proc. of Norwegian Artificial Intelligence Symposium*, pages 181–192, 2009.
18. J. Garcia-Rodriguez, F. Florez-Revuelta, and J. Garcia-Chamizo. Visual surveillance of objects motion using GNG. In S. Omatu, M. Rocha, J. Bravo, F. Fernndez, E. Corchado, A. Bustillo, and J. Corchado, Editors, *Distributed Computing, Artificial Intelligence, Bioinformatics, Soft Computing, and Ambient Assisted Living*, volume 5518 of *Lecture Notes in Computer Science*, pages 244–247. Springer, Berlin/Heidelberg, Germany, 2009.
19. R. M. Gaze. *The Formation of Nerve Connections*. Academic Press, London, United Kingdom, 1970.
20. G. Gemignani, L. Maddalena, and A. Petrosino. Real-time stopped object detection by neural dual background modeling. In *Lecture Notes in Computer Science*, 6586:327–334. Springer, Berlin/Heidelberg, Germany 2011.
21. D.O. Hebb. *The Organization of Behavior*. Wiley & Sons, New York, 1949.
22. Y.-L. Hou and G.K.H. Pang. People counting and human detection in a challenging situation. *IEEE Transactions on Systems, Man and Cybernetics, Part A: Systems and Humans*, 41(1):24–33, Jan. 2011.
23. J.C.S. Jacques Junior, S.R. Musse, and C.R. Jung. Crowd analysis using computer vision techniques. *IEEE Signal Processing Magazine*, 27(5):66–77, Sept. 2010.
24. T. Jan, M. Piccardi, and T. Hintz. Neural network classifiers for automated video surveillance. In *Proceedings of the IEEE 13th Workshop on Neural Networks for Signal Processing*, NNSP'03, pages 729–738, Sept. 2003.
25. Y. Jiang, K.-J. Chen, Z.-H. Zhou, Y. Jiang, K.-J. Chen, and Z.-H. Zhou. SOM based image segmentation. In *Lecture Notes in Artificial Intelligence*, 2639:640–643. Springer, 2003.
26. S. Khalid. Activity classification and anomaly detection using m-mediods based modelling of motion patterns. *Pattern Recognition*, 43(10):3636–3647,

2010.

27. T. Kohonen. Self-organised formation of topologically correct feature map. *Biol. Cybern.*, 43:56–69, 1982.
28. S. Kondra and V. Torre. Texture classification using three circular filters. In *Proceedings of the Sixth Indian Conference on Computer Vision, Graphics Image Processing*, ICVGIP '08, pages 429–434, Dec. 2008.
29. D. Kong, D. Gray, and H. Tao. A viewpoint invariant approach for crowd counting. In *Proceedings of the 18th International Conference on Pattern Recognition*, ICPR'06, 3:1187–1190, 2006.
30. E. Lopez-Rubio, R.M. Luque-Baena, and E. Dominguez. Foreground detection in video sequences with probabilistic self-organizing maps. *International Journal of Neural Systems*, 21(3):225–246, 2011.
31. R. M. Luque, J.M. Ortiz-De-Lazcano-Lobato, E. Lopez-Rubio, and E. J. Palomo. Object tracking in video sequences by unsupervised learning. In *Proceedings of the 13th International Conference on Computer Analysis of Images and Patterns*, CAIP '09, pages 1070–1077, Berlin, Heidelberg, 2009. Springer-Verlag.
32. L. Maddalena and A. Petrosino. A self-organizing approach to background subtraction for visual surveillance applications. *IEEE Transactions on Image Processing*, 17(7):1168–1177, July 2008.
33. L. Maddalena and A. Petrosino. 3D neural model-based stopped object detection. In *Proceedings of the 15th International Conference on Image Analysis and Processing*, ICIAP '09, pages 585–593. Springer-Verlag, 2009.
34. L. Maddalena and A. Petrosino. A fuzzy spatial coherence-based approach to background/foreground separation for moving object detection. *Neural Comput. Appl.*, 19:179–186, March 2010.
35. L. Maddalena and A. Petrosino. Bridging neural visual information processing and video surveillance. Springer-Verlag, 2012. To appear.
36. D. Marr. A theory of cerebellar cortex. *J. Physiol.*, 202(2):437–470, 1969.
37. M. K. Moghaddam and R. Safabakhsh. TASOM-based lip tracking using the color and geometry of the face. In M.A. Wani, M.G. Milanova, L.A. Kurgan, M. Reformat, and K. Hafeez, Editors, *Proc. of Fourth International Conference on Machine Learning and Applications*, ICMLA '05. IEEE Computer Society, 2005.
38. S. H. Ong, N. C. Yeo, K. H. Lee, Y. V. Venkatesh, and D. M. Cao. Segmentation of color images using a two-stage self-organizing network. *Image Vision Comput.*, pages 279–289, 2002.
39. J. Owens and A. Hunter. Application of the self-organising map to trajectory classification. In *Proceedings of the Third IEEE International Workshop on Visual Surveillance*, pages 77–83, 2000.
40. J. Owens, A. Hunter, and E. Fletcher. Novelty detection in video surveillance using hierarchical neural networks. In J. Dorronsoro, Editor, *Artificial Neural Networks ICANN 2002*, Vol. 2415 of *Lecture Notes in Computer Science*, page 140. Springer Berlin / Heidelberg, 2002.

41. C. Ozkurt and F. Camci. Automatic traffic density estimation and vehicle classification for traffic surveillance systems using neural networks. *Mathematical and Computational Applications*, 14(3):187–196, 2009.
42. G. Pajares. A Hopfield neural network for image change detection. *IEEE Transactions on Neural Networks*, 17(5):1250–1264, Sept. 2006.
43. S.K. Pal and S.C.K. Shiu. *Foundations of Case-Based Reasoning.* John Wiley & Sons, Inc., 2004.
44. C. Sacchi, G. Gera, L. Marcenaro, and C.S. Regazzoni. Advanced image-processing tools for counting people in tourist site-monitoring applications. *Signal Process.*, 81:1017–1040, May 2001.
45. C. Sacchi and C.S. Regazzoni. A distributed surveillance system for detection of abandoned objects in unmanned railway environments. *IEEE Transactions on Vehicular Technology*, 49(5):2013–2026, Sept. 2000.
46. C. Sacchi, C.S. Regazzoni, and G. Vernazza. A neural network-based image processing system for detection of vandal acts in unmanned railway environments. In *Proceedings of the 11th International Conference on Image Analysis and Processing*, pages 529–534, Sept. 2001.
47. A. J. Schofield, P. A. Mehta, and T. J. Stonham. A system for counting people in video images using neural networks to identify the background scene. *Pattern Recognition*, 29(8):1421–1428, 1996.
48. H. Shah-Hosseini and R. Safabakhsh. A TASOM-based algorithm for active contour modeling. *Pattern Recognition Letters*, 24(9-10):1361–1373, 2003.
49. D. Willshaw, O.P. Buneman, and H. Longnet-Higgins. Non-holographic associative memory. *Nature*, 222:960–962, 1969.
50. D. J. Willshaw and C. Von Der Malsburg. How patterned neural connections can be set up by self-organization. *Proc. Roy. Soc. London B*, 194(2):431–445, 1976.
51. M.-H. Yang, D.J. Kriegman, and N. Ahuja. Detecting faces in images: a survey. *IEEE Transactions on Pattern Analysis and Machine Intelligence*, 24(1):34–58, Jan. 2002.
52. W. Zhiming, Z. Li, and B. Hong. Pnn based motion detection with adaptive learning rate. In *Proceedings of the International Conference on Computational Intelligence and Security*, CIS '09, 1:301–306, Dec. 2009.

4

Video Summarization and Significance of Content: A Review

Rajarshi Pal
Indian Statistical Institute, Kolkata, India

Ashish Ghosh
Indian Statistical Institute, Kolkata, India

Sankar K. Pal
Indian Statistical Institute, Kolkata, India

4.1 Introduction

The need to monitor the activities in public places arises from the urge to secure the lives of general public and important installations. Surveillance systems installed at various places produce large volumes of video data to

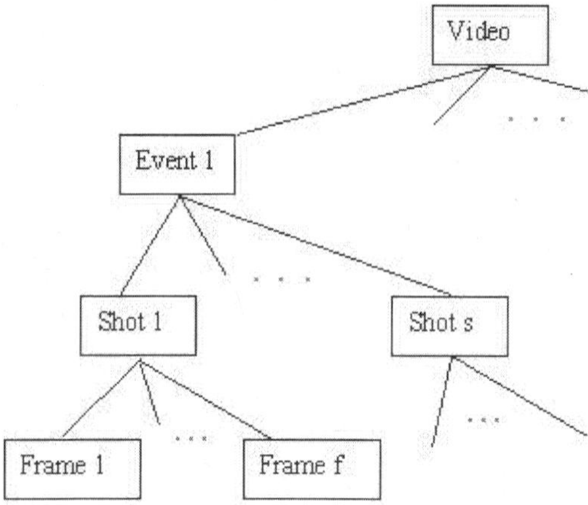

FIGURE 4.1 Various levels of units for video.

be scrutinized. Processing these huge chunks of video data to find out the possibility of any potential risk demands huge amounts of resources (in terms of time, man-power, and machine configurations). Video summarization plays an important role in this context. Moreover, summarization helps generate movie trailers, sports and news video highlights, and keeps records of interesting events for future inspection.

A video can be described as a time-ordered presentation of related events. An event is described in terms of interrelated objects and their activities. Each event may spread over one or more shots, which, in turn, is defined as a sequence of frames taken from one single camera in a continuous fashion and hence depicts a continuous action in time and space. A video can be analyzed using any of these levels of units. Various units are pictorially presented in Figure 4.1.

The summary of a video is a brief representation of it to convey the significant points in a concise form. Mathematically, a summary v of a video V can be defined with the help of units u_is as given in the following equations:

$$V = u_1 \oplus u_2 \oplus \ldots \oplus u_{n_V} \tag{4.1}$$

$$v = u'_{s_1} \oplus u'_{s_2} \oplus \ldots \oplus u'_{s_{n_v}}, \text{ where } n_v \leq n_V, \tag{4.2}$$

where u_i represents a unit in the original video and u'_i represents the same unit in the summarized video. u'_i is, at times (not always), a summarized version of the unit u_i. n_V and n_v denote the number of units at a particular level in the original video and its summary, respectively. $\oplus$ is the concatenation operator.

Two contradictory features — conciseness and high coverage value — characterize a good summary. It is concise enough to contain as small a content as possible. Moreover, its coverage is adequately high to retain all the significant contents. Therefore, judging the significance of the contents is a key challenge to generate a good video summary. In the absence of automatic semantic understanding of the contents of a video, researchers base their attempts on analyzing the contents on the basis of visual features, for example, color values of pixels, histogram of frames, motion vectors. More recently, few endeavors identify significant contents through attention modeling and semantics interpretation of events. As it is, in general, challenging to identify the semantic of a generic video, most of these approaches restrict themselves to a specific category — for example, sports video [4, 5, 8, 12], news video [6, 29, 54], and movie and drama [11, 49, 52, 57]. A few generic methods [45, 46, 61, 75] have been also proposed.

A survey on video summarization techniques can be found in [36]. It specifically focuses on summarizing videos of a specific genre, that is, movies. A systematic classification of video summarization techniques can also be found in [68]. This chapter reviews existing summarization techniques in the context of their relation to significant content identification. Suitability of these techniques for surveillance video summarization has been discussed. The issue of personalization is addressed. The chapter is organized as follows: Section 4.2 classifies the existing techniques depending on whether the concept of significant content has been utilized or not. Section 4.3 describes how several techniques produce summary despite skipping the significant content identification. Summarization schemes that apply knowledge about significant content are discussed in Section 4.4. Approaches to handling the personal nature of the definition of significance are mentioned in Section 4.5. Section 4.6 presents the future scope of research in the context of significant content identification for summarization. It also discusses several opportunities to exploit soft computing techniques for video summarization. Finally, Section 4.7 draws the conclusion.

4.2 Classification of Summarization Techniques

There may exist several criteria to classify existing video summarization techniques. In one such example [68], these techniques have been put into different categories depending on how the summaries are represented. Two main mechanisms to represent video summaries are key-frame based and skim based. Key-frame based techniques represent the summary as a collection of static frames and, hence, they cannot maintain the notion of continuity of the events in a video. On the contrary, video skims are just like a shorter version of the original video.

Our classification of the existing techniques is based on whether they depend on the detection of significant content within a video. A good summary retains the significant contents of a video. Identification of significant contents requires understanding the semantics of the video. It is a major challenge in automated analysis of video content. Though recently inroads have been made in domain-specific cases, the early approaches, even some recent ones too, are

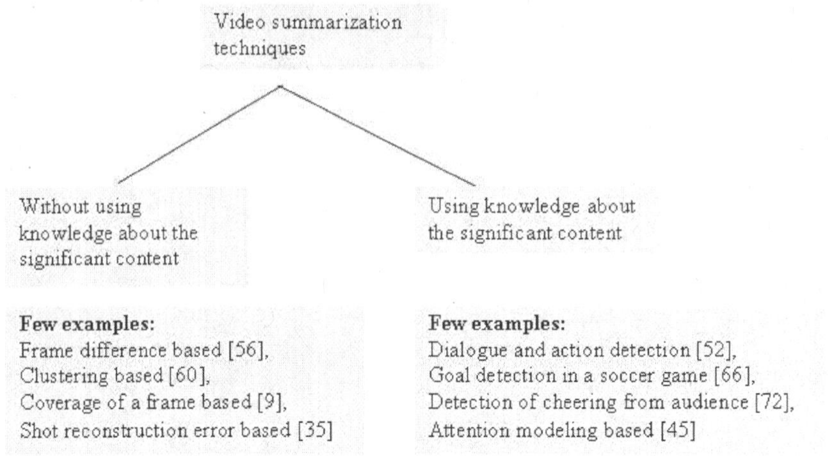

FIGURE 4.2 Categorization of video summarization techniques based on utilization of knowledge about significant content.

based on analysis of nonsemantic features of its content. From this perspective, video summarization techniques can be classified into two categories — techniques with and without knowledge of significant content (Figure 4.2).

Rudimentary techniques in this field involve segmentation of video into shots and the subsequent selection of a fixed set of key frames from each shot [62, 69]. Obviously, these techniques of the 1990's did not estimate the significance of the contents of a frame. Inter-frame difference-based techniques [23, 56], which capture intra-shot variation, have also been developed. Evolution from this category of techniques introduces clustering of similar frames/shots to select the representative frames/shots to be included in the summary [15, 26, 80]. Frames/shots are clustered on the basis of nonsemantic visual features like the histogram of a frame, average histogram of all frames in a shot. Some clustering-based approaches (for e.g., [74]) utilize the time-domain information along with visual features. Like clustering-based approaches, a few other summarization approaches also aim to reduce the redundancy by introducing concepts like coverage of a frame [9] and shot reconstruction error [35].

The second category of summarization techniques is based on the identification of significant content in the video. The definition of significance may vary across different domains. For example, dialogues and actions are assumed to be significant for movies [52]. Goal-scoring events are treated as significant in the case of a soccer video [66]. Therefore, the domain-dependent analysis of audio-visual features identifies the significant contents. Then these significant contents are retained in the summary and relatively less significant contents are discarded. There also exist few attempts to identify the significant content in a more generic way [45,61,75]. Few of these generic approaches employ attention models to understand parts of the video that draw human attention [45,61]. In some instances, the notion of significance may also vary across different persons. Few attempts on personalized abstracts [42, 48] have been referred in this context.

4.3 Summarization without Knowledge of Significant Content

Measuring the importance of the content in a video requires understanding the semantics. Difficulty in semantic analysis causes the evolution of various techniques that do not require such an important measure. These genres of summarization schemes are discussed in this section. These schemes can be categorized according to their characteristics as follows:

1. Techniques that select representative frames from every shot of a video
2. Techniques using redundancy removal
3. Techniques using motion analysis
4. Other techniques

4.3.1 Representative Frame Selection from Every Shot

In the absence of knowledge about significant shots, there have been several techniques that incorporate the representatives of all shots in the summary. These approaches are mentioned here.

Selecting a predefined set of frames

Early research in this category attempted to select a few fixed frames from each shot to generate the summary. For example, only the first frame [62], or the first and the last frames [69] of each shot are considered as the representative frame of the shot. Such frames may not necessarily reflect the content of a shot. For example, a shot having dynamic content is not well represented by these frames. Moreover, a frame lying close to the beginning or to the end of a shot may have reduced representative quality when it is a part of a dissolve effect at the shot boundary [22].

Capturing change of content using low-level visual features

The need to preserve the intra-shot variation in the constructed summary introduces few frame-difference-based techniques. These techniques can be placed into two groups. The first one requires estimation of differences between each pair of successive frames. Like shot boundary detection, a high difference value indicates a significant change in the content. The second frame in such a pair is considered the representative frame. The summarization technique in [56] belongs to this group. In a slight variation of this technique, in [23], the potential of a frame being a key frame depends on its difference with the previous frame and its difference with mean of the previous k frames. When both the differences are greater than a predefined threshold, the frame is considered a key frame. A histogram-based distance measure is adopted in [23], whereas [56] adopts a set of fuzzy geometrical and information measures as features of a frame.

In the second group of frame-difference-based techniques, the distance of each frame from the last selected key frame is examined to include the concerned frame in the summary [14, 78]. In [14], the distance between frames is measured by considering the difference between pixel intensities of the concerned frames. The method in [78] uses a histogram-based distance.

Both of these groups of techniques can act upon each individual shot as well as upon the entire video. An abrupt change of content in a frame with respect to the previous/reference frame indicates the start of a new event. These schemes can easily be utilized in surveillance applications, as demonstrated in [14].

Capturing change of content using object-based features

Apart from using low-level visual features (like color, intensity, and their histograms), object-based features (related to key video objects) are also utilized in inter-frame difference computations. In [32], video object planes (VOPs) are extracted for each frame. Connected component labeling is carried out on those VOPs. The number of regions in each successive frame is compared with that in the last selected key frame. A change in this number lets the concerned frame into the set of key frames. The same number of labeled regions in both cases demands a shape analysis of each labeled region. The shape of each labeled region is extracted using seven Hu moments [28]. Hu moments can describe shape in a translation- and scale-invariant manner. The distance of a region in a frame from its closest spatially labeled region in the previous key frame is estimated using the city block distance between their center points. The dissimilarity of the frame from its previous key frame is estimated by taking the maximum of distances of individual regions. If this frame dissimilarity exceeds a predefined threshold, the frame is also considered a key frame. In [33], time is considered another feature along with Hu's moments to estimate the dissimilarity of frames.

The work in [34] is an extension of that in [32] to examine the suitability of Fourier shape descriptors as compared with Hu's moment-based descriptors. Estimation of Fourier shape descriptors requires one to resample the contour of the video object at a fixed number of points. Then, the coordinate of each of these points is represented as a complex number. Fourier coefficients of these complex numbers are calculated. Experiments reveal that Hu's moment-based descriptors are more suitable as compared to Fourier descriptors. Contour-based descriptors are observed to be sensitive to a slight change in the object boundary.

Experiments are also reported in [16] where the shape dissimilarity between two key VOPs is estimated using the Hamming distance and Hausdorff distance. While generating the abstract, the initial VOP is always taken as the key VOP. For subsequent VOPs, those having significant shape dissimilarity with the previously selected key VOP are also considered the key VOP. The threshold of significance is proposed to vary with the activity level within the video and the size of the video object. Moreover, it is observed that a slight spatial shift between two similar shapes can cause a large Hamming distance. To cope with this problem, the mass centers of two comparing VOPs are aligned *a priori* if the Hamming distance is used in the computation.

Clustering frames within a shot

Clustering of frames of a shot and subsequent selection of representative frames from all the obtained clusters also capture intra-shot variability. The summarization scheme in [80] falls into this category. In this scheme, frames are inspected sequentially and placed into clusters. The similarity of a frame under observation with the existing cluster centroids determines whether the frame will be part of the existing clusters or not. If the similarity is estimated to be lower than a prespecified threshold, a new cluster is formed having the newly inspected frame. Similarly, clustering of frames within a shot is also performed in [17].

4.3.2 Redundancy Removal

The repeated appearance of similar content at several places in a video is reflected in a summary that has representative frames from every shot of the video. In some cases, this redundancy inadvertently increases the length of the summary. Summarization techniques that aim to remove the redundancy are categorized here based on their approaches.

Clustering

Clustering of frames followed by the selection of a few representative frames from each cluster reduces the redundancy. Frame-clustering-based summarization techniques have been proposed in [26, 60]. k-means and hierarchical clustering are performed in [26] and [60], respectively. As an alternative to clustering all the frames in a video, in [20], frames having high dissimilarities with their neighbors are selected at the first stage. In the second stage, these selected frames are further clustered. Moreover, it is ensured in [20] that a cluster represents at least one uninterrupted sequence of frames of a minimum duration. This ensures that the video artifacts and other nonsignificant events are not used as key frames.

Like frame-level analysis, clustering can also be applied for grouping visually similar shots [15, 21, 74]. Some of the distinguishing characteristics of these techniques are mentioned here. Construction of shot-level feature is the most notable characteristic of the technique in [15]. Color and depth-based image segmentation is carried out on the content of a frame using a multi-resolution implementation of the Recursive Shortest Spanning Tree (RSST) algorithm [50]. Then for each segment in a frame, the color, depth, size, and location of the segment center are considered features. A fuzzy feature vector is constructed by taking the above said features of all segments in the frame. A shot feature vector is derived from feature vectors of all frames in a shot. In [21], shots are clustered based on their visual similarities. The adopted clustering procedure groups the frames into the appropriate number of clusters while the maximum allowable radius of the resultant clusters is fixed. For each cluster, the shot with longest length is selected as the representative shot. Moreover, if a cluster's representative shot is shorter than 1.5 seconds, the cluster is ignored. A duration segment of 1.5 seconds is taken from each representative shot. These segments are then concatenated according to time order.

A time-constrained algorithm for video shot clustering is used in [74]. Shots are clustered according to their visual similarity and temporal proximity. Two shots may have similar visual characteristics but are located far apart in time. This distance in temporal dimension may reflect that these two shots arise in different contexts. The time-constrained clustering discriminates these shots despite their visual similarity to retain their contextual difference.

Coverage

Coverage of a frame is defined in [9] as a collection of all frames in the video sequence that are visually similar to it. The problem of summarization can be modeled as covering the entire video sequence with a minimum number of key frames. A greedy method for this is proposed in [9]. It selects the frames in descending order of their coverage, starting with the maximum one, and eliminates every frame in their coverage from the video sequence. The iteration is repeated until there is no frame left in the video sequence. In [73], the problem is formulated in a similar way. But the coverage of a frame is defined as the number of fixed-size excerpts (not individual frames) that contain at least one frame similar to the concerned frame. The dynamic programming procedure is adopted in this solution.

Shot reconstruction

Shot reconstruction degree or sequence reconstruction error is proposed in [35, 39, 40] to account for the capability of the selected key frame set to reconstruct the video shot/sequence. An interpolation function generates the remaining frames from the given set of key frames. In order to find an optimal set of key frames, the computations of break-points and key frames need to be integrated. On this note, [35] proposes an iterative scheme to select a predetermined number of key frames, while keeping the shot reconstruction error as small as possible. A similar solution is adopted in [39]. But to improve the efficiency, the latter scheme uses a heap-based structure. Assuming that each key frame only represents the frames following it, the dynamic programming-based approach in [37] generates an optimal solution. Moreover, assuming that the optimal solution is found for any size of the key frame set, a bisection search algorithm is proposed to find the optimal key frame set. It is to be noted that the methods discussed so far use one single key frame to interpolate the others frames. Alternatively, an inertia-based frame interpolation scheme is proposed in [40] that uses two adjacent key frames for interpolation.

Other redundancy removal approaches

Another redundancy elimination-based approach is found in [67]. The variance of features of a frame within a local window around a frame identifies the dynamic or highly active portions of a shot. Frames belonging to those highly active portions are selected as representative frames of the shot. Then redundancies in the sequence of selected frames are eliminated by considering both visual similarity and temporal coherence. Only a single frame among the key frames, which are visually similar as well as closely situated in temporal domain, is retained in the summary.

A two-level repetitive information detection-based video summarization technique is depicted in [19]. At first, redundant contents are removed using hierarchical agglomerative clustering in the shot level. To remove the redundancy further, the initial summarized video is again segmented. The Smith-Waterman [63] local alignment algorithm is used to identify similar segments. Thus, a two-level redundancy elimination approach generates the video abstract.

4.3.3 Motion Analysis

A few motion analysis-based video summarization techniques are discussed here. For example, trajectories of moving objects constitute the summary [30] for surveillance applications. On a lane-tracking system [38], a moving object is initially identified that maintains a uniform speed. All frames are aligned to fix the focused moving object in a fixed position. Then a synthesized frame hosts this reference object along with other moving objects whose positions are calculated according to the relative motion theory. Thus, this method produces a summary image containing all moving objects embedded with spatial and motion-related information.

In [52], a motion intensity index is proposed to indicate the temporal variation of activity. The time axis is partitioned into discrete regions of different sampling rates that are proportional to the value of the motion intensity index. Finally, the video sequence is sampled in each interval of time using the estimated sampling rate. But if the video abstract is played back at a fixed rate, due to smaller number of representative frames the static segments will be rendered at a faster speed compared to the dynamic segments. As a result, the temporal nature of the video will be distorted. As a remedy, a few intermediate frames are generated by simply copying the preceding frame sample as many times as desired.

In [71], the optical flow is computed for each frame. A motion metric evaluates the changes in the optical flow along the frame sequence. Key frames are then spotted at places where the metric, as a function of time, has its local minima. In another motion analysis-based approach [51], a set of key frames is obtained by combining motion analysis with a fast geometrical curve simplification algorithm.

4.3.4 Other Summarization Techniques

In [58], the entire video sequence is represented by a graph. Each frame is denoted by a node. An edge denotes correlation between two concerned frames. This correlation is estimated using the accumulative motion parameter vectors. The optimal key frame set is obtained by finding the shortest path from the first node to the last node. The A∗ search strategy is applied for this purpose.

A learning-based video preview generation is depicted in [65], where the user browsing behavior is learned through the use of Hidden Markov Model (HMM). Various potential browsing behavioral states are modeled as hidden states of HMM. This browsing pattern is defined as a transition of users through these states. The parameters of the HMM are estimated using a combination of both supervised and unsupervised learning. Another supervised learning-based approach to select the representative frames of a shot is depicted in [31]. At first, a shot is partitioned in subshots depending on camera motion. Then volunteers were asked to select one representative frame for each subshot.

The work in [64] segments the video into scenes, each of which is consistent in chromaticity, lighting, and sound. The Kolmogorov complexity of a shot that gives the minimum time required for its comprehension is measured. The starting and ending portions of a scene are selected as they incorporate the essential information of the scene according to the film grammar.

In a few other approaches [53, 59], video summarization has been posed as a multi-objective optimization problem. Genetic algorithm (GA) based video abstraction schemes are proposed to produce a meaningful summary in a search space of all video summaries.

4.3.5 Relevance to Surveillance Application

So far we have discussed the class of summarization schemes that do not utilize knowledge about significant contents. Here, we review the suitability of these techniques for surveillance applications. The task of surveillance can be of varied nature. A few applications only require us to identify the appearance of a new person or object and their disappearance. Keeping a record of persons who enter or exit a high security zone (for e.g., the President's residence) is one such example. Identification of activities performed by persons or objects is required in some other applications. Moreover, another kind of surveillance application needs to identify a few specific persons or situations. A surveillance system to identify notorious persons in busy public areas is one such example.

The rudimentary summarization schemes, such as the selection of one or more fixed sets of frames from each shot, have a high chance of missing the interesting events in a dynamic situation. For example, selection of only the first frame from each shot [62] fails to show the presence of any person or object that appears after the first frame of the shot.

Summarization techniques (such as [14, 23]) that capture the change of content using low-level visual features keep a record of arrival or departure of any new person or object from the scene. But summarization techniques (such as [32]) that capture change of object-level features are more suitable to represent different activities. If throughout the concerned video shot, a person is constantly present in all the frames, there may not be a significant change in low-level visual features of a frame throughout the shot. But if the person performs several activities in the concerned time, the shape of the VOP [32] changes across the frames in the shot.

Summarization using clustering of frames within a shot to select the representative frames of the shot [80] can capture the appearance or disappearance of persons/objects. The capability of identifying an activity depends on whether low-level or object-level features are used.

As discussed in Section 4.3.2, several summarization techniques aim to remove the redundant contents from the video. They eventually come up with a shorter representation of the summary as compared to shot-based techniques. Every change of event in the concerned video is not required for all applications. These redundancy removal-based techniques may be suitable in those cases. But again, there is a trade-off, that is, it may miss some significant information.

Therefore, the need to identify significant content is indispensable to produce a good summary. Moreover, surveillance applications sometimes identify some specific significant situations or persons. Several summarization schemes that use information about significant content are discussed in the next section.

4.4 Summarization Using Knowledge of Significant Content

Determination of important, interesting, or exciting events is an essential and perhaps the most challenging task in video summarization. Video clips corresponding to events having these characteristics are retained in the summary. Identification of the presence of this characteristic in video clips varies across various approaches. Most of these techniques exploit domain-specific knowledge to generate a compact video abstract. Examples include movie and drama [11, 49, 52, 57], sports [4, 5, 8, 12], news [6, 29, 54], medical video [44], home video [48], and audio-visual presentations [27]. Some generic approaches [45, 75] also exist. These are stated here in brief.

4.4.1 Movie and Drama Summarization

Actions and dialogues are considered the important parts of a movie or a drama [52,57,74]. In [74], identification of a dialogue is characterized by intense interactions between multiple parties. Therefore, there is repetition of similar-looking shots consisting of one or more people. On the contrary, an action event is characterized by the lack of repetition of similar shots due to the high dynamic nature of the event. Knowledge about repetition of a group of similar labeled shots differentiates these events. Therefore, action and dialogue events are identified using visual features of the shots.

On the contrary, analysis of both visual and auditory features is common for most of the movie summarization approaches. In [52], detection of emotional dialogues and action events helps to produce an event-based abstraction. Two acoustic features — average pitch frequency and temporal variation of speech signal intensity levels — are used in this work. Much higher values for these parameters indicate the presence of two emotions — anger and happiness. A k-means clustering (with $k = 2$) is carried out for this purpose. Detection of rapid movements by estimating spatio-temporal dynamic visual activities identifies violent and significant actions. Gunfire and explosions are also detected by spotting flames in the video. An increase in the number of bloody-colored pixels identifies violent actions causing bleeding. This has been done using pixel matching with a predefined color table. Moreover, analysis of the background sound-track distinguishes the violent and nonviolent scenes. Similarly, both auditory and visual features are used in [57] to find the activities with high motion and dialogues.

According to the television drama program summarization in [49], the psychological unfolding of events is more important than the behavior of objects. A set of heuristics based on audio and video channels is proposed to capture the psychological unfolding. Some of these are mentioned here. Frequent shot change signifies a tense situation. The beginning and end portions of the frequent part of a speech are also considered important. Moreover, the presence of special sound effects indicates the occurrence of important events.

A method for analyzing audio features is also proposed in [7] to distinguish among silence, environmental noise, speech, music, and speech with background music. This compartmentalization of the audio track helps in summarization.

In another approach [11], the structure of a story is understood according to concepts of characters, items, places, and time. The keywords for these concept entities are spotted from the text features extracted from video frames and speech transcripts. Low-level visual features of the video content are then mapped to high-level concept entities. A concept expansion method is formulated to establish the relationships among these entities. This facilitates summarization.

4.4.2 Sports Video Summarization

Content production style-based knowledge helps in sports video summarization. Generally, score captions convey information about the state of the game and those are placed in a particular place throughout the frames of the video. This knowledge has been exploited in [5] to track the state of the game as it progresses. The significance of an event in the game is estimated in terms of event rank, event occurrence time, and the profile. Event rank is based on whether it represents a state change event, exceptional score event, or not. Again, a score event occurring at the ending stage of a game is assumed to be of higher significance due to its influence on the final result of the game. Similarly, score captions are also analyzed in [55,66]. The presence of a replay sequence also indicates the occurrence of a significant event [55,66]. Similarly, the significant events are detected in [13] by locating the co-occurrence of keywords such as goal and touch-up with high audio energy.

Detection of a goal-scoring event [4] and detection of activities around the goal mouth [70] also help in summarizing a soccer video. As an example of more detailed semantic analysis, in [8], each shot of a baseball game is classified as pitch view, catch overview, catch closeup, running overview, running closeup, audience view, or touch-base closeup. Then HMM models have been developed to detect baseball events, like home run, catch, hit, and infield play and to include those events in the summary.

User excitation modeling-based highlight generation is proposed in [25], which is claimed to be suitable for sports video. Segments evoking high excitation are identified by analyzing the motion activity and the energy contained in the audio track of a video. While this is an interesting approach, its performance can be improved by finding new links between audio-visual stimuli and human responses. This will come in the domain of psycho-physiological research. Applause and cheering from the audience [72] and excited reactions from the commentator [12,47] also indicate some exciting events.

4.4.3 News Video Summarization

In [29], the hierarchical content structure of a news video helps generate the video skim. At first, the audio channel is analyzed to filter out commercials interweaved with the news program. Then, the presence of one or more anchor persons is detected. As the anchor persons present the news in the program, concatenation of all video segments containing those anchor persons generates the skim.

The repeated appearance of a footage in news programs indicates its importance [6]. The video abstraction scheme in [54] is able to extract interesting scenes in a video when few other interesting scenes have been specified. This was successfully applied to news video.

4.4.4 Other Domain-Specific Summarization Techniques

In a medical video summarization technique [43], the type of the video (e.g., lecture or surgery), salient objects, shot structure (the elements of the shot), and length of a principal video shot determine the weight of that shot. Shots are included in the summary in the order of their weights (starting from the shot with the highest weight) until their combined length reaches the desired length.

Salient objects in medical education videos are extracted in [44]. Using the features of these salient objects, SVM is trained to classify the video shots into a set of semantic classes. This classification results help in the process of concept-oriented video skimming.

The psychological response of observers is used to find out the entertaining portions in a home video summarization scheme [48]. This psychological response is guessed through electro-dermal response, heart rate, blood volume pulse, respiration rate, and respiration amplitude.

Abstracts for online audio-video presentations are generated in [27] based on pitch and pause in audio signals and slide transition points in the presentation.

4.4.5 Generic Approaches

Here we describe some of the domain-independent approaches to identify significant content. Various psychological phenomena are modeled in this context. The summarization techniques in [45, 46] assign a score to each frame of a video to indicate its capability to draw human attention. The time series of these attention values of each frame in a video produces an attention curve. Zero-crossing points from positive to negative on the derivative of this attention curve indicate the locations of peaks in this attention curve. Frames at those peaks are the key frames. Thus, all the key frames can be generated without shot-boundary detection. Even this method can be combined with a shot-based approach. Key frames in a shot are selected by identifying the peaks within the shot. If no peak exists in a shot, the middle frame of that shot is selected as the key frame. Video skims are also generated around each key frame while maintaining some conditions. In another attention model-based approach [61], video summarization is modeled as a 0–1 knapsack optimization problem. A greedy approach balances the video importance gain and the duration as cost.

The method in [75] is based on the idea that motion, contrast, face, and caption influence human perception. These features are used to generate a perception curve that models the perception. Frames that correspond to the peak points in the curve are then extracted to build the summary.

In [41], it is assumed that a dramatic change in motion speed occurs due to some interesting turns in action. Therefore, a video shot is partitioned into subshots of consecutive motion patterns in terms of acceleration and deceleration. The representative frame is selected where the motion turns from acceleration to deceleration. They have developed a triangular model of motion energy, and the frame corresponding to the top vertex of a triangle is selected. In a slightly different approach, the model in [24] depicts the camera motion using a 2D curve. It considers the direction of motion along with its magnitude. Key frames are extracted at the curvature points.

Studying viewing patterns of users [76] and capturing psychological patterns of the filming person by high magnitude of α-brainwaves [2] also indicate the interesting events.

4.4.6 Relevance to Surveillance Application

Identification of significant content is crucial for video summarization. As discussed in this section, a variety of ideas and subsequent technological research to implement those has enriched our knowledge. These inventions can be utilized to summarize surveillance videos too. A few such examples are mentioned here. The methods of detecting the violent and significant actions by estimating spatio-temporal visual activities [52], and those to detect gunfire and explosion by spotting flames in the video [52], can be mentioned in this context. A variant of the technique in [11], which comprehends the story structure of a movie in terms of the concepts of characters, items, and time, can also be applied to surveillance video summarization. Analysis of the motion and the energy contained in audio track (similar as in [25]) can identify the exciting events that occur in the area under observation. In accordance with the attention models in [45, 46], a surveillance video-specific model can also be developed to monitor significant events.

4.5 Personalized Summaries

The explanation of significance, at times, may depend on the individual who is going to use the summary. Thus, personalization of significance is important in this context. Personalization is more important in the context of surveillance activities where systems need to be configured to seek some specific objects or persons. Few research attempts have been found for generating the personalized summaries.

The individual user is required to inform about her bias toward a set of candidate features to indicate her preference. This feature set is also designed to be domain specific. For example, in [42], candidate features consist of human face, speech, camera zoom, and caption text in the case of news video. For movies, the candidate features are gunshot, explosion, loud voice, face, and fire color. An approach designed to summarize a home video [79] used speech, laughter, song, applause, scream, motion, and blur as features.

Contrary to explicit specification by users, there are models to predict biasness depending on several physical or characteristic parameters of the user. According to [48], psychological response is obtained using electro-dermal response, heart rate, blood volume pulse, respiration rate, and respiration amplitude. Analysis of these recorded data serves as an information source for personalized video summarization. In another technique [1], a personalized abstract is achieved by considering personality and other issues related to the user, for example, gender, introvert or extrovert.

The APIDIS (Autonomous Production of Images based on Distributed and Intelligent Sensing) project [3] is another fine initiative for producing personalized summaries applicable to surveillance and sports events. In the context of two-team sports videos [10], users specify the extent to which they prefer important portions of the game and close-up views of players to catch a glimpse of emotions. Users also can control the redundancy brought by replays and close-up views. Similar efforts of personalization can also be made in the context of surveillance video summarization.

4.6 Challenges

As discussed in the previous sections, several attempts have been made to detect few significant or interesting events. Examples include goal scoring events in a soccer video, and dialogue and action events in a movie. But a complete semantic understanding of a video in an automated way is still a challenge. Inventions in different, varied disciplines will enable the summarization techniques to flourish. A few of these are listed below:

1. Improvement in object recognition technique (especially in a view-invariant way, in different illumination conditions, and in the presence of obstacles) will immensely help video summarization.

2. Improvement in natural language processing models will also help video summarization.

3. Working with large variety of objects and vocabularies is challenging. This explains why most of the existing semantic analysis-based approaches focus on a specific domain.

4. Automated categorization of video domains may also lead to automatic selection of a domain specific technique.

5. Approaches in the generic category of significant event based summarization are mainly limited to attention modeling. In the last few decades,

extensive researches have been carried out on visual attention modeling. These are mainly based on low-level visual features. Among the high-level semantic features, the presence of face has been mostly exploited. But study on the effects of other semantic concepts in drawing attention is still unexplored.

6. Attention modeling in auditory channel is also an emerging area. This includes attention models in speech, music, or any audible sound in general. Research in this topic is in nascent stage.

7. Moreover, multi-view video summarization [18] has appeared as a new avenue of research.

4.6.1 Scope of Soft Computing Methodologies

Soft computing techniques can be utilized in summarization. Few approaches exist in the literature exploiting the concepts of fuzziness [56, 77] and genetic algorithm [53, 59]. But further exploration along this avenue is needed to achieve a larger benefit. A few potential scopes are mentioned here.

Detection of shot boundary is a prerequisite for many video summarization techniques. But a gradual transitions from one shot to another produces vagueness in terms of locating the boundary. Concepts of fuzzy set theories can be utilized in this context. Moreover, features (both geometrical and spatio-temporal) of such a shot may also be fuzzy. This opens up opportunities to experiment with a large set of fuzzy techniques. Rough set theory can also be exploited to handle the impreciseness of information arising due to gradual shot transition effects. Similarly, neural computation can be utilized for efficient learning of user preferences to produce personalized summaries.

Moreover, the summarization process generates a concise representation of the original video. On the other hand, all significant contents must be covered. As a result, summarization can be modeled as a multi-objective optimization problem. Therefore, a genetic algorithm can be utilized to produce an acceptable summary.

4.7 Conclusion

This chapter reviewed several interesting methods of video summarization. It classifies them in the context of whether they use the knowledge about significant content or not. The suitability of these research efforts for surveillance video summarization is also discussed. Several methods of generating personalized summaries were mentioned. Some of the challenging issues, including the relevance of soft computing, were addressed. Thus, by putting a huge chunk of reference materials under one umbrella, this chapter will help present and future researchers significantly. Achievements being put together, limitations and potential research avenues have also come into focus.

Acknowledgments

The work was performed when Prof. Sankar K. Pal held a J. C. Bose National Fellowship of the Government of India.

References

1. L. Agnihotri, J. Kender, N. Dimitrova, and J. Zimmerman, Framework for personalized multimedia summarization. In *Proceedings of the 7th ACM SIGMM International Workshop on Multimedia Information Retrieval*, pages 81–88, New York, USA, November 2005.
2. K. Aizawa, K. Ishijima, and M. Shina, Summarizing wearable video. In *Proceedings of the IEEE International Conference on Image Processing*, pages 398-401, Thessaloniki, Greece, October 2001.
3. APIDIS Project, `http://cordis.europa.eu/fp7/ict/content-knowledge/projects-apidis_en.html`
4. J. Assfalg, M. Bertini, C. Colombo, A. D. Bimbo, and W. Nunziati, Semantic annotation of soccer videos: Automatic highlights identification, *Computer Vision and Image Understanding*, 92(2-3): 285-305, 2003.
5. N. Babaguchi, Y. Kawai, T. Ogura, and T. Kitahashi, Personalized abstraction of broadcast American football video by highlight selection, *IEEE Transactions on Multimedia*, 6(4): 575–586, 2004.
6. A. Bagga, J. Hu, J. Zhong, and G. Ramesh, Multi-Source combined-media video tracking for summarization. In *Proceedings of the IEEE International Conference on Pattern Recognition*, pages 818-821, Quebec City, Canada, August 2002.
7. R. B. Bhatt, P. Krishnamoorthy, and S. Kumar, Efficient general genre video abstraction scheme for embedded devices using pure audio cues. In *Proceedings of 7th International Conference on ICT and Knowledge Engineering*, pages 63–67, Bangkok, Thailand, December 2009.
8. P. Chang, M. Han, and Y. Gong, Extract highlights from baseball game video with hidden markov models. In *Proceedings of the IEEE International Conference on Image Processing*, pages I-609–612, Rochester, New York, USA, September 2002.
9. H. S. Chang, S. Sull, and S. U. Lee, Efficient video indexing scheme for content-based retrieval, *IEEE Transactions on Circuits and Systems for Video Technology*, 9(8): 1269-1279, 1999.
10. F. Chen and C. D. Vleeschouwer, Formulating team-sport video summarization as a resource allocation problem, *IEEE Transactions on Circuits and Systems for Video Technology*, 21(2): 193–205, 2011.
11. B.-W. Chen, J.-C. Wang, and J.-F. Wang, A novel video summarization based on mining the story-structure and semantic relations among concept entities, *IEEE Transactions on Multimedia*, 11(2): 295–312, 2009.
12. F. Coldefy and P. Bouthemy, Unsupervised soccer video abstraction based on

pitch, dominant color and camera motion analysis. In *Proceedings of the ACM International Conference on Multimedia*, pages 268-271, New York, USA, October 2004.

13. S. Dagtas and M. Abdel-Mottaleb, Multimodal detection of highlights for multimedia content, *Multimedia Systems*, 9(6): 586-593, 2004.
14. U. Damnjanovic, V. Fernandez, E. Izquierdo, and J. M. Martinez, Event detection and clustering for surveillance video summarization. In *Proceedings of the Ninth International Workshop on Image Analysis for Multimedia Interactive Services*, pages 63–66, Klagenfurt, Austria, May 2008.
15. N. D. Doulamis, A. D. Doulamis, Y. S. Avrithis, K. S. Ntalianis, and S. D. Kollias, Efficient summarization of stereoscopic video sequences, *IEEE Transactions on Circuits and Systems for Video Technology*, 10(4): 501–517, 2000.
16. B. Erol and F. Kossentini, Video object summarization in the mpeg-4 compressed domain. In *Proceedings of the IEEE International Conference on Acoustics, Speech, and Signal Processing*, pages 2027–2030, Istanbul, Turkey, June 2000.
17. A. M. Ferman and A. M. Tekalp, Multiscale content extraction and representation for video indexing. In *Proceedings of SPIE Multimedia Storage and Archiving Systems*, vol. 3229, Dallas, TX, USA, November 1997.
18. Y. Fu, Y. Guo, Y. Zhu, F. Liu, C. Song, and Z.-H. Zhou, Multi-view video summarization, *IEEE Transactions on Multimedia*, 12(7): 717–729, 2010.
19. Y. Gao, W.-B. Wang, and J.-H. Yong, A video summarization tool using two-level redundancy detection for personal video recorders, *IEEE Transactions on Consumer Electronics*, 54(2): 521–526, 2008.
20. A. Girgensohn and J. Boreczky, Time-Constrained Keyframe Selection Technique. In *Proceedings of the IEEE International Conference on Multimedia Computing and Systems*, pages 756–761, Florence, Italy, June 1999.
21. Y. Gong and X. Liu, Video summarization with minimal visual content redundancies. In *Proceedings of the IEEE International Conference on Image Processing*, pages 362–365, Thessaloniki, Greece, October, 2001.
22. P. Gresle and T. S. Huang, Video sequence segmentation and key frames selection using temporal averaging and relative activity measure. In *Proceedings of the 2nd International Conference on Visual Information Systems*, San Diego, CA, USA, December 1997.
23. B. Gunsel and A. M. Tekalp, Content-based video abstraction. In *Proceedings of the IEEE International Conference on Image Processing*, pages 128–132, Chicago, IL, USA, October 1998.
24. S.-H. Han and I.-S. Kweon, Scalable temporal interest points for abstraction and classification of video events. In *Proceedings of the IEEE International Conference on Multimedia and Expo*, Amsterdam, The Netherlands, July 2005.
25. A. Hanjalic, Adaptive extraction of highlights from a sports video based on

excitement modeling, *IEEE Transactions on Multimedia*, 7(6): 1114–1122, 2005.

26. A. Hanjalic and H. Zhang, An integrated scheme for automated video abstraction based on unsupervised cluster-validity analysis, *IEEE Transactions on Circuits and Systems for Video Technology*, 9(8): 1280–1289, 1999.
27. L. He, E. Sanocki, A. Gupta, and J. Grudin, Auto-summarization of audio-video presentations. In *Proceedings of the 7th ACM International Conference on Multimedia*, pages 489–498, Orlando, FL, USA, October 1999.
28. M. Hu, Visual pattern recognition by moment invariants, *IRE Transactions on Information Theory*, 8(2): 179–182, 1962.
29. Q. Huang, Z. Lou, A. Rosenberg, D. Gibbon, and B. Shahraray, Automated generation of news content hierarchy by integrating audio, video, and text information. In *Proceedings of IEEE International Conference on Acoustics, Speech, and Signal Processing*, pages 3025–3028, Arizona, USA, March 1999.
30. Z. Ji, Y. Su, R. Qian, and J. Ma, Surveillance video summarization based on moving object detection and trajectory extraction. In *Proceedings of the 2nd International Conference on Signal Processing Systems*, pages V2-250–253, Dalian, China, July 2010.
31. H.-W. Kang and X.-S. Hua, To learn representativeness of video frames. In *Proceedings of the 13th Annual ACM International Conference on Multimedia*, pages 423–426, Singapore, November 2005.
32. C. Kim and J.-N. Hwang, An integrated scheme for object-based video abstraction. In *Proceedings of the Eighth ACM International Conference on Multimedia*, pages 303–311, Los Angeles, CA, USA, October 2000.
33. C. Kim and J.-N. Hwang, Object-based video abstraction using cluster analysis. In *Proceedings of the International Conference on Image Processing*, pages 657–660, Thessaloniki, Greece, October 2001.
34. C. Kim and J.-N. Hwang, Object-based video abstraction for video surveillance systems, *IEEE Transactions on Circuits and Systems for Video Technology*, 12(12): 1128–1138, 2002.
35. H.-C. Lee and S.-D. Kim, Iterative key frame selection in the rate-constraint environment, *Signal Processing: Image Communication*, 18(1): 1-15, 2003.
36. Y. Li, S.-H. Lee, C.-H. Yeh, and C.-C. J. Kuo, Techniques for movie content analysis and skimming: Tutorial and overview on video abstraction techniques, *IEEE Signal Processing Magazine*, 23(2): 79–89, 2006.
37. Z. Li, G. Schuster, A. K. Katsaggelos, and B. Gandhi, Optimal video summarization with a bit budget constraint. In *Proceedings of the IEEE International Conference on Image Processing*, pages 617–620, Singapore, October 2004.
38. C. Li, Y.-T. Wu, S.-S. Yu, and T. Chen, Motion-focusing key frame extraction and video summarization for lane surveillance system. In *Proceedings*

of 16th IEEE International Conference on Image Processing, pages 4329–4332, Cairo, Egypt, November 2009.

39. T. Liu and J. R. Kender, An efficient error-minimizing algorithm for variable-rate temporal video sampling. In *Proceedings of the International Conference on Multimedia and Expo*, pages 413–416, Lausanne, Switzerland, 2002.
40. T. Liu, X. Zhang, J. Feng, and K. Lo, Shot reconstruction degree: A novel criterion for keyframe selection, *Pattern Recognition Letters*, 25(12): 1451-1457, 2004.
41. T. Liu, H.-J. Zhang, and F. Qi, A novel video key-frame extraction algorithm based on perceived motion energy model, *IEEE Transactions on Circuits and Systems for Video Technology*, 13(10): 1006-1013, 2003.
42. S. Lu, I. King, and M. Lyu, Video summarization using greedy method in a constraint satisfaction framework. In *Proceedings of the 9th International Conference on Distributed Multimedia Systems*, pages 456-461, Miami, FL, USA, 2003.
43. H. Luo and J. Fan, Concept-oriented video skimming and adaptation via semantic classification. In *Proceedings of 6th ACM SIGMM International Workshop Multimedia Information Retrieval*, pages 213–220, New York, USA, October 2004.
44. H. Luo, Y. Gao, X. Xue, J. Peng, and J. Fan, Incorporating feature hierarchy and boosting to achieve more effective classifier training and concept-oriented video summarization and skimming, *ACM Transactions on Multimedia Computing, Communications, and Applications*, 4(1): 1–25, 2008.
45. Y.-F. Ma, X.-S. Hua, L. Lu, and H.-J. Zhang, A generic framework for user attention model and its application in video summarization, *IEEE Transactions on Multimedia*, 7(5): 907–919, 2005.
46. Y.-F. Ma, L. Lu, H.-J. Zhang, and M. Li, A user attention model for video summarization. In *Proceedings of the tenth ACM International Conference on Multimedia*, pages 533–542, Juan les Pins, France, December 2002.
47. S. Marlow, D. Sadlier, N. O'Connor, and N. Murphy, Audio processing for automatic TV sports program highlights detection. In *Proceedings of the Irish Signals and Systems Conference*, Cork, Ireland, June 2002.
48. A. G. Money and H. Agius, ELVIS: Entertainment-led video summaries, *ACM Transactions on Multimedia Computing, Communications, and Applications*, 6(3): 1–30, 2010.
49. T. Moriyama and M. Sakauchi, Video summarization based on the psychological content in the track structure. In *Proceedings of the ACM workshops on Multimedia*, pages 191–194, Los Angeles, CA, USA, November 2000.
50. O. J. Morris, M. J. Lee, and A. G. Constantinides, Graph theory for image analysis: An approach based on the shortest spanning tree, *IEE Proceedings*, 133(2): 146-152, 1986.

51. M. Mrak, J. Calic, G. Cordara, and A. Kondoz, Hierarchical motion analysis for fast summarization of scalable coded video. In *Proceedings of the IEEE International Conference on Multimedia and Expo*, pages 1309–1312, Hannover, Germany, June 2008.
52. J. Nam and A. H. Tewfik, Dynamic video summarization and visualization. In *Proceedings of the Seventh ACM International Conference on Multimedia*, pages 53–56, Orlando, FL, USA, October 1999.
53. H. Narasimhan, S. Satheesh, and D. Sriram, Automatic summarization of cricket video events using genetic algorithm. In *Proceedings of the 12th Annual Conference Companion on Genetic and Evolutionary Computation*, pages 2051–2054, Portland, OR, USA, July 2010.
54. J. H. Oh and K. A. Hua, An efficient technique for summarizing videos using visual contents. In *Proceedings of the IEEE International Conference on Multimedia and Expo*, pages 1167–1170, New York, USA, July-August 2000.
55. H. Pan, P. Beek, and M. Sezan, Detection of slow-motion replay segments in sports video for highlights generation. In *Proceedings of the International Conference on Acoustic, Speech, and Signal Processing Conference*, pages 1649–1652, Salt Lake City, UT, USA, May 2001.
56. S. K. Pal and A. B. Leigh, Motion frame analysis and scene abstraction: discrimination ability of fuzziness measures, *Journal of Intelligent and Fuzzy Systems*, 3: 247–256, 1995.
57. S. Pfeiffer, R. Lienhart, S. Fischer, and W. Effelsberg, Abstracting digital movies automatically, *Journal of Visual Communication and Image Representation*, 7(4): 345-353, 1996.
58. S. V. Porter, M. Mirmehdi, and B. T. Thomas, A shortest path representation for video summarisation. In *Proceedings of the 12th IEEE International Conference on Image Analysis and Processing*, pages 460-465, Mantova, Italy, September 2003.
59. D. K. A. Raju and C. S. Velayutham, A study on genetic algorithm based video abstraction system. In *Proceedings of the IEEE World Congress on Nature and Biologically Inspired Computing*, pages 878–883, Coimbatore, India, December 2009.
60. K. Ratakonda, M. I. Sezan, and R. Crinon, Hierarchical video summarization. In *Proceedings of the SPIE Visual Communications and Image Processing*, 3653: 1531-1541, January 1999.
61. R. Ren, P. P. Swamy, J. Urban, and J. Jose, Attention based video summarisation in rushes collection. In *Proceedings of the ACM International Workshop on TRECVID Video Summarization*, pages 89–93, Bavaria, Germany, September 2007.
62. B. Shahraray and D. C. Gibbon, Automatic generation of pictorial transcripts of video programs. In *Proceedings of the SPIE Multimedia Computing and Networking*, pages 512-519, San Jose, CA, USA, February 1995.
63. T. F. Smith and M. S. Waterman, Identification of common molecular subsequences, *Journal of Molecular Biology*, 147: 195–197, 1981.

64. H. Sundaram and S.-F. Chang, Condensing computable scenes using visual complexity and film syntax analysis. In *Proceedings of the IEEE International Conference on Multimedia and Expo*, pages 389–392, Tokyo, Japan, August 2001.
65. T. Syeda-Mahmood and D. Ponceleon, Learning video browsing behavior and its application in the generation of video previews. In *Proceedings of the 9th ACM International Conference on Multimedia*, pages 119–128, Ottawa, Ontario, Canada, September-October, 2001.
66. D. W. Tjondronegoro, Y.-P. P. Chen, and B. Pham, Classification of self-consumable highlights for soccer video summaries. In *Proceedings of the IEEE International Conference on Multimedia and Expo*, pages 579–582, Taipei, Taiwan, June 2004.
67. P. Toharia, O. D. Robles, L. Pastor, and A. Rodriguez, Combining activity and temporal coherence with low-level information for summarization of video rushes. In *Proceedings of the 2nd ACM TRECVid Video Summarization Workshop*, pages 70–74, Vancouver, Canada, October 2008.
68. B. T. Truong and S. Venkatesh, Video abstraction: a systematic review and classification, *ACM Transactions on Multimedia Computing, Communications, and Applications*, 3(1), 2007.
69. H. Ueda, T. Miyatake, and S. Yoshizawa, IMPACT: An interactive natural-motion-picture dedicated multimedia authoring system. In *Proceedings of the SIGCHI Conference on Human Factors in Computing Systems*, pages 343–350, New Orleans, USA, April-May, 1991.
70. K. Wan and C. Xu, Efficient multimodal features for automatic soccer highlight generation. In *Proceedings of the 17th IEEE International Conference on Pattern Recognition*, pages 973–976, Cambridge, UK, August 2004.
71. W. Wolf, Key frame selection by motion analysis. In *Proceedings of the IEEE International Conference on Acoustics, Speech, and Signal Processing*, pages 1228-1231, Atlanta, GA, USA, May 1996.
72. Z. Xiong, R. Radhakrishnan, and A. Divakaran, Generation of sports highlights using motion activity in combination with a common audio feature extraction framework. In *Proceedings of the IEEE International Conference on Image Processing*, pages 5-8, Barcelona, Catalonia, Spain, September 2003.
73. I. Yahiaoui, B. Merialdo, and B. Huet, Automatic video summarization. In *Proceedings of the International Conference on Context Based Multimedia Information Retrieval Conference*, 2001.
74. M. M. Yeung and B.-L. Yeo, Video visualization for compact presentation and fast browsing of pictorial content, *IEEE Transactions on Circuits and Systems for Video Technology*, 7(5): 771–785, 1997.
75. J. You, G. Liu, L. Sun, and H. Li, A multiple visual models based perspective analysis framework for multilevel video summarization, *IEEE Transactions on Circuits and Systems for Video Technology*, 17(3): 273–285, 2007.

76. B. Yu, W.-Y. Ma, K. Nahrstedt, and H.-J. Zhang, Video summarization based on user log enhanced link analysis. In *Proceedings of the ACM International Conference on Multimedia*, pages 382–391, Berkeley, CA, USA, November 2003.
77. X.-D. Yu, L. Wang, Q. Tian, and P. Xue, Multi-level video representation with application to keyframe extraction. In *Proceedings of the International Multimedia Modeling Conference*, pages 117-121, Melbourne, Australia, January 2004.
78. H. Zhang, J. Wu, D. Zhong, and S. W. Smoliar, An integrated system for content-based video retrieval and browsing, *Pattern Recognition*, 30(4): 643-658, 1997.
79. M. Zhao, J. Bu, and C. Chen, Audio and video combined for home video abstraction. In *Proceedings of the IEEE International Conference on Acoustic, Speech and Signal Processing*, pages 620–623, Hong Kong, April 2003.
80. Y. Zhuang, Y. Rui, T. S. Huang, and S. Mehrotra, Adaptive key frame extraction using unsupervised clustering. In *Proceedings of the IEEE International Conference on Image Processing*, pages 866–870, Chicago, IL, USA, October 1998.

5

Background Subtraction for Visual Surveillance: A Fuzzy Approach

Thierry Bouwmans
University of La Rochelle, La Rochelle, France

5.1 Introduction

Analysis and understanding of video sequences is an active research field. Many applications in this research area (video surveillance [18], optical motion capture [2], multimedia application [16]) need in the first step to detect the moving objects in the scene. So, the basic operation needed is the separation of the moving objects, called the foreground, from the static information, called the background. The process mainly used is background subtraction [12, 14, 24]. The simplest way to model the background is to acquire a background image that does not include any moving object. In some environments, the background is not available and can always be changed under critical situations like illumination changes, or objects being introduced or removed from the scene. So, the background representation model must be more robust and adaptive. The different background representation models can be classified into the following categories:

- ***Basic background modeling:*** In this case, the background is modeled using an average [11], a median [42], or a histogram analysis over time [64]. Once the model is computed, pixels of the current image are classified as foreground by thresholding the difference between the background image and the current frame as follows:

$$d\left(I_t(x,y), B_{t-1}(x,y)\right) > T. \tag{5.1}$$

  Otherwise, pixels are classified as background. T is a constant threshold; $I_t(x,y)$, $B_{t-1}(x,y)$ are respectively the current image at time t; and the background image at time $t-1$. $d(.,.)$ is a distance measure that is usually the absolute difference between the current and the background images.

- ***Statistical background modeling:*** The background is represented using a single Gaussian [61], a Mixture of Gaussians [57], or a Kernel Density Estimation [23]. Statistical variables are used to classify the pixels as foreground or background.

- ***Background modeling based on clusters:*** The background model supposes that each pixel in the frame can be represented temporally by clusters. Incoming pixels are matched against the corresponding cluster group and are classified according to whether the matching cluster is considered part of the background. The clustering approach consists of using the K-mean algorithm [15] or using Codebook [34].

- ***Neural network background modeling:*** The background is represented by the mean of the weights of a neural network suitably trained on N clean frames. The network learns how to classify each pixel as background or foreground [19, 38].

- ***Background estimation:*** The background is estimated using a filter. Any pixel of the current image that deviates significantly from its predicted value is declared foreground. This filter may be a Wiener filter [60], a Kalman filter [43], or a Tchebychev filter [17].

All these different models present the same following steps and issues:

- Background modeling, which describes the kind of model used to represent the background
- Background initialization, which concerns the initialization of the model
- Background maintenance, which relies on the mechanism used for adapting the model to the changes occurring in the scene over time

- Foreground detection which consists of the classification of the pixel as a background or as a foreground pixel
- Choice of the picture's element, which is used in the previous steps; this element may be a pixel [57], a block [25, 46], or a cluster [9, 10]
- Choice of the features which characterize the picture's element. In the literature, there are five features commonly used: color features, edge features, stereo features, motion features, and texture features. In [37], these features are classified as spectral features (color features), spatial features (edge features, texture features), and temporal features (motion features). These features have different properties that allow one to handle differently the critical situations (illumination changes, motion changes, structure background changes)

Figure 5.1 shows an overview of a background subtraction process.

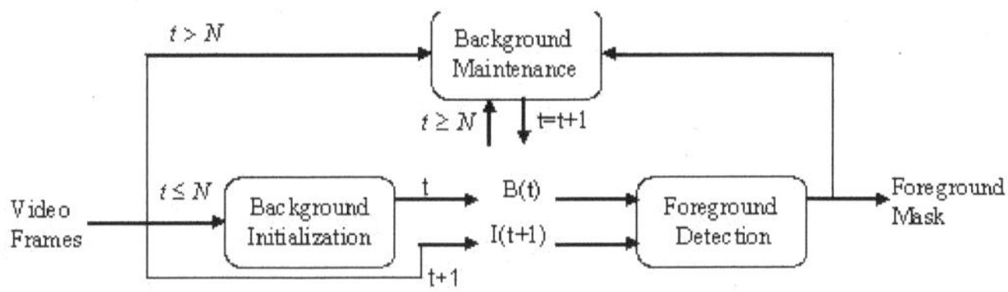

FIGURE 5.1 Background subtraction process.

In developing a background subtraction method, researchers must design each step and choose the features in relation to the critical situations [60] they want to handle: noise image due to a poor quality image source, camera jitter, camera automatic adjustments, time of the day, light switch, bootstrapping, camouflage, foreground aperture, moved background objects, inserted background objects, multimodal background, waking foreground object, sleeping foreground object, and shadows. These critical situations have different spatial and temporal properties. The main difficulties come from the illumination changes and dynamic backgrounds:

- Illumination changes appear in indoor and outdoors scenes. Figure 5.2 shows an indoor scene that presents a gradual illumination change. It causes false detections in several parts of the foreground mask as can be seen in Figure 5.2-d). Figure 5.3 shows the case of a sudden illumination change due to a light on/off. As all the pixels are affected by this change, a large number of false detections are generated (see Figure 5.3-c)).
- Dynamic backgrounds appear in outdoor scenes. Figure 5.4 shows three main types of dynamic backgrounds: waving trees, water rippling, and water surface. The first row shows the original images and the second row shows the foreground mask obtained by the GMM [57]. In each case, there are a large number of false detections.

FIGURE 5.2 a) An indoor scene with low illumination; b) the same scene with a moderate illumination; c) the same scene with a high illumination; d) the foreground mask obtained with GMM [57]. This sequence, called "Time of Day", comes from the Wallflower dataset [60].

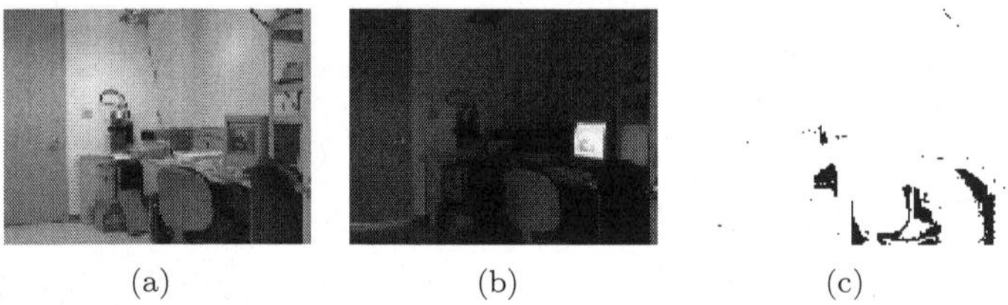

FIGURE 5.3 a) An indoor scene with light-on; b) the same scene with light-off; c) the foreground mask obtained with GMM [57]. This sequence, called "Light Switch", comes from the Wallflower dataset [60].

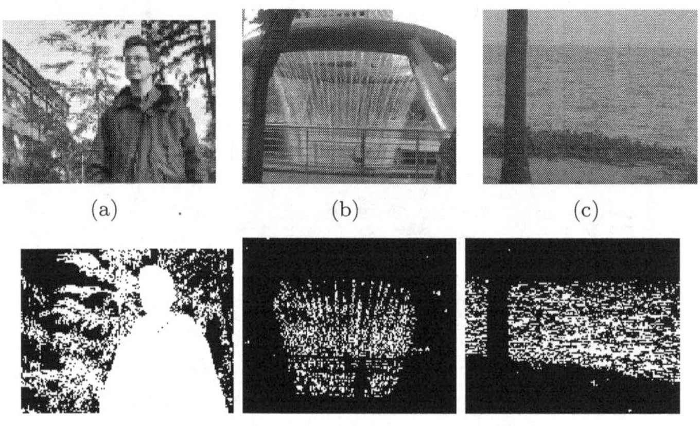

FIGURE 5.4 The first row presents original scenes containing dynamic backgrounds: a) waving trees [60], b) fountain [37] and c) water surface [37]. The second row shows the corresponding foreground masks obtained by the GMM [57].

All these critical situations generate imprecision and uncertainties in the whole process of background subtraction. Therefore, some authors have recently introduced fuzzy concepts in the different steps of background subtraction as follows:

- ***Fuzzy Background Modeling:*** The main challenge consists of modeling a multimodal background. The algorithm usually used is the Gaussian Mixture Models [61]. The parameters are determined using a training sequence which contains insufficient or noisy data. So, the parameters are not well determined. In this context, Type-2 Fuzzy Gaussian Mixture Models [6,7,13] are used to model uncertainties when dynamic backgrounds occur.

- ***Fuzzy Foreground Detection:*** In this case, a saturating linear function is used to avoid crisp decision in the classification of the pixels as background or foreground. The background model can be unimodal, such as the running average in [54,55], or multi-modal, such as the background modeling with confidence measure proposed in [50]. Another approach consists of aggregating different features such as color and texture features using the Sugeno integral [63] or the Choquet integral [3–5]. Fuzzy foreground detection is more robust to illumination changes and shadows than crisp foreground detection.

- ***Fuzzy Background Maintenance:*** The idea is to update the background following the membership of the pixel at the class background or foreground. This membership comes from the fuzzy foreground detection. This fuzzy adaptive background maintenance allows one to deal robustly with illumination changes and shadows.

- ***Fuzzy Post-Processing:*** Fuzzy inference can be used between the previous and the current foreground masks to perform detection of the moving objects, as developed recently by Sivabalakrishnan and Manjula [56].

TABLE 5.1 Fuzzy background subtraction: An overview.

Background Subtraction Steps	Algorithm	Authors - Dates
Background Modeling	Type-2 Fuzzy GMM (3)	El Baf et al. (ISVC 2008) [6]
Foreground Detection	Saturating Linear Function(2)	Sigari et al. (IJCSNS 2008) [55]
	Saturating Linear Function(2)	Shakeri et al. (ICSP 2008) [52]
	Saturating Linear Function(1)	Rossel et al. (ICPR 2010) [50]
	Sugeno Integral (1)	Zhang and Xu (FSKD 2006) [63]
	Choquet Integral (3)	El Baf et al. (WIAMIS 2008) [3]
	Choquet Integral (1)	Ding et al. (ComSIS 2010) [22]
	Choquet Integral (1)	Azab et al. (ICIP 2010) [1]
Background Maintenance	Fuzzy Learning Rate (2)	Sigari et al.(IJCSNS 2008) [55]
	Fuzzy Learning Rate (3)	Maddalena et al.(WILF 2009) [39]
	Fuzzy Maintenance Rule (1)	El Baf et al.(ICIP 2008) [8]
	Fuzzy Maintenance Rule (1)	Kashani et al. (IJCTE 2010) [33]

Table 5.1 shows an overview of the fuzzy concepts used in the background subtraction. The first column indicates the background subtraction steps. The second column indicates the fuzzy concepts used with the papers counted for each method in the parenthesis. The third column gives the name of the authors and the date of their first publication that use the corresponding fuzzy concept.

The rest of this chapter is organized as follows. In the Section 5.2, we present the fuzzy background modeling method using Type-2 Fuzzy Gaussian Mixture Models. Then, in Section 5.3, fuzzy foreground detection methods are described. In Section 5.4, we present how some authors used the results of the fuzzy foreground detection to update the background. Finally, we present several experimental results that show the performance of the fuzzy methods versus the traditional methods.

5.2 Background Modeling by Type-2 Fuzzy Gaussian Mixture Models

Gaussian Mixture Models (GMMs) [61] are the most popular techniques to deal with dynamic backgrounds but present some limitations when some rapid dynamic changes occur, like camera jitter, illuminations changes, and movement in the background. Furthermore, the GMMs are initialized by an expectation-maximization (EM) algorithm that allows one to estimate GMMs parameters from a training sequence according to the maximum-likelihood (ML) criterion. The GMMs are completely certain once their parameters are specified. However, because of insufficient or noisy data in the training sequence, the GMMs may not accurately reflect the underlying distribution of the observations according to the ML estimation. It may seem problematical to use likelihoods that are themselves precise real numbers to evaluate GMMs with uncertain parameters. To solve this problem, Type-2 Fuzzy Gaussian Mixture Models (T2-FGMMs) were developed by Zeng et al. [62] to introduce descriptions of uncertain parameters in the GMMs. T2-FGMMs have proved their superiority in pattern classification [62]. Recently, El Baf et al. [6, 7, 13] have proposed to model the background by using T2-FGMMs. Therefore, each pixel can be characterized by its intensity in the RGB color space. So, the observation o is a vector X_t in the RGB space with three components, that is $d = 3$. Then, the GMMs are composed of K mixture components of multivariate Gaussian as follows:

$$P(X_t) = \sum_{i=1}^{K} \omega_{i,t}\eta\left(X_t, \mu_{i,t}, \sum\nolimits_{i,t}\right), \tag{5.2}$$

where K is the number of distributions, $\omega_{i,t}$ is a weight associated with the ith Gaussian at time t with mean $\mu_{i,t}$ and standard deviation $\sum_{i,t}$. η is a Gaussian probability density function:

$$\eta(X_t, \mu, \textstyle\sum) = \frac{1}{(2\pi)^{\frac{3}{2}}\left|\sum\right|^{\frac{1}{2}}} exp\left(-\frac{1}{2}(X_t - \mu)^T \textstyle\sum^{-1}(X_t - \mu)\right). \tag{5.3}$$

For the T2-FGMM-UM, the multivariate Gaussian with uncertain mean vector is:

$$\eta\left(X_t, \tilde{\mu}, \sum\right) = \frac{1}{(2\pi)^{\frac{3}{2}}\left|\sum\right|^{\frac{1}{2}}} \prod exp\left[-\frac{1}{2}\left(\frac{X_{t,c}-\mu_c}{\sigma_c}\right)^2\right] \quad (5.4)$$

with $\mu_c \in \left[\underline{\mu}_c, \overline{\mu}_c\right]$ and $c \in \{R, G, B\}$.
For the T2-FGMM-UV, the multivariate Gaussian with uncertain variance vector is:

$$\eta\left(X_t, \mu, \tilde{\sum}\right) = \frac{1}{(2\pi)^{\frac{3}{2}}\left|\sum\right|^{\frac{1}{2}}} \prod exp\left[-\frac{1}{2}\left(\frac{X_{t,c}-\mu_c}{\sigma_c}\right)^2\right], \quad (5.5)$$

where $\sigma_c \in [\underline{\sigma}_c, \overline{\sigma}_c]$ and $c \in \{R, G, B\}$.

$\tilde{\mu}$ and $\tilde{\sigma}$ denote the uncertain mean vector and covariance matrix, respectively. Because there is no prior knowledge about the parameter uncertainty, practically Zeng et al. [62] assume that the mean and standard deviation vary within intervals with uniform possibilities, that is, $\mu \in \left[\underline{\mu}, \overline{\mu}\right]$ or $\sigma \in [\underline{\sigma}, \overline{\sigma}]$. Each exponential component in Equation (5.3) and Equation (5.4) is the Gaussian primary membership function (MF) with uncertain mean or standard deviation, as shown in Figure 5.5. The shaded region is the footprint of uncertainty (FOU). The thick solid and dashed lines denote the lower and upper MFs. In the Gaussian primary MF with uncertain mean, the upper MF is:

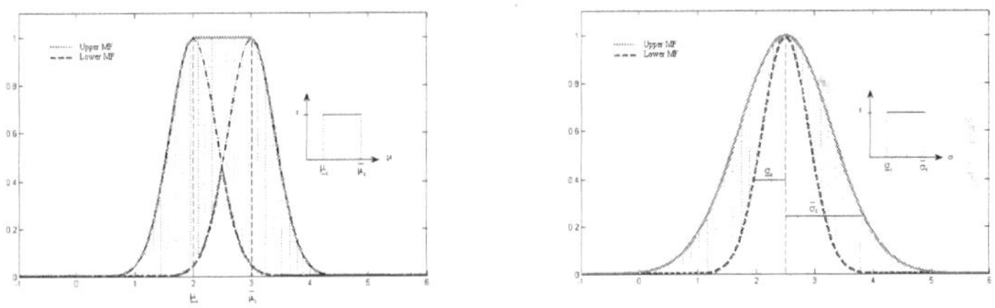

(a) (b)

FIGURE 5.5 a) The Gaussian primary MF with uncertain mean; b) the Gaussian primary MF with uncertain std having uniform possibilities. The shaded region is the FOU. The thick solid and dashed lines denote the lower and upper MFs.

$$\overline{h}(o) = \begin{cases} f\left(o; \underline{\mu}, \sigma\right), & \text{if } o < \underline{\mu} \\ 1, & \text{if } \underline{\mu} \leq o < \overline{\mu} \\ f\left(o; \overline{\mu}, \sigma\right), & \text{if } o > \overline{\mu}, \end{cases} \quad (5.6)$$

where $f\left(o; \underline{\mu}, \sigma\right) = exp\left[-\frac{1}{2}\left(\frac{o-\underline{\mu}}{\sigma}\right)^2\right]$ and $f\left(o; \overline{\mu}, \sigma\right) = exp\left[-\frac{1}{2}\left(\frac{o-\overline{\mu}}{\sigma}\right)^2\right]$.

The lower MF is:

$$\underline{h}(o) = \begin{cases} f(o; \overline{\mu}, \sigma), & \text{if } o \leq \frac{\mu+\overline{\mu}}{2} \\ f(o; \underline{\mu}, \sigma), & \text{if } o > \frac{\mu+\overline{\mu}}{2}, \end{cases} \tag{5.7}$$

In the Gaussian primary MF with uncertain standard deviation, the upper MF is $\underline{h}(o) = f(o; \mu, \overline{\sigma})$ and the lower MF is $\overline{h}(o) = f(o; \mu, \underline{\sigma})$.

The factors k_m and k_ν control the intervals in which the parameter varies as follows:

$$\underline{\mu} = \mu - k_m\sigma, \quad \overline{\mu} = \mu + k_m\sigma, \quad k_m \in [0,3], \tag{5.8}$$

$$\underline{\sigma} = k_\nu\sigma, \quad \overline{\sigma} = \frac{1}{k_\nu}\sigma, \quad k_\nu \in [0.3, 1]. \tag{5.9}$$

Because a one-dimensional gaussian has 99.7% of its probability mass in the range of $[\mu - 3\sigma, \mu + 3\sigma]$, Zeng et al. [62] constrain $k_m \in [0,3]$ and $k_\nu \in [0.3, 1]$. These factors also control the area of the FOU. The bigger k_m or the smaller k_ν, the larger the FOU, which implies the greater uncertainty.

Both T2-FGMM-UM and T2-FGMM-UV can be used to model the background, and we can expect that T2-FGMM-UM will be more robust than T2-FGMM-UV. Indeed, only the means are estimated and tracked correctly over time in the GMM maintenance. The variance and the weights are unstable and unreliable, as explained by Greiffenhagen et al. [28].

Training a T2-FGMM consists of estimating the parameters μ, $\sum$ and the factor k_m or k_ν. Zeng et al. [62] set the factors k_m and k_ν as constants according to prior knowledge. In background modeling, they are fixed depending on the video. Thus, parameter estimation of T2-FGMM includes three steps:

- Step 1: Choose K between 3 and 5.
- Step 2: Estimate GMM parameters by an EM algorithm.
- Step 3: Add the factor k_m or k_ν to GMM to produce T2-FGMM-UM or T2-FGMM-UV.

Once the training is made, a first foreground detection can be processed. By using the ratio $= r_j = \omega_j/\sigma_j$, we first order the K Gaussians as in [61]. This ordering supposes that a background pixel corresponds to a high weight with a weak variance due to the fact that the background is more present than moving objects and that its value is practically constant. The first B Gaussian distributions that exceed certain threshold T are retained for a background distribution:

$$B = argmin_b \left(\sum_{i=1}^{b} \omega_{i,t} > T \right). \tag{5.10}$$

The other distributions are considered to represent a foreground distribution. When the new frame arrives at times $t+1$, a match test is made for each pixel. For this, we use the log-likelihood, and thus we are only concerned with the length between two bounds of the log-likelihood interval, that is, $H(X_t) = \left| ln\left(\underline{h}(X_t)\right) - ln\left(\overline{h}(X_t)\right)\right|$. In Figure 5.5-a), the Gaussian primary MF with uncertain mean has:

$$H(X_t) = \begin{cases} \frac{2k_m|X_t-\mu|}{\sigma}, & \text{if } X_t \leq \mu - k_m\sigma \text{ or } X_t \geq \mu + k_m\sigma \\ \\ \frac{|X_t-\mu|}{2\sigma^2} + \frac{k_m|X_t-\mu|}{\sigma} + \frac{k_m^2}{2}, & \text{if } \mu - k_m\sigma < X_t < \mu + k_m\sigma. \end{cases} \quad (5.11)$$

In Figure 5.5-b), the Gaussian primary MF with uncertain standard deviation has

$$H(X_t) = \left(\frac{1}{1/k_\nu^2 - k_\nu^2}\right)\frac{|X_t-\mu|^2}{2\sigma^2}. \quad (5.12)$$

μ and σ are the mean and the std of the original certain T1 MF without uncertainty. Both Equations (5.11) and (5.12) are increasing functions in terms of the deviation $|X_t - \mu|$. For example, given a fixed k_m, the farther the X_t deviates from μ, the larger $H(X_t)$ is in Equation (5.11), which reflects a higher extent of the likelihood uncertainty. This relationship agrees with the outlier analysis. If the outlier X_t deviates farther from the center of the class-conditional distribution, it has a larger $H(X_t)$, showing its greater uncertainty to the class model. So, a pixel is ascribed to a Gaussian if

$$H(X_t) < k\sigma, \quad (5.13)$$

where k is a constant threshold determined experimentally and is equal to 2.5. Then, two cases can occur: (1) A match is found with one of the K Gaussians. In this case, if the Gaussian distribution is identified as a background one, the pixel is classified as background, or else the pixel is classified as foreground. (2) No match is found with any of the K Gaussians. In this case, the pixel is classified as foreground. At this step, a binary mask is obtained. Then, to make the next foreground detection, the parameters must be updated.

T2-FGMM Maintenance is made as in the original GMM [61] as follows:

- Case 1: A match is found with one of the K Gaussians. For the matched component, the update is done as follows:

$$\omega_{i,t+1} = (1-\alpha)\,\omega_{i,t} + \alpha, \quad (5.14)$$

  where α is a constant learning rate.

$$\mu_{i,t+1} = (1-\rho)\,\mu_{i,t} + \rho X_{t+1} \quad (5.15)$$

$$\sigma_{i,t+1}^2 = (1-\rho)\,\sigma_{i,t}^2 + \rho\,(X_{t+1} - \mu_{i,t+1})\,(X_{t+1} - \mu_{i,t+1})^T, \quad (5.16)$$

  where $\rho = \alpha\eta\left(X_{t+1}, \mu_i, \sum_i\right)$.
  For the unmatched components, μ and σ are unchanged, only the weight is replaced by $\omega_{j,t+1} = (1-\alpha)\,\omega_{j,t}$.

- Case 2: No match is found with any of the K Gaussians. In this case, the least probable distribution k is replaced by a new one with parameters:

$$\omega_{k,t+1} = \text{Low Prior Weight} \tag{5.17}$$
$$\mu_{k,t+1} = X_{t+1} \tag{5.18}$$
$$\sigma^2_{k,t+1} = \text{Large Initial Variance} \tag{5.19}$$

Once a background maintenance is made, another foreground detection can be processed, and so on.

Results presented in Section 5.5.1 show the relevance of T2-FGMM in the presence of camera jitter, waving trees, and water rippling.

5.3 Foreground Detection

Foreground detection consists of classifying pixels as background and foreground by comparing the background and the current images. In general, a simple subtraction is made between these two images to detect regions corresponding to foreground. Another way to establish this comparison consists of defining a similarity measure between pixels in current and background images. In this case, pixels corresponding to background should be similar in the two images, while pixels corresponding to foreground should not be similar.

Another issue in the foreground detection is the choice of features. Color features are the main features used because colors are often very discriminative features of objects, but they do have several limitations in the presence of some critical situations: illumination changes, camouflage, and shadows. To solve these problems, some authors have proposed to use other features like edge [31], texture [30], and stereo features [29], in addition to the color features. The stereo features deal with the camouflage but two cameras are needed. The edge feature allows us to deal with the local illumination changes and the ghost left when waking foreground objects begin to move [2]. The texture features are appropriate for both of illumination changes and shadows.

In the literature, fuzzy concepts have been used in foreground detection in two ways: The first interest [55] consists of avoiding a crisp decision when the comparison is made for the classification and the second one [3, 63] is to aggregate different features using fuzzy integrals. The idea is to be more robust to illumination changes and shadows.

5.3.1 Saturating Linear Function

In standard foreground detection, a crisp limiter function is used to classify pixels, as background or foreground:

$$F_t(x,y) = \begin{cases} 1, & \text{if } d\left(I_t(x,y), B_{t-1}(x,y)\right) > T \\ 0, & \text{otherwise,} \end{cases} \tag{5.20}$$

where $I_t(x, y)$, $B_{t-1}(x, y)$ and $F_t(x, y)$ are respectively the current image at time t, the background images at time $t-1$, and the foreground mask at time t. $B_{t-1}(x, y)$ can be obtained by any background modeling method. Using a running average as the background model, Sigari et al. [55] proposed the use of a saturating linear function instead of a crisp limiter as follows:

$$F_{f,t}(x,y) = \begin{cases} 1, & \text{if } d\left(I_t(x,y), B_{t-1}(x,y)\right) > T \\ \frac{|I_t(x,y)-B_{t-1}(x,y)|}{T}, & \text{otherwise,} \end{cases} \tag{5.21}$$

So, the result of fuzzy foreground detection is a real value in the range [0,1]. To obtain a binary foreground mask, Sigari et al. [55] applied a lowpass filter (LPF). In simple mode, a 3*3 mean filter is used. Then, the binary foreground mask is computed as follows:

$$F_t(x,y) = \begin{cases} 1, & \text{if } |LPF(F_{f,t}(x,y)| > T_f \\ 0, & \text{otherwise,} \end{cases} \tag{5.22}$$

where T_f is a threshold.

Sigari et al. [54] applied this fuzzy foreground detection in vehicle detection. The background model is a running average and the background maintenance is a fuzzy adaptive one (see Section 5.4.1). Results [54] show the pertinence of this approach in the case of camouflage.

Furthermore, Shakeri et al. [51, 52] have adapted this method in cellular automata for urban traffic applications. Each frame sequence is modeled by a cellular automata, and specific cellular automata rules are applied on pixels. Computation is done independently in all cells. Instead of using a trial and error procedure [55], this approach uses a specific threshold value that is obtained from the following equation:

$$F_t(x,y) = \begin{cases} 1, & \text{if } d\left(I_t(x,y), B_{t-1}(x,y)\right)/255 > exp(-(3+m)/2) * k \\ 0, & \text{otherwise,} \end{cases} \tag{5.23}$$

where $I_t(x, y)$, $B_{t-1}(x, y)$, and $F_t(x, y)$ are respectively the current image, the background images, and the foreground mask at time t. $B_{t-1}(x, y)$ can be obtained by any background modeling method. Note that dividing the distance between the background and current image assumes a number in the range [0,1]. In this equation, m is the number of steps that foreground detection procedure has been performed on this frame. Also, k is obtained by fuzzy sets.The initial values of m and k are respectively set to 0 and 1. Experimental results [52] show better performance for the fuzzy cellular foreground detection against the fuzzy foreground detection [55].

The background model used in Sigari et al. [55] and Shakeri et al. [52] is the running average and thus is unimodal. This fuzzy scheme can be used for multi-modal background as developed by Rossel-Ortega et al. [50]. The background model is a multi-modal such as in [48]. Foreground detection is made by thresholding the background membership value. Experiments [50] show the performance on the Wallflower dataset against the crisp multi-modal [48] and the crisp unimodal [49] background model with confidence measure.

This fuzzy foreground detection based on a saturating linear function allows us to avoid crisp decisions and enhance the robustness to some critical situations such as camouflage in a traffic scene.

5.3.2 Fuzzy Integrals

Another way to perform the foreground detection is to use another feature with the color features and to aggregate the result by a fuzzy integral. As seen in the Section 5.1, the choice of the feature is essential and using more than one feature allows to be more robust to illumination changes and shadows. In this context, Zhang and Xu [63] have used texture and color features to compute similarity measures between current and background pixels. Then, these similarity measures are aggregated by applying the Sugeno integral. The assumption made by the authors reflects that the scale is ordinal. The moving objects are detected by thresholding the results of the Sugeno integral. Recently, El Baf et al. [3–5] have shown that the scale is continuum in the foreground detection. Therefore, they used the same features with the Choquet integral instead of the Sugeno integral. Ding et al. [22] used the Choquet integral too but they change the similarity measures. Azab et al. [1] have aggregated recently three features, i.e color, edge and texture. The Figure 5.6 shows the fuzzy foreground detection process with the color and texture features. The fuzzy integral can be the Sugeno integral [63] or the Choquet

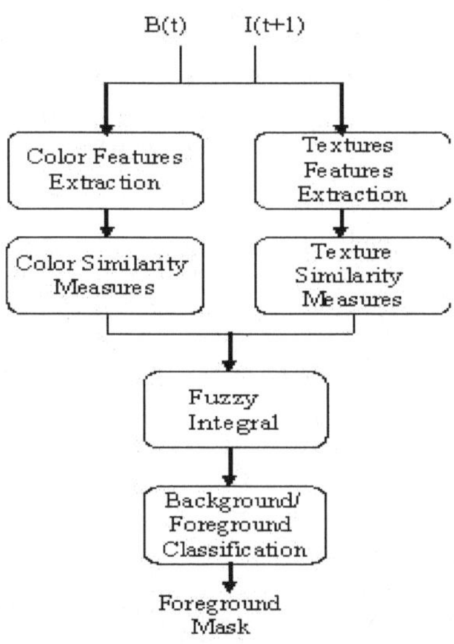

FIGURE 5.6 Fuzzy foreground detection process.

integral [3]. We describe in the following each step of this process with the Choquet integral. Furthermore, the process is similar for more than two features.

Color and textures features

- *Color Features:* The selection of the color space, such as color features, is one of the key factors for efficient color information extraction. In foreground detection, the most commonly used is the RGB space, because it is the one directly available from the sensor or the camera. The RGB color space has an important drawback: their three components are dependent which increase its sensitivity to illumination changes. For example, if a background point is covered by the shadow, the three component values at this point could be affected because the brightness and the chromaticity information are not separated. A number of color space comparisons are presented in the literature [32,35,47] and usually YCrCb is selected as the most appropriate color space. But first, let us define the different color spaces (the Ohta, HSV, and YCrCb) tested that separate the luminance and the chrominance channels. The axes of the Ohta space [45] are the three largest eigenvectors of RGB space, found from the principal components analysis of a large selection of natural images. This color space is a linear transformation of RGB and has three components I_1,I_2, and I_3. Equation (5.24) shows the relationship between RGB to the Ohta space:

$$\begin{aligned} I_1 &= (R+G+B)/3 \\ I_2 &= (R-B)/2 \ (or \ (B-R)/2) \\ I_3 &= (2G-R-B)/4 \end{aligned} \tag{5.24}$$

  HSV and YCrCb are closer to human interpretation of colors in the sense that brightness, or intensity, is separated from the base color. YCrCb uses cartesian coordinates to describe the base color while HSV uses polar coordinates. For HSV, the color information improves the discrimination between shadow and object, classifying as shadows those pixels having approximately the same hue and saturation values compared to the background, but lower luminosity. Equations (5.25) and (5.26) below show the relationships between RGB and HSV, then YCrCb color spaces:

$$\begin{aligned} H &= \begin{cases} 60\,(G-B)/\Delta & \text{if } max\,(R,G,B)=R \\ 60\,(B-R)/(\Delta+120) & \text{if } max\,(R,G,B)=G \\ 60\,(R-G)/(\Delta+240) & \text{if } max\,(R,G,B)=B \end{cases} \\ S &= \Delta/max\,(R,G,B) \\ V &= max\,(R,G,B), \end{aligned} \tag{5.25}$$

  where $\Delta = max\,(R,G,B) - min\,(R,G,B)$.

$$\begin{aligned} Y &= 0.25R + 0.504G + 0.098B + 16 \\ Cr &= 0.439R - 0.368G - 0.071B + 128 \\ Cb &= -0.148R - 0.291G + 0.439B + 128 \end{aligned} \tag{5.26}$$

For each color space, two components are chosen according to their sensitivity to illumination changes. For example, the components Cr and Cb are retained as the luminance Y is very sensitive to illumination changes.

- *Texture Feature:* The texture feature used is the Local Binary Pattern (LBP), which was developed by Heikkila and Pietikinen [30]. The LBP is invariant to monotonic changes in gray scale, which makes it robust against illumination changes. The operator labels the pixels of an image block by thresholding the neighborhood of each pixel with the centre value and considering the result as a binary number:

$$LBP(x, y) = \sum_{i=0}^{N-1} s(g_i - g)\, 2^i,$$

where g corresponds to the gray value of the center pixel (x, y) and g_i to the gray values of the N neighborhood pixels. The function s is defined as follows:

$$s(z) = \begin{cases} 1 & if \quad z \geq 0 \\ 0 & if \quad z < 0. \end{cases}$$

The original LBP operator worked with the 3×3 neighbourhood of a pixel.

Once the features are chosen, we can compare the background and current images using similarity measures that have to be defined.

Similarity measures

The comparison is made using a similarity measure between pixels in the current and background images. We define in the following the similarity measures for the color and the texture features as they were developed in [5] but other similarity measures can be found in [1, 22].

- *Color Similarity Measures:* We describe the similarity measure in a general way, that is, the color features may be any color space with three components noted I_1, I_2 and I_3. Then, the color similarity measure $S_k^C(x, y)$ at the pixel (x, y) is computed as in [63]:

$$S_k^C(x, y) = \begin{cases} \frac{I_k^C(x,y)}{I_k^B(x,y)} & if \quad I_k^C(x, y) < I_k^B(x, y) \\ 1 & if \quad I_k^C(x, y) = I_k^B(x, y) \\ \frac{I_k^B(x,y)}{I_k^C(x,y)} & if \quad I_k^C(x, y) > I_k^B(x, y), \end{cases}$$

where $k \in \{1, 2, 3\}$ is one of the three color features. B and C represent respectively the background and the current images at time t. B can be obtained using any of the background modelling method. Note that $S_k^C(x, y)$ is between 0 and 1. Furthermore, $S_k^C(x, y)$ is closed to 1 if $I_k^C(x, y)$ and $I_k^B(x, y)$ are very similar.

- *Texture Similarity Measure:* The texture similarity measure $S^T(x, y)$ at the pixel (x, y) is computed as follows:

$$S^T(x, y) = \begin{cases} \frac{L^C(x,y)}{L^B(x,y)} & if \quad L^C(x, y) < L^B(x, y) \\ 1 & if \quad L^C(x, y) = L^B(x, y) \\ \frac{L^B(x,y)}{L^C(x,y)} & if \quad L^C(x, y) > L^B(x, y), \end{cases}$$

where $L^B(x, y)$ and $L^C(x, y)$ are respectively the texture LBP of pixel (x, y) in the background and current images at time t. Note that $S^T(x, y)$ is between 0 and 1. Furthermore, $S^T(x, y)$ is close to 1 if $L^B(x, y)$ and $L^C(x, y)$ are very similar.

Once these two similarity measures are computed, they can be aggregated by a fuzzy integral. We remind in the following some concepts on fuzzy integrals and then we explain the aggregation of the similarity measures with the Choquet integral.

Fuzzy integrals: A brief background

There are several mathematical operators used for the aggregation of different measures. In the literature [21], we find the basic ones like the average, the median, the minimum, and the maximum, as well as some generalizations like the Ordered Weighted Average (OWA) having the minimum, and the maximum as particular cases and k-order statistics. Then, the family of fuzzy integrals has presented through its discrete version a generalization of OWA or the weighted average using the Choquet integral, as well as the minimum and the maximum using the Sugeno integral. The advantage of fuzzy integrals is that they take into account the importance of the coalition of any subset of criteria. We briefly summarize the necessary concepts around fuzzy integrals (Sugeno and Choquet).
Let μ be a fuzzy measure on a finite set X of criteria and $h : X \to [0, 1]$ be a fuzzy subset of X.

Definition 1 *The Sugeno integral of h with respect to μ is defined by*

$$S_\mu = Max\left(Min\left(h\left(x_{\sigma(i)}\right), \mu\left(A_{\sigma(i)}\right)\right)\right), \tag{5.27}$$

where σ is a permutation of the indices such that
$h\left(x_{\sigma(1)}\right) \leq \ldots \leq h\left(x_{\sigma(n)}\right)$ *and* $A_{\sigma(i)} = \{\sigma(1), \ldots, \sigma(n)\}$.

Definition 2 *The Choquet integral of h with respect to μ is defined by*

$$C_\mu = \sum_{i=0}^{n} h\left(x_{\sigma(i)}\right)\left(\mu\left(A_{\sigma(i)}\right) - \mu\left(A_{\sigma(i+1)}\right)\right) \tag{5.28}$$

with the same notations as above.

An interesting interpretation of the fuzzy integral arises in the context of the source fusion. The measure μ can be viewed as the factor that describes the relevance of the sources of information, where h denotes the values that the criteria have reported. The fuzzy integrals then aggregate nonlinearly the outcomes of all criteria. The Choquet integral is adapted for cardinal aggregation while the Sugeno integral is more suitable for ordinal aggregation. More details can be found in [44, 58, 59].

In the fusion of different criteria or sources, the fuzzy measures take on an interesting interpretation. A pixel can be evaluated based on criteria or sources providing information about the state of the pixel, whether the pixel corresponds to background or foreground. The more the criteria provide information about the pixel, the more relevant the decision of the pixel's state. Let $X = \{x_1, x_2, x_3\}$, with each criterion, we associate a fuzzy measure $\mu(x_1) = \mu(\{x_1\})$, $\mu(x_2) = \mu(\{x_2\})$ and $\mu(x_3) = \mu(\{x_3\})$ such that the higher the $\mu(x_i)$, the more important the corresponding criterion in the decision. To compute the fuzzy measure of the union of any two disjoint sets whose fuzzy measures are given, we use an operational version proposed by Sugeno that is called the λ-*fuzzy measure*. To avoid excessive notation, let us denote this measure by μ_λ-*fuzzy measure*, where λ is a paramater of the fuzzy measure used to describe an interaction between the criteria that are combined. Its value can be determined through the boundary condition, that is, $\mu(X) = \mu(\{x_1, x_2, x_3\}) = 1$. The fuzzy density values over a given set $K \subset X$ is computed as

$$\mu_\lambda(K) = \frac{1}{\lambda}\left[\prod_{x_i \in K}\left(1 + \lambda\mu_\lambda(x_i)\right) - 1\right]. \tag{5.29}$$

Aggregation of color and texture similarity measures by a fuzzy integral

As defined above, the computed measures are obtained by dividing the intensity values in the background and current images by endpoints denoted by 0 and 1, where 0 means that the pixels at the same location in background and current images respectively are not similar, and 1 means that these pixels are similar, that is, the pixel corresponding to background. In such a case, the scale is a continuum and is constructed as a cardinal one where the distances or the differences between values can be defined. For example, the distance between 0.1 and 0.2 is the same as the distance between 0.8 and 0.9, because numbers have a real meaning. In the case of an ordinal scale, the numbers correspond to modalities when an order relation on the scale should be defined. A typical example of the former is when we define a scale [a, b, c, d, e] to evaluate the level of some students, where "a" corresponds to "excellent" and "e" to "very bad." So that, the difference between "b" (very good) and "c" (good) is not necessarily the same as the difference between "c" (good) and "d" (bad). Hence, operations other than comparison on a cardinal scale can be allowed such as standard arithmetic operations, typically addition and multiplication. In this sense, the Choquet integral is considered more suitable than the Sugeno integral because of its ability to aggregate well the features on a cardinal scale and to use such arithmetic operations. So, for each pixel, color and texture similarity measures are computed, as explained previously from the background and the current frame. We define the set of criteria $X = \{x_1, x_2, x_3\}$ with, (x_1, x_2) = two components color features of the chosen color space (i.e. Ohta, HSV, YCrCb, etc..) and x_3 = texture feature obtained by the LBP.

For each x_i, let $\mu(x_i)$ be the degree of importance of the feature x_i in the decision whether the pixel corresponds to background or foreground. The fuzzy functions $h(x_i)$ are defined in $[0, 1]$ so that, $h(x_1) = S_1^C(x, y)$, $h(x_2) = S_2^C(x, y)$ and $h(x_3) = S^T(x, y)$. To compute the value of the Choquet integral for each pixel, we need first to rearrange the features x_i in the set X with respect to the order: $h(x_1) \geq h(x_2) \geq h(x_2)$.

The pixel at position (x, y) is considered foreground if its Choquet integral value is less than a certain constant threshold Th:

$$if \quad C_\mu(x, y) < Th \quad then \quad (x, y) \ is \ foreground,$$

which means that pixels at the same position in the background and the current images are not similar. Th is a constant value depending on each video dataset.

Results presented in Section 5.5.2 show the relevance of the fuzzy integrals in the presence of illuminations changes and shadows.

5.4 Background Maintenance

The background maintenance determines how the background will adapt itself to take into account the critical situations that can occur. In the literature, there are two maintenance schemes: the blind one and the selective one.

Blind background maintenance consists of updating all the pixels with the same rules, which is usually an IIR filter, as follows:

$$B_t(x, y) = (1 - \alpha) B_{t-1}(x, y) + \alpha I_t(x, y), \tag{5.30}$$

where α is a learning rate that is a constant in the interval $[0, 1]$.

The disadvantage of this scheme is that the value of pixels classified as foreground are taken into account in the computation of the new background and thus pollute the background image. To solve this problem, some authors use a selective maintenance which consists of computing the new background image with a different learning rate following its previous classification into foreground or background as follows:

$$\begin{aligned} B_t(x,y) &= (1-\alpha)B_{t-1}(x,y) + \alpha I_t(x,y) \\ & \quad if\ \ (x,y)\ \ is\ background \\ B_t(x,y) &= (1-\beta)B_{t-1}(x,y) + \beta I_t(x,y) \\ & \quad if\ \ (x,y)\ \ is\ foreground. \end{aligned} \tag{5.31}$$

Here, the idea is to adapt very quickly a pixel classified as background and very slowly a pixel classified as foreground. For this reason, $\beta << \alpha$ and usually $\beta = 0$. So the second expression in Equation (5.31) becomes

$$B_t(x,y) = B_{t-1}(x,y). \tag{5.32}$$

In this crisp decision, there are two learning rates: one learning rate for pixels classified as background and another for pixels classified as foreground. This crisp decision generates the fact that erroneous classification results may make a permanent incorrect background model. To avoid this problem, an adaptive learning rate that integrates the membership of the pixels to one of the class background or foreground seems to be better. So, the adaptive background maintenance consists of the following:

$$B_t(x,y) = (1-\alpha_t(x,y))B_{t-1}(x,y) + \alpha_t(x,y)I_t(x,y), \tag{5.33}$$

where $\alpha_t(x,y)$ is an adaptive learning rate that is a variant in the interval $[0,1]$ and depends on the membership of the pixels in the class background or foreground. Different approaches were proposed in the literature to compute $\alpha_t(x,y)$.

5.4.1 Fuzzy Learning Rates

Sigari el al. [54, 55] proposed to compute an adaptive learning rate $\alpha_t(x,y)$ at each pixel following the fuzzy membership value obtained by each pixel in fuzzy foreground detection (see Section 5.3.1). Therefore, $\alpha_t(x,y)$ is computed as follows:

$$\alpha_t(x,y) = (1-\alpha_{min})exp(-5 * F_{f,t}(x,y)), \tag{5.34}$$

where $F_{f,t}(x,y)$ is the fuzzy foreground mask obtained by the saturating linear function (see Section 5.3.1) and α_{min} is the minimum value for α. In Sigari et al. [54, 55], α_{min} is set to 0.9.

In the same idea to adapt the background, Maddalena and Petrosino [39] proposed an adaptive learning rate in a neural network background modeling. In this case, $\alpha_t\,(x, y)$ is computed as follows:

$$\alpha_t\,(x, y) = (1 - F_t(x, y)) * \alpha_t * w_{x,y}, \tag{5.35}$$

where $w_{x,y}$ are Gaussians weights in the n*n neighborhood, α_t represents the learning factor that is the same for each pixel of the t-th sequence frame and depends on the scene variability, and $F_t(x, y)$ is a saturating linear function given by:

$$F_t(x, y) = \begin{cases} \frac{d\,(c_m(x,y), p_t(x,y))}{\epsilon}, & \text{if } d\,(c_m(x, y), p_t(x, y)) < \epsilon \\ 1, & \text{otherwise,} \end{cases} \tag{5.36}$$

where $c_m(x, y)$ is a weight vector coming from the neural network and corresponds to the background value at the pixel (x,y). $p_t(x, y)$ is the current value of the pixel (x,y), and ϵ is a constant threshold. $F_t(x, y)$ can be interpreted as the membership value of the pixel (x,y) to the background model. So, the closer the current pixel (x,y) is to the background model, the more it can contribute to the maintenance.

Recently, Maddalena and Petrosino [40,41] performed the adaptivity by introducing spatial coherence information. In this scheme, $\alpha_t\,(x, y)$ is computed as follows:

$$\alpha_t\,(x, y) = (1 - F_{1,t}(x, y) * F_{2,t}(x, y)) * \alpha_t * w_{x,y}, \tag{5.37}$$

where the function $F_{1,t}$ is the adaptive fuzzy factor and $F_{2,t}$ is the fuzzy coherence factor. So, $F_{1,t}(x, y)$ is equal to $F_t(x, y)$, which is defined in Equation (5.36). $F_{2,t}$ is defined as follows:

$$F_{2,t}(x, y) = \begin{cases} 2NCF(p_t(x, y)) - 1, & \text{if } NCF(p_t(x, y)) > 0.5 \\ 1, & \text{otherwise,} \end{cases} \tag{5.38}$$

where $NCF(p_t(x, y))$ is a neighborhood coherence factor defined in [40]. This factor is based on the fact that the greater $NCF(p_t(x, y))$ is, the greater is the number of the pixels in the neigborhood of the pixel (x,y) that represented the background model and then the better the pixel (x,y) represents the background.

Comparative results between the adaptive learning rate [55] and the adaptive learning rate with spatial coherence [40] are presented qualitatively in Section 5.5.3 and quantitatively in Section 5.5.4.

5.4.2 Fuzzy Maintenance Rules

One disadvantage of selective maintenance is mainly due to the crisp decision that attributes a different rule following the classification in background or foreground. To solve this problem, El Baf et al. [8] proposed to take into account the uncertainty of the classification. This can be made by graduating the update rule using the result of the Choquet integral as follows:

$$\begin{aligned} B_t(x,y) &= \mu_F B_{t-1} + \\ & (1-\mu_F)\left((1-\alpha) B_{t-1}(x,y) + \alpha I_t(x,y)\right), \end{aligned} \tag{5.39}$$

where $\mu_F = 1 - \mu_B$. μ_F and μ_B are respectively the fuzzy membership values of the pixel (x, y) to the class foreground and background. μ_B is a function of $C_\mu(x,y)$ such as $\mu_B = 1$ for $Max\left(C_\mu(x,y)\right)$ and $\mu_B = 0$ for $Min\left(C_\mu(x,y)\right)$.

We can note that adaptive maintenance is a generalized version of selective maintenance. Indeed, if the pixel is classified as background with the Choquet integral value equal to 1, we retrieve the first expression of Equation (5.31), and if the pixel is classified as foreground with the Choquet integral value equal to 0, Equation (5.40) is equal to Equation (5.32). This fuzzy adaptive scheme is tested on the Wallflower test in Section 5.5.3.

5.5 Experimental Results

We have evaluated the different existing fuzzy approaches. First, we compare the fuzzy method with their corresponding crisp method. Then, we evaluate them on the same dataset [53].

5.5.1 Fuzzy Background Modeling

In this section, the experiments are conducted to compare the results of T2-FGMM [6] with the crisp GMM [57]. So, we have applied the GMM, T2-FGMM-UM, and T2-FGMM-UV algorithms to indoor and outdoor videos where different critical situations occur, such as camera jitter, movement in the background, illuminations changes, and shadows. For T2-FGMM, the best results were obtained with the values 2 and 0.9, respectively, for the factors k_m and k_ν.

Indoor scene videos

The PETS 2006 dataset [27] provides several video presenting indoor sequences in a video surveillance context. In these video sequences, there are illumination changes and shadows. Figure 5.7 presents the results obtained by GMM [57], T2-FGMM-UM, and T2-FGMM-UV. It is noticed that the results obtained using T2-FGMM-UM and T2-FGMM-UV are better than using the crisp GMM. The silhouettes are well detected with the T2-FGMM-UM. T2-FGMM-UV is more sensitive because the variance is more unstable over time.

a) Current Image b) GMM c) T2-FGMM-UM d) T2-FGMM-UV

FIGURE 5.7 Background subtraction with illumination changes (PETS 2006 dataset). From left to right: a) The original image (Frame 165), b) the mask obtained by the GMM, c) the mask obtained using T2-FGMM-UM, d) the mask obtained using T2-FGMM-UV [6]. (©2008 Springer Verlag)

Outdoor scene videos

We have chosen three videos presenting different dynamic backgrounds: camera jitter, waving trees, and water rippling. The first outdoor sequence that was tested involved a camera mounted on a tall tripod and comes from [37]. The wind caused the tripod to sway back and forth, causing nominal motion in the scene. In Figure 5.8, the first row shows different current images. The second row shows the results obtained by the GMM proposed in [57]. It is evident that the motion causes substantial degradation in performance. The third and fourth rows show respectively the results obtained by T2-FGMM-UM and T2-FGMM-UV. As for the indoor scene, T2-FGMM-UM and T2-FGMM-UV give better results than the crisp GMM.

We have also tested our method for the sequences Campus and Water Surface that come from [36]. Figure 5.9 shows the robustness of T2-FGMM-UM against waving trees and water rippling.

5.5.2 Fuzzy Foreground Detection

We have applied the Sugeno and Choquet integrals to different datasets: the first one is the PETS dataset. We have chosen particularly the PETS 2000 [26] and PETS 2006 [27] datasets used in video surveillance. The output images of these two datasets are respectively 768×576 and 720×576 pixels. The second one is our Aqu@theque dataset used in a multimedia application [2], where the output images are 384×288 pixels. The results are obtained without post processing and the threshold for each algorithm is optimized to give the best results.

PETS 2000 and 2006 dataset

We first tested the Sugeno and Choquet integrals on the PETS 2000 and 2006 benchmark data indoor and outdoor sequences in video surveillance context. The goal is to detect moving persons and/or vehicules. The use of the Choquet integral [3] with YCrCb color space shows more robustness to illumination

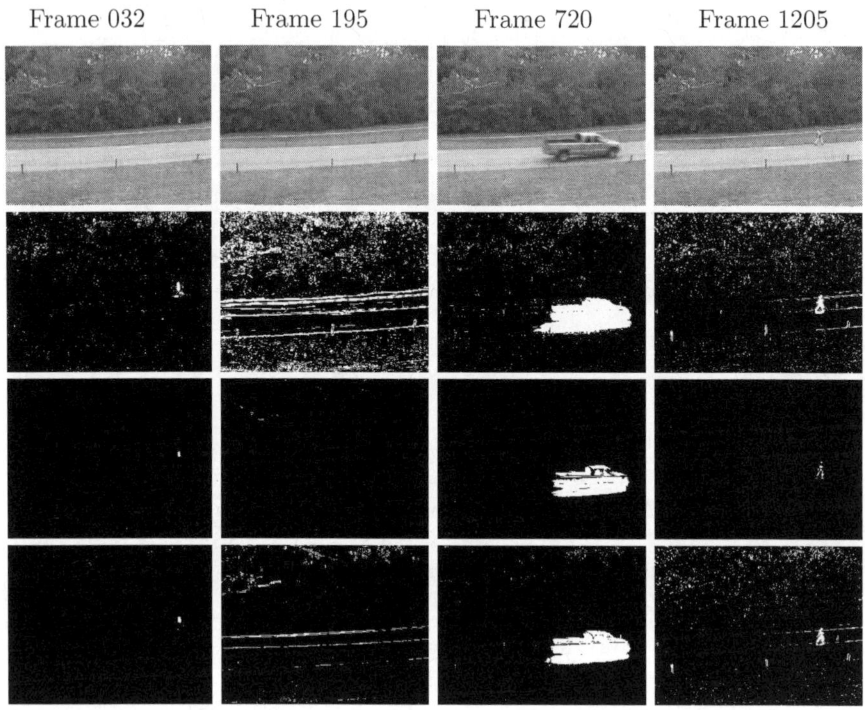

FIGURE 5.8 Fuzzy background modeling: The first row shows the original frames for the sequence Camera Jitter. The second row presents the segmented images obtained by GMM. The third and fourth rows illustrate the result obtained using T2-FGMM-UM and T2-FGMM-UV, respectively [6]. (©2008 Springer Verlag.)

FIGURE 5.9 Fuzzy background modeling: The first row shows the original frames for Campus and Water Surface sequences. The second row presents the segmented images obtained by GMM. The third and fourth rows illustrate the result obtained using T2-FGMM-UM and T2-FGMM-UV, respectively [6]. (©2008 Springer Verlag.)

changes and shadows than the Sugeno integral [63], as we can see qualitatively in the Figures 5.10 and 5.11.

a) Current Image b) Sugeno-Ohta c) Choquet-YCrCb

FIGURE 5.10 From left to right: a) The current image, b) the mask obtained by the Sugeno integral using the Ohta space, c) the mask obtained by the Choquet integral using the YCrCb space [5]. (©2008 IEEE.)

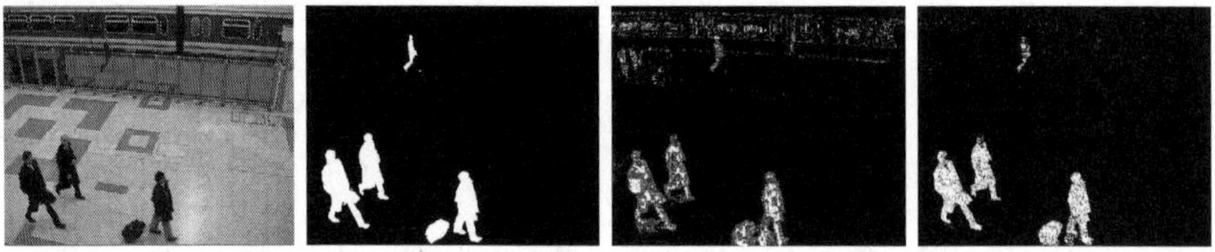

a) Current Image b) Ground Truth c) Sugeno-Ohta d) Choquet-YCrCb

FIGURE 5.11 From left to right: a) The current image, b) the ground truth, c) the mask obtained by the Sugeno integral using the Ohta space, d) the mask obtained by the Choquet integral using the YCrCb space [5]. (©2008 IEEE.)

Aqu@theque dataset

This dataset contains several video sequences presenting fish in a tank. The goal of the application Aqu@theque [2] is to detect fish and identify them. In these aquatic video sequences, there are many critical situations. For example, there are illumination changes due to the ambient light, the spotlights that light the tank from the inside and from the outside, the movement of the water due to fish, and the continuous renewal of the water. These illumination changes can be local or global, following their origin. Furthermore, the constitution of the aquarium (rocks, algae) and the texture of the fish amplify the consequences of the brilliant variation. Figure 5.12 shows the experiments made on one sequence. In Table 5.2, we show the fuzzy density values that we have tested. The best results are obtained with $\{0.53, 0.0.34, 0.13\}$. It is noticed that the results obtained using the Choquet integral [3] are better than those using the Sugeno integral [63] with the same color space, that is, Ohta. The results obtained with the Choquet integral using other color spaces, that

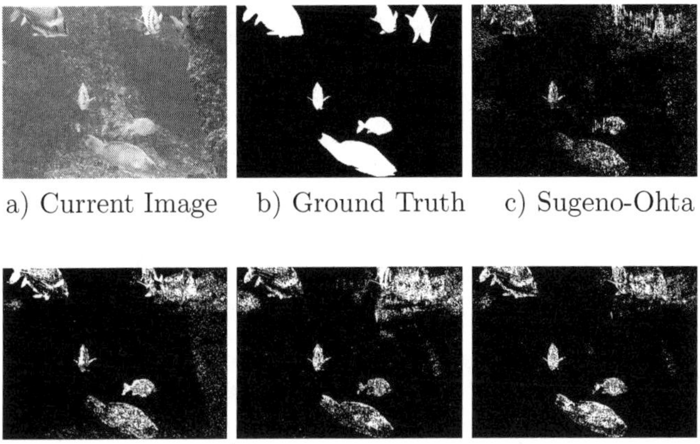

a) Current Image b) Ground Truth c) Sugeno-Ohta

d) Choquet-Ohta e) Choquet-HSV f) Choquet-YCrCb

FIGURE 5.12 First row: The current image, the ground truth, Sugeno-Ohta. Second row: Choquet-Ohta, Choquet-HSV, and Choquet-YCrCb [5]. (©2008 IEEE.)

is, the HSV and YCrCb, confirmed that optimum results are obtained using the Choquet integral with the YCrCb color features.

TABLE 5.2 Fuzzy measure values.

$\{x_1\}$	$\{x_2\}$	$\{x_3\}$	$\{x_1, x_2\}$	$\{x_1, x_3\}$	$\{x_1, x_3\}$	$\{X\}$
0.6	0.3	0.1	0.9	0.7	0.4	1
0.5	0.4	0.1	0.9	0.6	0.5	1
0.5	0.3	0.2	0.8	0.7	0.5	1
0.50	0.39	0.11	0.89	0.61	0.5	1
0.53	0.34	0.13	0.87	0.66	0.47	1

The quantitative evaluation was done first using the similarity measure derived by Li [37]. Let A be a detected region and B be the corresponding ground truth, the similarity between A and B can be defined as

$$S(A, B) = \frac{A \cap B}{A \cup B}. \tag{5.40}$$

If A and B are the same, $S(A, B)$ approaches 1, otherwise 0; that is, A and B have the least similarity. The ground truth is marked manually. Table 5.3 shows the similarity value obtained for the previous experiments. It is

TABLE 5.3 Similarity measure.

Integral Color Space	Sugeno Ohta	Choquet Ohta	Choquet HSV	Choquet YCrCb
$S(A, B)$	0.27	0.44	0.34	0.46

well identified that optimum results are obtained by the Choquet integral.

Furthermore, the Ohta and the YCrCb spaces give almost similar results ($S_{Ohta} = 0.44, S_{YCrCb} = 0.46$), when the HSV space registers ($S_{HSV} = 0.34$). When observing the effect of YCrCb and Ohta spaces on the images, we have noticed that the YCrCb is slightly better than the Ohta space. To see the progression of the performance of each algorithm, we use the ROC curves [20]. For that, we compute the false positive rate (FPR) and the true positive rate (TPR) as follows:

$$FPR = \frac{FP}{FP + TN}, \qquad TPR = \frac{TP}{TP + FN},$$

where TP is the total of true positives, TN is the total of true negatives, FP is the total of false positives, and FN is the total of false negatives. The FPR is the proportion of background pixels that was erroneously reported as being moving object pixels. And the TPR is the proportion of moving object pixels that was correctly classified among all positive samples. Figure 5.13 represents the ROC curves for the Sugeno and the Choquet integrals with the Ohta color space. These curves confirm that the Choquet integral outperforms the Sugeno one using the Ohta space.

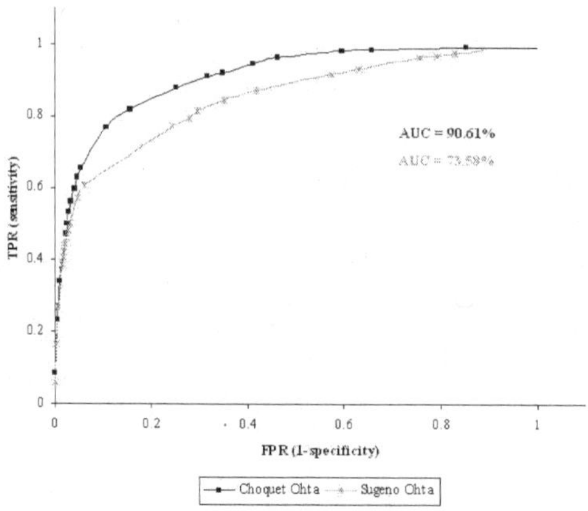

FIGURE 5.13 ROC curve: Comparison of the two detection algorithms using respectively the Sugeno and the Choquet integrals in the Ohta color space [5].(©2008 IEEE.)

Then, we compared the previous results with other color spaces. Figure 5.14 shows the ROC curves for the Choquet integral with the Ohta, HSV, and YCrCb color spaces. Once again, the curves confirm the previous conclusion. Indeed, the Areas Under Curve (AUCs) are very similar for YCrCb and Ohta spaces.

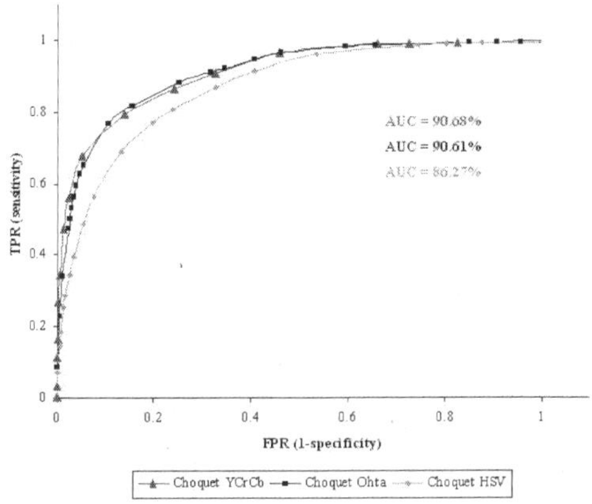

FIGURE 5.14 ROC curve: Evaluation of the effect of different color spaces on the detection algorithm using the Choquet integral [5].(©2008 IEEE.)

5.5.3 Fuzzy Background Maintenance

This section presents tests of the fuzzy learning rates and the fuzzy maintenance rules on the Wallflower dataset [60], which contains seven real video sequences. Each one of them presents typical critical situations.

Fuzzy learning rates

This section presents the results obtained by the fuzzy learning rates from [55] and the fuzzy learning rate with spatial coherence from [40, 41] on the sequence called "Waving Trees". Figure 5.15 shows the original image, the ground truth, the foreground mask obtained by [55] and the foreground mask obtained by [40]. The optimum results are obtained by the fuzzy learning rate with spatial coherence.

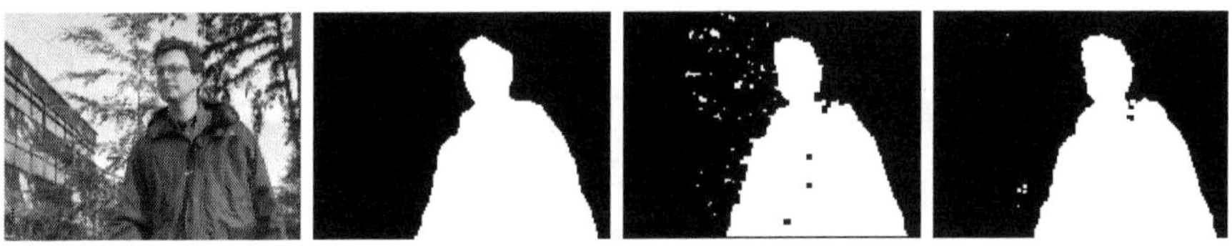

a) Current Image b) Ground Truth c) FBGS [55] d) SOBS-CF [40]

FIGURE 5.15 From left to right: a) The original image, b) the ground truth, c) the foreground mask obtained by [55] and d) the foreground mask obtained by [40,41]. (©2010 Springer Verlag.)

Fuzzy maintenance rules

We have tested the fuzzy maintenance rules [8] on the sequence called "Time of Day". We choose this sequence because it presents gradual illumination changes that alter the background and thus permit us to show the performance of the fuzzy maintenance rules. Figure 5.16 shows the original image, the ground truth, the final foreground mask obtained by the blind scheme, the selective one, and the fuzzy one. Table 5.4 shows the similarity value obtained

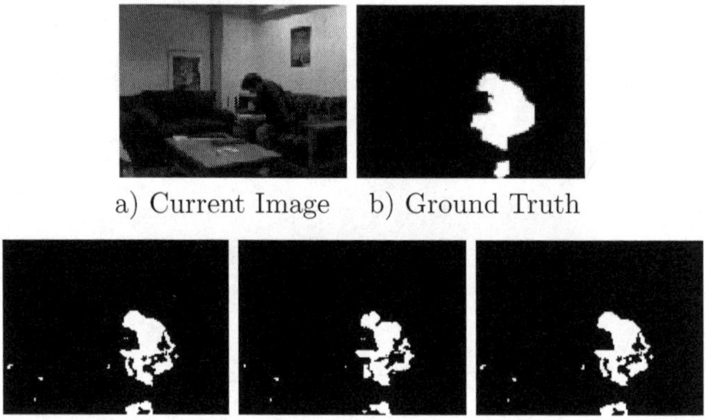

a) Current Image b) Ground Truth

c) Blind d) Selective e) Fuzzy Adaptive

FIGURE 5.16 First row: a) the current image, b) the ground truth. Second row: c) blind maintenance, d) selective maintenance, e) fuzzy adaptive maintenance [8]. (©2008 IEEE.)

for the previous experiments. The optimum results are obtained by the fuzzy adaptive scheme.

TABLE 5.4 Similarity measure.

Maintenance Scheme	Blind Maintenance	Selective Maintenance	Adaptive Maintenance
$S(A,B)$ %	58.40	57.08	58.96

5.5.4 Comparison

In this section, we compare, on the same dataset, the following fuzzy approaches: background modeling by Type-2 Fuzzy GMMs [6], foreground detection by a linear saturation function [55], foreground detection by the Sugeno integral [63], foreground detection by the Choquet integral [3], and the background maintenance with the fuzzy spatial-coherence learning rate [41]. All the methods are compared in their original version and their parameters have been adapted until the results seem optimal over the entire sequence:

- ***Type-2 Fuzzy Gaussian Mixture Models (Section 5.2):*** The background initialization was made on 100 frames. The factors k_m and k_ν were set respectively to the values 2 and 0.9.
- ***Linear Saturation Function (Section 5.3.1):*** The background model is based on the running average. The method is called FBGS as in [41].
- ***Sugeno Integral and Choquet Integral (Section 5.3.2):*** The background model used is the running average applied on 100 frames. For the Sugeno integral, the color space is the Ohta space, while for the Choquet integral it is the YCrCb. The fuzzy measures values are the ones indicated in Table 5.2.
- ***Fuzzy spatial-coherence learning rate(Section 5.4.1):*** The background model is a self-organization through the artificial neural network. This method is called SOBS-CF as in [41].

This dataset comes from [53] and consists of a sequence of 500 frames of 360*240 pixels with ground truth masks. The camera is mounted on a tall tripod and the wind caused it to sway back and forth, causing nominal motion in the scene. This scene consists of a street crossing, where several people and cars pass by. Figure 5.18 shows the original test images in the first row and the corresponding ground truth in the second row while the corresponding results obtained by the fuzzy approaches, that is, T2-FGMM-UM, T2-FGMM-UV, Sugeno, Choquet, FBGS, and SOBS-CF, are reported in this order in the following rows.

We used ground truth-based metrics computed from the true positives (TP), true negatives (TN), false positives (FP), and false negatives (FN) as defined in Section 5.5.2. Table 5.5 shows the FP and FN for the different fuzzy approaches on the frames 271, 373, 410, and 465.

TABLE 5.5 Performance analysis.

Method	Error Type	Frame 271	Frame 373	Frame 410	Frame 465	Total Error
GMM	FN	0	1120	4818	2050	
	FP	2093	4124	2782	1589	18576
T2-FGMM-UM	FN	0	1414	6043	2520	
	FP	203	153	252	46	10631
T2-FGMM-UV	FN	0	957	2217	1069	
	FP	3069	1081	1119	1158	10670
Sugeno Ohta-LBP	FN	0	1852	7370	3210	
	FP	210	146	203	222	13213
Choquet YCrCb-LBP	FN	0	1243	5871	2592	
	FP	54	58	137	40	9995
FBGS	FN	0	431	814	249	
	FP	2034	288	400	684	4900
SOBS-CF	FN	0	147	298	91	
	FP	0	125	292	179	1132

Then, we computed the following metrics: detection rate,precision and F-measure. Detection rate gives the percentage of corrected pixels classified as background when compared with the total number of background pixels in the ground truth:

$$DR = \frac{TP}{TP + FN}. \tag{5.41}$$

Precision gives the percentage of corrected pixels classified as background as compared to the total pixels classified as background by the method:

$$Precision = \frac{TP}{TP + FP}. \tag{5.42}$$

A good performance is obtained when the detection rate is high without altering the precision. We also computed the F-measure used in [41] as follows:

$$F = \frac{2 * DR * Precision}{DR + Precision}. \tag{5.43}$$

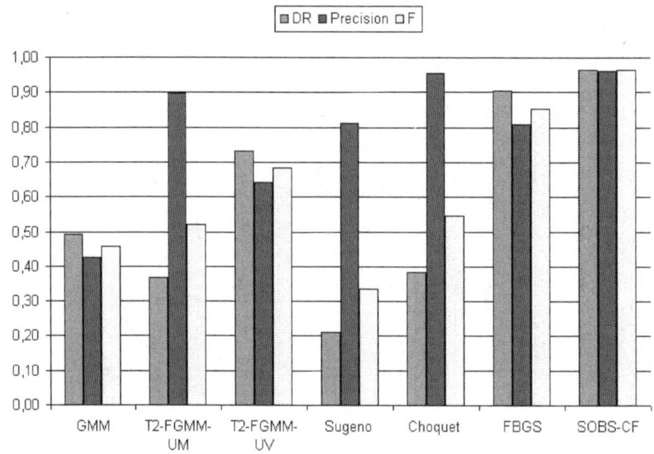

FIGURE 5.17 Overall performance.

Figure 5.17 shows the detection rate, precision and F-measure obtained for each fuzzy approach. We can observe that T2-FGMM-UM and T2-FGMM-UV outperform the crisp GMM. Furthermore, the Choquet integral shows better performance than the Sugeno integral. It confirms that the Choquet integral is more appropriate for foreground detection. Finally, the SOBS-CF algorithm performs better than all other methods due to the background model and the spatial information taken in the adaptive learning rate.

5.6 Conclusion

This chapter attempted to provide a comprehensive survey of research on fuzzy background subtraction. Thus, we reviewed the contribution of the fuzzy

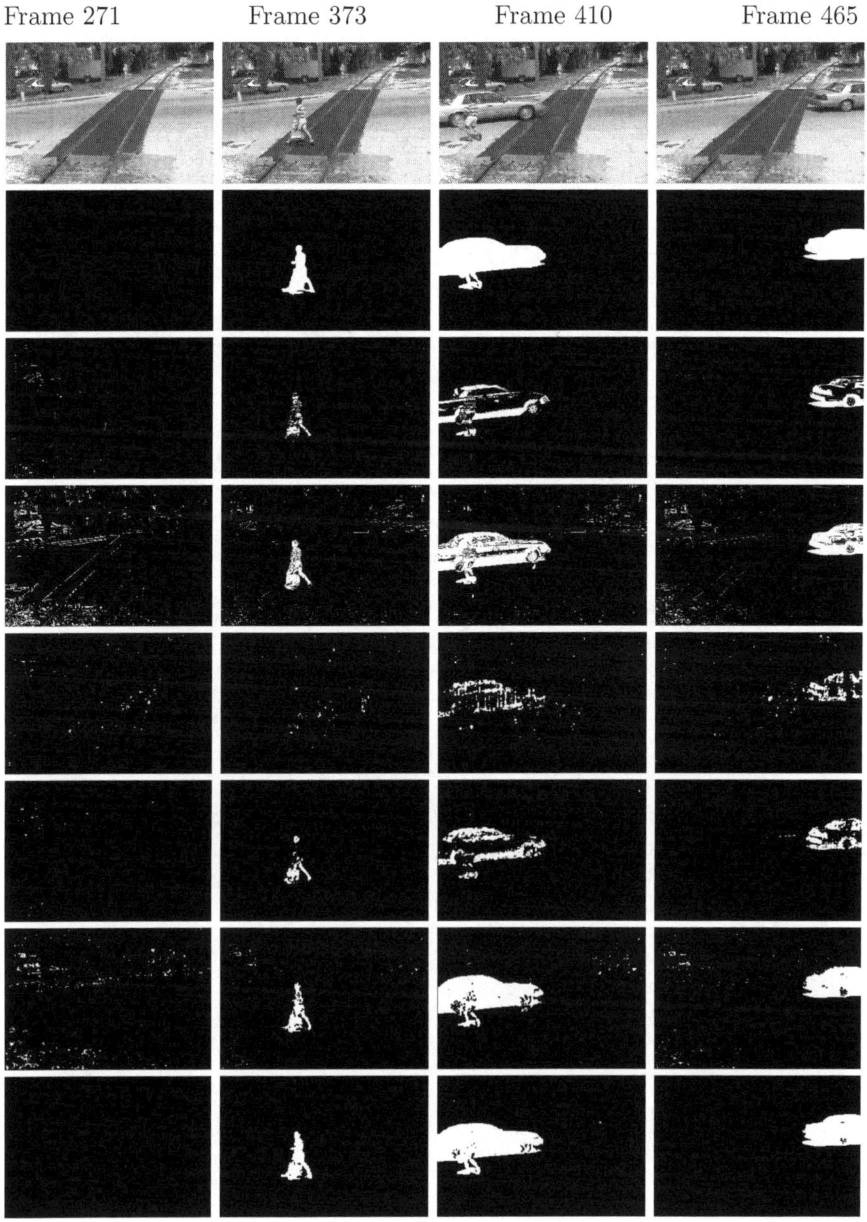

FIGURE 5.18 The first row contains the original images, the second row contains the ground truth images, the third row contains the results obtained using T2-FGMM-UM, and the fourth contains the results obtained by using T2-FGMM-UV, the fifth row contains the results obtained by using FBGS, the sixth and the seventh rows contain respectively the results obtained by using the Sugeno and the Choquet integrals, and the last row contains the results obtained by SOBS-CF. Images from [6] (©2008 Springer Verlag) and [41] (©2010 Springer Verlag).

concepts in each step. The Type-2 Fuzzy Gaussian Mixture Models allow us to model robustly dynamic backgrounds. For foreground detection, the use of a linear saturation function avoids crisp decision in the classification. The fuzzy integrals (Sugeno and Choquet integrals) aggregate adequately the color and texture features to deal with illumination changes and shadows. Fuzzy adaptive learning rates permit us to adapt robustly the background to the changes that occur in videos. Future investigations may concern other issues of background subtraction:

- Fuzzy background modeling: T2-FGMMs allow us to deal with dynamic backgrounds but it is a parametric method, as are the crisp GMMs. It will be interesting to develop a fuzzy nonparametric method to model multimodal background. Furthermore, it would be interesting to use spatial information in the T2-FGMMs.

- Fuzzy foreground detection: Foreground detection is considered as a classification process with two classes, that is, background and foreground. Maybe it will be pertinent to add a third class for the shadows. Furthermore, feature selection using the fuzzy support vector machine can be applied to select the more discriminant features.

Acknowledgments

The author wants to acknowledge Lucia Maddalena (ICAR, National Research Council, Italy) and Alfredo Petrosino (DSA, University Parthenope of Naples, Italy) for providing the comparative results on the adaptive learning rate [55] and their algorithm [40, 41]. The author wants to thank Fida El Baf (Laboratoire MIA, University of La Rochelle, France) for providing her results on the Sugeno and Choquet integrals [3–5].

References

1. M. Azab, H. Shedeed, and A. Hussein. A new technique for background modeling and subtraction for motion detection in real-time videos. In *International Conference on Image Processing*, ICIP 2010, pages 3453–3456, September 2010.
2. F. El Baf and T. Bouwmans. Comparison of background subtraction methods for a multimedia learning space. In *International Conference on Signal Processing and Multimedia*, SIGMAP'07, July 2007.
3. F. El Baf, T. Bouwmans, and B. Vachon. Foreground detection using the Choquet integral. In *International Workshop on Image Analysis for Multimedia Interactive Integral*, WIAMIS 2008, pages 187–190, May 2008.
4. F. El Baf, T. Bouwmans, and B. Vachon. Fuzzy foreground detection for infrared videos. In *International Workshop on Object Tracking and Clas-*

sification in and beyond the Visible Spectrum, OTCBVS 2008, pages 1–6, June 2008.

5. F. El Baf, T. Bouwmans, and B. Vachon. Fuzzy integral for moving object detection. In *International Conference on Fuzzy Systems*, FUZZ-IEEE 2008, pages 1729–1736, June 2008.
6. F. El Baf, T. Bouwmans, and B. Vachon. Type-2 fuzzy mixture of Gaussians model: Application to background modeling. In *International Symposium on Visual Computing*, ISVC 2008, pages 772–781, December 2008.
7. F. El Baf, T. Bouwmans, and B. Vachon. Fuzzy statistical modeling of dynamic backgrounds for moving object detection in infrared videos. In *OTCBVS 2009*, pages 60–65, June 2009.
8. F. El Baf, T. Bouwmans, and B. Vachon. A fuzzy approach for background subtraction. In *International Conference on Image Processing*, ICIP 2008, pages 2648–2651, October 2008.
9. H. Bhaskar, L. Mihaylova, and A. Achim. Video foreground detection based on symmetric alpha-stable mixture models. *IEEE Transactions on Circuits, Systems and Video Technology*, March 2010.
10. H. Bhaskar, L. Mihaylova, and S. Maskell. Automatic target detection based on background modeling using adaptive cluster density estimation. *LNCS from the 3rd German Workshop on Sensor Data Fusion: Trends, Solutions, Applications*, September 2007.
11. B.Lee and M. Hedley. Background estimation for video surveillance. In *Image and Vision Computing New Zealand*, IVCNZ, pages 315–320, 2002.
12. T. Bouwmans. Subspace learning for background modeling: A survey. *RPCS*, 2(3):223–234, November 2009.
13. T. Bouwmans and F. El Baf. Modeling of dynamic backgrounds by type-2 fuzzy Gaussians mixture models. *MASAUM Journal of Basics and Applied Sciences*, 1(2):265–277, September 2009.
14. T. Bouwmans, F. El-Baf, and B. Vachon. Background modeling using mixture of Gaussians for foreground detection: A survey. *RPCS*, 1(3):219–237, November 2008.
15. D. Butler, V. Bove, and S. Shridharan. Real time adaptive foreground/background segmentation. *EURASIP*, pages 2292–2304, 2005.
16. J. Carranza, C. Theobalt, M. Magnor, and H. Seidel. Free-viewpoint video of human actors. *ACM Transactions on Graphics*, 22(3):569–577, 2003.
17. T. Chang, T. Ghandi, and M. Trivedi. Vision modules for a multi sensory bridge monitoring approach. In *ITSC 2004*, pages 971–976, 2004.
18. S. Cheung and C. Kamath. Robust background subtraction with foreground validation for urban traffic video. *Journal of Applied Signal Processing, EURASIP*, 2005.
19. D. Culbrik, O. Marques, D. Socek, H. Kalva, and B. Furht. Neural network approach to background modeling for video object segmentation. *IEEE Transaction on Neural Networks*, 18(6):1614–1627, 2007.
20. J. Davis and M. Goadrich. The relationship between precision-recall and roc curves. In *International Conference on Machine Learning*, ICML 2006,

pages 233–240, 2006.
21. M. Detyniecki. Fundamentals on aggregation operators. In *AGOP*, 2001.
22. Y. Ding, W. Li, T. Fan, and H. Yang. Robust moving object detection under complex background. *Computer Science and Information Systems*, 7(1), February 2010.
23. A. Elgammal and L. Davis. Non-parametric model for background subtraction. In *6th European Conference on Computer Vision, ECCV 2000*, pages 751–767, June 2000.
24. S. Elhabian, K. El-Sayed, and S. Ahmed. Moving object detection in spatial domain using background removal techniques — State-of-art. *RPCS*, 1(1):32–54, January 2008.
25. X. Fang, W. Xiong, B. Hu, and L. Wang. A moving object detection algorithm based on color information. In *International Symposium on Instrumentation Science and Technology*, 48:384–387, 2006.
26. J. Ferryman. http://www.cvg.rdg.ac.uk/pets2000/data.html. *PETS 2000*, 2000.
27. J. Ferryman. http://www.cvg.rdg.ac.uk/pets2006/data.html. *PETS 2006*, 2006.
28. M. Greiffenhagen, V. Ramesh, and H. Niemann. The systematic design and analysis cycle of a vision system: A case study in video surveillance. In *CVPR 2001*, 1(2):704, 2001.
29. M. Harville, G. Gordon, and J. Woodfill. Adaptive background subtraction using color and depth. In *IEEE International Conference on Image Processing*, ICIP 2001, October 2001.
30. M. Heikkila and M. Pietikinen. A texture-based method for modeling the background and detecting moving objects. *IEEE Transactions on Pattern Analysis and Machine Intelligence*, 28(4):657–662, 2006.
31. O. Javed, K. Shafique, and M. Shah. A hierarchical approach to robust background subtraction using color and gradient information. In *IEEE Workshop on Motion and Video Computing*, WMVC 2002, December 2002.
32. S. Kanpracha and S. Tangkawanit. Performance of RGB and HSV color systems in object detection applications under different illumination intensities. In *International Multi Conference of Engineers and Computer Scientists*, 2:1943–1948, March 2007.
33. H. Kashani, S. Seyedin, and H. Yazdi. A novel approach in video scene background estimation. *International Journal of Computer Theory and Engineering*, 2(2):274–282, April 2010.
34. K. Kim, T. Chalidabhongse, D. Harwood, and L. Davis. Real time foreground background segmentation using codebook model. *Real Time Imaging*, 11(3):167–256, 2005.
35. F. Kristensen, P. Nilsson, and V. Wall. Background segmentation beyond RGB. In *ACCV 2006*, pages 602–612, 2006.
36. L. Li and W. Huang. http://perception.i2r.astar.edu.sg/bkmodel/bkindex.html.
37. L. Li and W. Huang. Statistical modeling of complex background for foreground object detection. *IEEE Transaction on Image Processing*,

13(11):1459–1472, November 2004.
38. L. Maddalena and A. Petrosino. A self organizing approach to background subtraction for visual surveillance applications. *IEEE Transactions on Image Processing*, 17(7):1729–1736, 2008.
39. L. Maddalena and A. Petrosino. Multivalued background/foreground separation for moving object detection. In *International Workshop on Fuzzy Logic and Applications*, WILF 2009, 5571:263–270, June 2009.
40. L. Maddalena and A. Petrosino. Self organizing and fuzzy modelling for parked vehicles detection. In *Advanced Concepts for Intelligent Vision Systems*, ACIVS 2009, LNCS 5807, pages 422–433, 2009.
41. L. Maddalena and A. Petrosino. A fuzzy spatial coherence-based approach to background/foreground separation for moving object detection. *Neural Computing and Applications*, Springer London, 19:179–186, 2010.
42. N McFarlane, C. Schofield. Segmentation and tracking of piglets in images. *British Machine Vision and Applications*, pages 187–193, 1995.
43. S. Messelodi, C. Modena, N. Segata, and M. Zanin. A Kalman filter based background updating algorithm robust to sharp illumination changes. In *ICIAP 2005*, 3617:163–170, September 2005.
44. Y. Narukawa and T. Murofushi. Decision modelling using the Choquet integral. *Modeling Decisions for Artificial Intelligence*, 3131:183–193, 2004.
45. Y. Ohta, T. Kanade, and T. Sakai. Color information for region segmentation. *Computer Graphics and Image Processing*, 13(3):222–241, 1980.
46. D. Pokrajac and L. Latecki. Spatiotemporal blocks-based moving objects identification and tracking. *IEEE Visual Surveillance and Performance Evaluation of Tracking and Surveillance*, pages 70–77, October 2003.
47. H. Ribeiro and A. Gonzaga. Hand image segmentation in video sequence by GMM: A comparative analysis. In *XIX Brazilian Symposium on Computer Graphics and Image Processing*, SIBGRAPI 2006, pages 357–364, 2006.
48. J. Rossel-Ortega, G. Andrieu, F. Lopez-Garcia, and V. Atienza-Vanacloig. Background modeling with motion criterion and multi-modal support. In *International Conference on Computer Vision Theory and Application*, VISAPP 2010, May 2010.
49. J. Rossel-Ortega, G. Andrieu, A. Rodas-Jorda, and V. Atienza-Vanacloig. Background modeling in demanding situations with confidence measure. In *International Conference on Computer Vision*, ICPR 2008, December 2008.
50. J. Rossel-Ortega, G. Andrieu, A. Rodas-Jorda, and V. Atienza-Vanacloig. A combined self-configuring method for object tracking in colour video. In *International Conference on Computer Vision*, ICPR 2010, August 2010.
51. M. Shakeri and H. Deldari. Fuzzy-cellular background subtraction technique for urban traffic applications. *World Applied Sciences Journal*, 5(1), 2008.
52. M. Shakeri, H. Deldari, H. Foroughi, A. Saberi, and A. Naseri. A novel fuzzy

background subtraction method based on cellular automata for urban traffic applications. In *9th International Conference on Signal Processing*, ICSP 2008, pages 899–902, October 2008.
53. Y. Sheikh. http://www.cs.ucf.edu/yaser/backgroundsub.htm.
54. M. Sigari. Fuzzy background modeling/subtraction and its application in vehicle detection. In *World Congress on Engineering and Computer Science*, WCECS 2008, October 2008.
55. M. Sigari, N. Mozayani, and H. Pourreza. Fuzzy running average and fuzzy background subtraction: Concepts and application. *International Journal of Computer Science and Network Security*, 8(2):138–143, 2008.
56. M. Sivabalakrishnan and D. Manjula. Adaptive background subtraction in dynamic environments using fuzzy logic. *International Journal of Image Processing*, 4(1), 2010.
57. C. Stauffer and E. Grimson. Adaptive background mixture models for real-time tracking. In *IEEE Conference on Computer Vision and Pattern Recognition*, pages 246–252, 1999.
58. M. Sugeno and S. Kwon. A new approach to time series modeling with fuzzy measures and the Choquet integral. In *IEEE International Conference on Fuzzy Systems*, pages 799–804, March 1995.
59. H. Tahani and J. Keller. Information fusion in computer vision using the fuzzy integral. In *IEEE Transaction on Systems, Man, and Cybernetics*, 20(3):733–741, 1990.
60. K. Toyama, J. Krumm, B. Brumiit, and B. Meyers. Wallflower: Principles and practice of background maintenance. In *International Conference on Computer Vision*, pages 255–261, September 1999.
61. C. Wren and A. Azarbayejani. Pfinder: Real-Time Tracking of the Human Body. *IEEE Transactions on Pattern Analysis and Machine Intelligence*, 19(7):780–785, July 1997.
62. J. Zeng, L. Xie, and Z. Liu. Type-2 fuzzy Gaussian mixture. *Pattern Recognition*, 41(2):3636–3643, September 2009.
63. H. Zhang and D. Xu. Fusing color and texture features for background model. In *Third International Conference on Fuzzy Systems and Knowledge Discovery*, FSKD'06, 4223(7):887–893, September 2006.
64. J. Zheng, Y. Wang, N. Nihan, and E. Hallenbeck. Extracting roadway background image: A mode based approach. *Journal of Transportation Research Report*, 1944:82–88, 2006.

6

Sensor and Data Fusion: Taxonomy, Challenges, and Applications

Lawrence A. Klein
Consultant, Santa Ana, California, USA

Lyudmila Mihaylova
Lancaster University, Lancaster, United Kingdom

Nour-Eddin El Faouzi
LICIT IFSTTAR – ENTPE, Cedex, France

6.1 Introduction

Sensor and data fusion is a process of paramount importance for many domains and applications. Its potential for rapid data and information processing is of primary importance for surveillance, security, intelligent transportation systems, navigation, and communications. Effective use of the data requires the sensor and context data to be aggregated or "fused" in such a way that high-quality information results and serves as a basis for decision support. Data fusion encompasses groups of methods for merging various types of data and information [35]. This process is especially important for tracking systems.

Reliable tracking methods enable surveillance systems to remotely monitor activity across a variety of environments such as: 1) transport systems (railway transportation, airports, urban and motorway road networks, and maritime transportation); 2) banks, shopping malls, car parks, and public buildings; 3) industrial areas, and 4) government facilities (military bases, prisons, strategic infrastructures, radar centers, and hospitals). As an integral part of surveillance systems, tracking algorithms can enhance situation awareness across multiple scales of space and time. They can assist a surveillance analyst who needs to track people and vehicles in a space (identity tracking), know where the people are in a space (location tracking), and know what the people, vehicles, and objects in a space are doing (activity tracking). Multiple-sensor systems can provide surveillance coverage across a wide area, ensuring constant object visibility.

This chapter describes processes and methods that support sensor and data fusion, its challenges and some applications to tracking objects in video sequences. The chapter is organized as follows. Section 6.2 describes the JDL data fusion model with its six levels of fusion from 0 to 5. Section 6.3 discusses basic sensor architectures while Sections 6.4 and 6.5 describe taxonomies for detection, classification, and identification algorithms and state estimation and tracking algorithms, respectively. An introduction to Bayesian inference and filtering is found in Section 6.6. Section 6.7 presents the Dempster-Shafer theory. The application of sensor and data fusion to video-based tracking is found in Sections 6.9 and 6.10. Finally, conclusions are given in Section 6.11.

6.2 Data Fusion Model

The U.S. Department of Defense Joint Directors of Laboratories Data Fusion Subpanel (JDL DFS) model divides data fusion into low-level and high-level processes. Low-level fusion supports preprocessing of data and target detection, classification, identification, and state estimation, including tracking. High-level functions consist of situation and threat (or impact) assessment, fusion process refinement, and user refinement [17, 27]. The definition of data fusion formulated by the panel has been modified over the years and in one version exists as [80, 84]:

"A multilevel, multifaceted process dealing with the automatic detection, association, correlation, estimation, and combination of data and information from single and multiple sources to achieve refined position and identity estimates, and complete and timely assessments of situations and threats and their significance."

The IEEE Geoscience and Remote Sensing Society Data Fusion Technical Committee produced an alternative definition of data fusion as:

"The process of combining spatially and temporally indexed data provided by different instruments and sources in order to improve the processing and interpretation of these data."

The goals of data fusion are realized through a six-level hierarchy of processing described as follows:

- Level 0 processing: encompasses source preprocessing to address estimation, computational, and scheduling requirements through normalization, formatting, ordering, batching, and compression of input data.

- Level 1 processing: achieves refined state and identity estimates by fusing individual sensor position, velocity, acceleration, and identity estimates. Includes analyzing data from all appropriate sources, for example, point and wide-area sensors whether land, sea, air, or space based; probes (e.g., cellular telephones, Personal Digital Assistants (PDAs), toll payment devices, and GPS equipped vehicles); and emergency call boxes found on highways, malls, and some universities.

- Level 2 processing: assists in the complete and timely assessment of the probable situation causing the observed data and events by incorporating relations among the entities of interest. The relations may be physical, organizational, informational, or perceptual, as appropriate to the need. In a defense application, Level 2 fusion utilizes data gathered in Level 1 to gain insights into prescribed event and activity sequences, force structures, and the overall battle environmental factors. In a traffic management application, Level 2 processing integrates information from external sources and databases, including law enforcement agency reports and databases, architectural or roadway configuration drawings, weather reports, anticipated traffic mix, time-of-day and seasonal traffic patterns, construction and special event schedules, and police action reports. In a video surveillance application, integration of information could occur from sources such as visible spectrum and infrared cameras and scheduled events.

- Level 3 processing: a prediction function that assists in the complete and timely assessment of the impact or significance of the situation on the system or organization using inferences drawn from Level 2 associations. Level 3 fusion estimates the outcome of various plans as they interact with one another and with the surroundings. In the defense application, Level 3 processing develops a threat-oriented data perspective to estimate enemy capabilities, identify threat opportunities, estimate enemy intent, and determine levels of danger. In a traffic management application, Level 3 processing assesses the impact of traffic flow patterns and other data on the likely occurrence of an incident, travel time delay, or other event and the effect of that event on traffic flow. In video surveillance, Level 3 processing ascertains the impact of the fused real-time data and relationships gleaned from informational databases on the operation and effectiveness of the organization.

- Level 4 processing: improves results of the fusion process by continuously refining estimates and assessments through planning and control, which includes evaluating the need for additional sources of information, assigning tasks to available resources, or modifying the fusion process itself.

- Level 5 processing: user refinement focuses on issues related to human processing of fused information, for example, when automatic object recognition or other computerized analyses are not paramount. Level 5 fusion addresses adaptive determination about (1) who queries and has access to information and (2) which data is retrieved and displayed to support cognitive decision making and action taking [8]. A significant amount of information from external databases is usually needed to support the Level 2 and 3 fusion processes. Interrelationships in Levels 1 through 3 fusion processes are illustrated in Figure 6.1 [46, 84]. Target detection, classification, and tracking occur simultaneously in some applications such as surveillance, rather than in separate paths as displayed in the figure.

6.3 Sensor and Data Fusion Architectures

Categorization of data fusion architectures occurs in several ways. In one approach, the architecture is defined by the extent of the data processing that occurs in each sensor, the information products produced by the individual

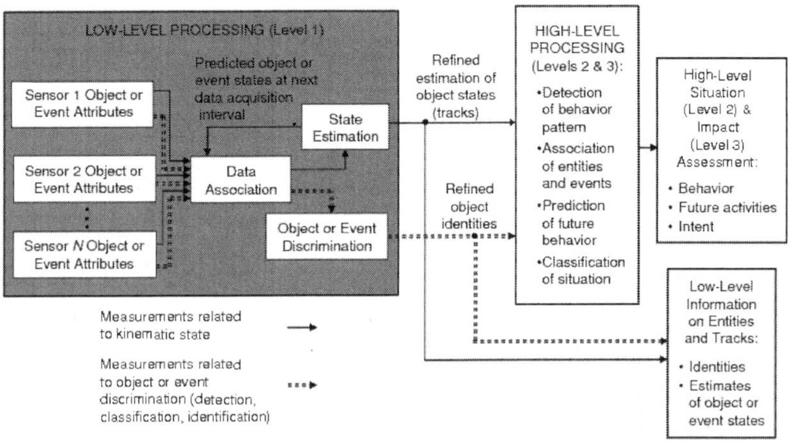

FIGURE 6.1 Data fusion processing Levels 1, 2, and 3.

sensors, and the location of the fusion processes. For example, sensors supplying information to detection, classification, and identification fusion algorithms may use complex processing techniques to provide the object class and a measure of the confidence of that decision to a fusion algorithm for further refinement. Alternatively, the sensors may simply provide filtered signals or features to a fusion algorithm, where the signals or features are analyzed in conjunction with those from other sensors to determine the object class. By way of contrast, sensors supplying information to state estimation and tracking algorithms may provide either measurement data, that is, reports that contain the position and velocity of objects, or tracks of the objects. Current values of measurement data may be combined with previously obtained data to generate new tracks, or the current data may be utilized to update preexisting tracks using Kalman filtering [42, 44]. These processes can occur in the individual sensors or at the central processing node. If the sensors supply tracks, the tracks can be correlated with preexisting tracks residing in individual sensors or at a central processing node [46].

The terms that describe data fusion architectures based on the extent of the data processing, data product types, and fusion location are sensor-level fusion (also referred to as autonomous fusion and distributed fusion), central-level fusion (also referred to as centralized fusion), and hybrid fusion, which uses combinations of the sensor-level and central-level approaches [4, 16, 40]. The resolution of the data and the extent of processing by each sensor may also be employed to define another fusion architecture lexicon, namely pixel-level, feature-level, and decision-level fusion.

6.3.1 Sensor-Level Fusion

In sensor-level fusion, illustrated in Figure 6.2, each sensor detects, classifies, identifies, and estimates the trajectories of objects of interest before sending its results to the fusion processor. The fusion processor combines the information from the individual sensors to improve the classification, identification, or state estimate of the objects.

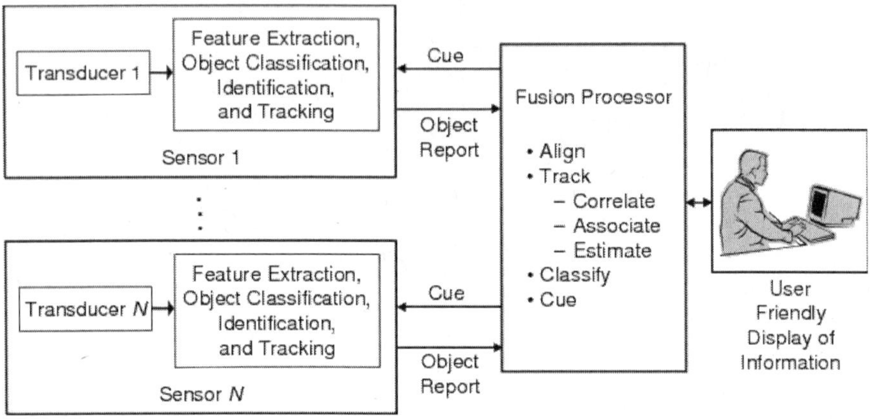

FIGURE 6.2 Sensor-level fusion architecture.

Functions typically performed by a fusion processor include [18, 46, 54]:

- Alignment: referencing of sensor data to common time and spatial origins;
- Correlation: using a metric to identify new measurements or tracks closest to existing tracks as candidates for the association process;
- Association: decision to use one specific measurement or track to update an existing track;
- Estimation: calculation of an object's future position by updating the state vector and error covariance matrix using the results of the association process;
- Classification: assessing the tracks and object discrimination data to determine object type, impact on its surroundings, and action priority;
- Cueing: feedback of threshold, integration time, and other signal processing parameters or information about areas over which to conduct a more detailed search, based on the results of the fusion process. For example, if a highly cluttered region is found, a command may be sent to the appropriate sensor to increase the threshold setting. Alternatively, when the fusion processing identifies a decoy, a message describing the decoy's location is sent to minimize object search-related signal processing in this region. Another

application of cueing is to initiate a search of a small, but high-interest region using a sensor of limited field of regard having high resolution, such as laser radar.

Sensor-level fusion is as optimal (based on Bayesian decision logic) for detecting and classifying objects as central-level fusion (described below) when the sensors rely on independent signature-generation phenomena to develop information about the identity of objects in the field of regard, that is, they derive object signatures from different physical processes and generally do not false alarm on the same artifacts [70]. The sensor footprints must be registered with respect to each other to ensure that the sensor signatures are characteristic of events or objects at the same spatial locations. Additional information about signature generation phenomena and how they are influenced by background, sensor type, and sensor design parameters is found in [46].

6.3.2 Central-Level Fusion

Figure 6.3 depicts the central-level fusion architecture. Here each sensor provides minimally processed measurement data, although sensor-generated tracks may also be sent to the fusion processor when state estimation is the desired outcome of the fusion process. Minimal processing includes operations such as filtering and baseline estimation.

Central-level fusion algorithms are generally more complex and must process data at higher rates than in sensor-level fusion, because the centralized architecture is designed to operate on the minimally analyzed data from each sensor. Central-level fusion is optimal for tracking objects, as it is more effective than sensor-level fusion in estimating or predicting the future position of the object due to a combination of effects: (1) processing all the data in one place; (2) forming the initial tracks based on observations from more than one sensor, thus eliminating tracks established from partial data received by individual sensors; (3) processing sensor measurement data directly, eliminating difficulties associated with combining the sensor-level tracks produced by individual sensors; and (4) facilitating multiple hypothesis tracking by having all data available in a central processor [40]. Deficiencies of the method are reflected in the large amount of data that must be transferred in a timely manner and then processed by the central processor(s).

6.3.3 Hybrid Fusion

In hybrid fusion, shown in Figure 6.4, the central-level fusion process is supplemented by individual sensor signal processing algorithms that may, in turn, provide inputs to a sensor-level fusion algorithm. Hybrid fusion allows the tracking benefits of central-level fusion to be realized utilizing sensor measurement data and, in addition, allows sensor-level fusion of target tracks as computed by individual sensors. Global track formation that combines the central- and sensor-level fusion tracks occurs in the central-level processor. Hybrid fusion can also support target attribute classification when the signature data is not truly generated by independent phenomena. In this case, minimally processed data is sent to a central processor where it is combined using a fusion algorithm that detects and classifies objects in the field of regard of the sensors. The disadvantages of hybrid fusion are the increased processing complexity and possibly increased data transmission rates.

Hybrid fusion can manifest itself in the form of hierarchical and distributed architectures [57,83]. Hierarchical architectures contain fusion nodes arranged such that the lowest-level nodes process sensor data and send the results to higher-level nodes to be combined. Distributed-fusion architectures have benefits and drawbacks that are summarized in [46]. Some of the issues involve sharing of fusion responsibility among nodes, bandwidth required of the communications system to transfer data and fused products among nodes, and

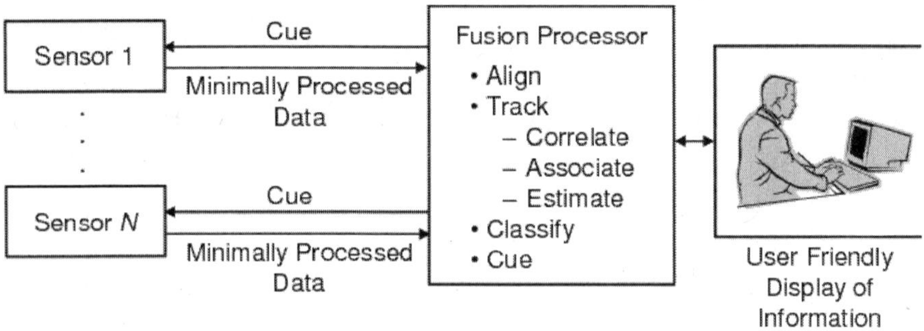

FIGURE 6.3 Central-level fusion architecture.

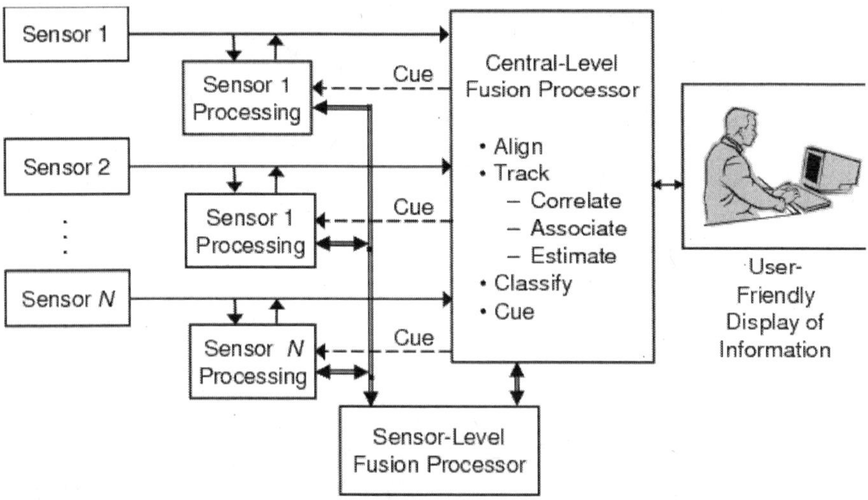

FIGURE 6.4 Hybrid fusion architecture.

timing of user access to the fusion products. Challenges of sensor data fusion for intelligent transportation systems are summarized in [25].

Many hybrid architectures are application specific. One hybrid architecture employs multiple classifiers in parallel to operate on the same set of input features. The outputs of the classifiers are then processed through a series of data fusion algorithms, which produce the final result [66]. In another hybrid architecture, the input features are again input to multiple classifiers in parallel, but this time the classifier outputs are subject to a reliability test. For example, if the classifier uses fuzzy logic, the output is deemed reliable if one class has a high membership value in a fuzzy set and the others membership values close to zero. These results are weighted further by using prior knowledge about the performance of each classifier in the scenario under consideration. Finally, the classifier results are combined using a fusion rule, which may be conjunctive (i.e., intersection or minimum operator)), disjunctive (i.e., union or maximum operator), or a compromise (i.e., one that lies between the minimum and maximum operators) [26].

6.4 Detection, Classification, and Identification Algorithm Taxonomy

The fusion of video imagery concerns Level 1 fusion. Therefore, we concentrate on the algorithms that apply to this case. Reference [46] contains a detailed discussion of Level 2 and 3 processing. Figure 6.5 shows a taxonomy for detection, classification, and identification algorithms used in Level 1 processing [17, 36–38, 84]. The major algorithm categories are physical models, feature-based inference techniques, and cognitive-based models.

6.4.1 Physical Models

Physical models replicate object discriminators that are easily and accurately observable or calculable. Examples of discriminators are radar cross-section as a function of aspect angle; infrared emissions as a function of object type or surface characteristics such as roughness, emissivity, and temperature; multispectral signatures; edges; line relationships; and height profile images. A list of feature categories used in developing physical models, and representative physical features and other attributes of the categories are found in [36, 46].

Physical modeling techniques include simulation, estimation, and syntactic methods. Physical models estimate the classification and identity of an object by matching modeled or prestored object signatures to observed data. The signature or imagery gathered by a sensor is analyzed for pre-established physical characteristics or attributes, which are input into an identity declaration process. Here the characteristics identified by the analysis are compared with stored physical models or signatures of objects of interest. The stored model or signature having the closest match to the real-time sensor data is

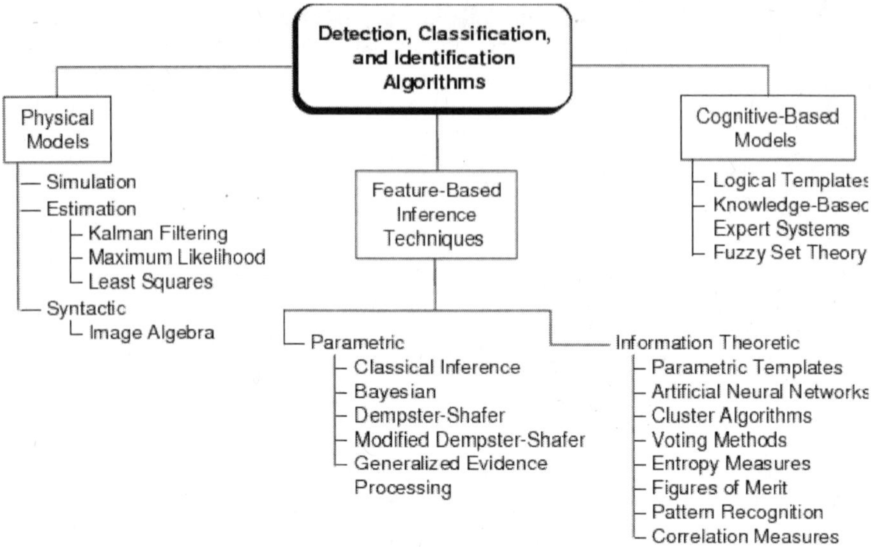

FIGURE 6.5 Taxonomy of detection, classification, and identification algorithms.

declared the correct identity of the object.

6.4.2 Feature-Based Inference Techniques

Feature-based inference techniques perform classification or identification by mapping data, such as statistical knowledge about an object or recognition of object features, into a declaration of identity. Feature-based algorithms may be further divided into parametric and information theoretic techniques (i.e., algorithms that have some commonality with information theory) as depicted in Figure 6.5.

Parametric classification directly maps parametric data (e.g., features) into a declaration of identity. Stochastic properties may be modeled, but physical properties are not. Parametric techniques include classical inference, Bayesian inference, Dempster-Shafer evidential theory, modified Dempster-Shafer methods, and generalized evidence processing.

Information theoretic techniques transform or map parametric data into an identity declaration. All these methods share a common concept, namely, that similarity in identity is reflected in the similarity in observable parameters. No attempt is made to directly model the stochastic aspects of the observables. The techniques included under this category are parametric templates, artificial neural networks, cluster algorithms, voting methods, entropy-measuring techniques, figures of merit, pattern recognition, and correlation measures.

6.4.3 Cognitive-Based Models

Cognitive-based models, including logical templates, knowledge-based systems, and fuzzy set theory, attempt to emulate and automate the decision-making processes used by human analysts.

Templating allows predetermined and stored patterns to be matched against observed data, thereby inferring the identity of an object or assessing a situation. Parametric templates that compare real-time patterns with stored ones can be combined with logical templates derived, for example, from Boolean relationships [80]. Fuzzy logic may also be applied to the pattern matching technique to account for uncertainty in either the observed data or the logical relationships used to define a pattern.

Knowledge-based systems incorporate rules and other knowledge from known experts to automate the object identification process. They retain the expert knowledge for use at a time when the human inference source is no longer available.

Fuzzy set theory opens the world of imprecise knowledge or indistinct boundary definition to mathematical treatment. It facilitates the mapping of system state variable data into control, classification, or other outputs. The three major elements of a fuzzy system are: fuzzy sets, membership functions, and production rules. Fuzzy sets are the state variables defined in imprecise terms. Membership functions are the graphical representation of the boundary between fuzzy sets. Production rules, also known as fuzzy associative memory, are the constructs that specify the membership of a state variable in a given fuzzy set. Membership value can range from 0 (definitely not a member) to 1 (definitely a member). The production rules, which govern the behavior of the system, are in the form of IF-THEN statements. Defuzzification is the process that converts the result of the application of the production rules into a crisp output value, which is used to control the system.

6.5 Taxonomy of State Estimation and Tracking Algorithms

Figure 6.6 contains a taxonomy for state estimation and tracking algorithms used in Level 1 processing [17, 36–38, 84]. The processes required to perform the tracking function are represented at the top level by algorithms that (1) determine the search direction and (2) correlate and associate data and tracks. Correlation and association are further separated into data alignment, data and track association; and position, kinematic, and attribute estimation. Each of these subgroups has procedures and algorithms coupled to them as depicted in the figure and explained below.

Direction tracking systems can be sensor (data) driven or target (goal) driven. In sensor-driven systems, measurement data (consisting of combinations of range, azimuth, elevation, and range-rate sensor data) initiate a search

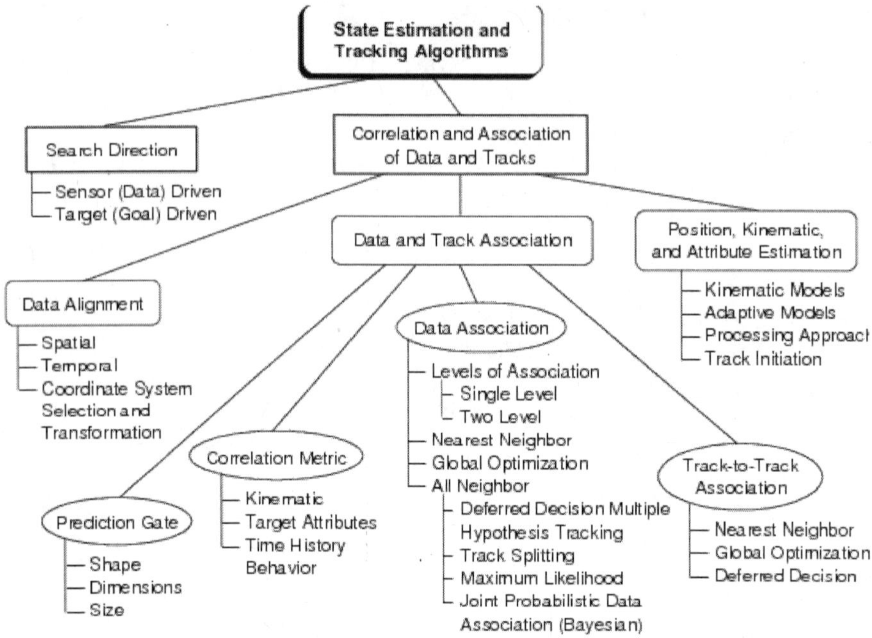

FIGURE 6.6 Taxonomy of state estimation and tracking algorithms.

through the tracks for those that can be associated with the data. Target-driven systems use a primary sensor for tracking and use the target track to direct other sensors to acquire data or search databases for data that can be associated with particular tracks.

The proper correlation and association of measurement data and tracks from multisensor inputs ultimately generate optimal central track files. Each file ideally represents a unique physical object or entity. Correlation and association require algorithms that define data alignment, prediction gates, correlation metrics, data and track association, and position, kinematic, and attribute estimation.

6.5.1 Data Alignment

Data alignment is performed through spatial and temporal reference adjustments and coordinate system selection and transformations that establish a common space-time reference for fusion processing. Errors introduced by measurement accuracies, coordinate transformations, and unknown object motion are accounted for through the data alignment process.

6.5.2 Data and Track Association

Data and track association consist of processes that establish the prediction gate, define the correlation metric, perform data association, and perform track-to-track association.

Prediction gate

Prediction gates control the correlation of datasets into one of two categories, namely candidates for track update or initial observations for forming a new tentative track. Data that was originally categorized for track update may later be used to initiate new tracks if it is not ultimately assigned to an existing track. The size of the gates reflects the calculated or otherwise anticipated object position and velocity errors associated with their calculation, sensor measurement errors, and desired probability of correct association.

Correlation metrics

These metrics quantify the closeness of new measurements or tracks to existing tracks, and thus identify candidates for the association process. They are also used in track-to-track correlation to assist in associating tracks produced by different sensors. The metric can be based on spatial distance (e.g., Euclidean distance) or statistical measures of correlation between observations and predictions (e.g., Mahalanobis or Bhattacharyya distance) [36], heuristic functions such as figures-of-merit that use the kinematic and object attribute information, and measures that quantify the realism of an observation or track based on prior assumptions such as track lengths, object densities, or track behavior.

Data association

In a multiple object and sensor scenario, data association refers to the statistical decision process that links sets of measurement data from overlapping gates, multiple returns (hits) in a gate, clutter in a gate, and new objects that appear in a gate on successive scans. Thus, data association partitions the measurements into sets that could have originated from the same objects [22].

Track-to-track association

Track-to-track association is used to merge sensor-level tracks to obtain a central track file. Tracks can be characterized by position, velocity, covariance, and other features. In order to associate the sensor-level tracks, they first are transformed into a common coordinate system and time aligned. Gates are then formed and a metric is chosen to evaluate the track association process. Many of the methods discussed for data association can be used to perform track-to-track association. These include nearest neighbor, global optimization, and deferred decision. The latter operates on tracks obtained over several future scans. After the track associations are made, the object position and covariance matrix corresponding to the input tracks are combined to form a new object position and covariance for the fused track. If the states observed by the various sensors are not identical, then only those that are common are used in the association process. The remaining states are augmented to the track and carried along.

6.5.3 Position, Kinematic, and Attribute Estimation

These processes optimally combine multiple observations to obtain improved estimates of the position, velocity, and attributes (e.g., size, temperature, and shape) of an object. Estimates of updated object parameters are provided by a tracking filter. The filter uses algorithms that operate on time sequences of associated measurements to develop predictions of object state and attributes. Kinematic and adaptive models of object motion and sequential or batch processing (i.e., where all data is processed simultaneously) are used to support the estimation process. A priori models of track dynamics and observations are also used as estimators to refine the state estimate and to predict the state at the next observation interval for gating.

Even with a priori knowledge, the object may maneuver. If this occurs, the state of the tracking filter must be changed to accommodate the maneuver. This can be accomplished in several ways. The first method, used with track splitting, augments the state of the parent track to include the maneuver. The second method, called the multiple-model maneuver, parameterizes the range of the expected maneuver and constructs tracking filters for each set of parameter values [9].

In the next section we present the framework of the Bayesian inference and filtering.

6.6 Bayesian Inference and Filtering

Bayesian probability theory is traditionally used to model uncertainty in several disciplines [52]. These methods provided the first successful data fusion implementation and are still widely used in many applications, such as image processing. Their popularity rests, in part, on their ability to incorporate multi-source data (often called evidence) and a priori knowledge about the states of the underlying system. By interpreting the conditional probability as a measure of uncertainty, it is one of the classical methods for solving inverse problems through the utilization of Bayesian statistics based on *Bayes' rule*.

Bayes' rule

In Bayesian inference, all the uncertainties (including statements, parameters that are either time-varying or fixed but unknown) are treated as random variables and inference is based on probability distributions and prior information.

Bayes' rule allows probability distributions to be combined. It estimates the probability of occurrence of a future event by observing the occurrence of similar events in the past. A probability distribution that describes the occurrence of a past event is called the *prior probability.* If a new experiment is performed, the resulting observations modify this probability and produce the *posterior probability.* Bayes' rule evaluates the posterior probability $P(H_i|y_j)$ of a given hypothesis H_i, knowing the measurements y_j, the likelihood function $P(y_j|H_i)$, and the prior probability $P(H_i)$ of H_i, as

$$P(H_i|y_j) = \frac{P(H_i)P(y_j|H_i)}{\sum_{k=1}^{n} P(y_j|H_k)P(H_k)} \tag{6.1}$$

Thus, given the a priori probability distribution for the hypotheses H_i, one can determine the posterior distribution for the hypothesis H_i knowing the likelihood function of y_j.

There are three central considerations related to the application of Bayesian inference:

- *Normalization*: Given the prior probability $P(H)$ and probability $P(y|H)$, the posterior probability $P(H|y)$ can be calculated as the product of the conditional probability and the prior probability divided by a normalization factor:

$$P(H|y) = \frac{P(y|H)P(H)}{\int_H P(y|h)P(h)dh}. \tag{6.2}$$

- *Marginalization*: Given the joint probability of (H, z) conditioned on the occurrence of event A, the marginal posterior is found as

$$P(x|y) = \int P(H, z|A)dz. \tag{6.3}$$

- *Computation of average value*: Given the density function for the conditional probability, the statistical average can be calculated as

$$E\big[P(x|y)\big] = \int P(x|y)dP_X = \int P(x|y)f(x)dx, \tag{6.4}$$

where $f(x)$ is the probability density function of x.

In a multi-source configuration where the vectors of observations are derived from k sources, the above equations can be generalized in a form that is given in the next subsection. As has been emphasized in [23], the fusion does not appear in this formula. It operates upstream to derive the likelihood functions associated with each source and possibly other information sources on dependency and their respective reliabilities.

The Bayesian inference fusion process

Most Bayesian fusion applications rely on the maximum posterior probability (MAP) decision rule to select the hypothesis that is most likely giving rise to the observations. This criterion requires choosing the hypothesis with the maximum posterior probability. Other decision rules can be used, such as the maximum likelihood estimator or the least squares estimator, to choose the most likely event or hypothesis that represents the observed data or information.

The probability distributions are derived either by using parametric models or learning techniques from a finite set. However, in the latter approach, additional conditions (consistency and exhaustivity) must be observed by the learning sample.

As an example of multisource data fusion, we consider the estimation of an unknown parameter θ given outputs X_s of the different information sources or sensors $s = 1, \ldots, \ell$ at hand. We usually assume that these different sources exhibit conditional independence, that is, that the sources are independent for a given value of θ [33]. Therefore, the likelihood function $P(X|\theta)$ is given by

$$P(X|\theta) = \prod_{s=1}^{\ell} P(X_s|\theta). \tag{6.5}$$

Using the *maximum likelihood* principle, we choose the value of θ that maximizes the likelihood function. The derived estimator $\hat{\theta}$ of θ, is given by

$$\hat{\theta} = \max_{\theta}\left(\sum_{s=1}^{\ell} \log P(X_s|\theta)\right) \tag{6.6}$$

To illustrate the implementation of such fusion, let us consider k Gaussian sensor measurement errors. Here, the likelihood function $P(X_s|\theta)$ is given by

$$P(X_s|\theta) = |2\pi\Omega_s|^{-1/2} \exp\left(-\frac{1}{2}(X_s - \theta)^T \Omega_s^{-1}(X_s - \theta)\right), \tag{6.7}$$

where Ω_s is the measurement error covariance matrix and $|\Omega_s|$ denotes its determinant.

The logarithm of the likelihood function is then given by

$$\begin{aligned} L(\theta) &= \sum_{s=1}^{\ell} \log\Big(P(X_s|\theta)\Big) = \\ &\sum_{s=1}^{\ell}\left(-\frac{1}{2}\log((2\pi)^{\ell}|\Omega_s|) - \frac{1}{2}(X_s - \theta)^T \Omega_s^{-1}(X_s - \theta)\right), \end{aligned} \tag{6.8}$$

where T is the transpose operation. The maximum likelihood estimate of θ is obtained by solving the normal equations (partial derivative of L with respect to θ is zero), giving

$$\hat{\theta} = \frac{\sum_{s=1}^{\ell} \Omega_s^{-1} X_s}{\sum_{s=1}^{\ell} \Omega_s^{-1}}. \tag{6.9}$$

For example, if there are two sensors ($\ell = 2$) and each provides a measurement of θ, X_i, $i = 1, 2$, then from the above equation, the maximum likelihood estimate of θ is given by

$$\hat{\theta} = \frac{\sigma_1^2}{\sigma_1^2 + \sigma_2^2} X_1 + \frac{\sigma_2^2}{\sigma_1^2 + \sigma_2^2} X_2, \tag{6.10}$$

where σ_1 and σ_2 are the standard deviations of the measurement errors, respectively, from sensor 1 and 2 [24].

This result highlights the two ways in which the Bayesian approach can be applied to data fusion: either merging information from multiple sources at a given time or merging the information from a *same source* but at different times. In the latter case, new measures may lead to revisions of the probability of a state that corresponds to the Kalman-Bucy filter [42, 44, 45].

Bayesian inference suffers from some drawbacks that prompted researchers to develop other techniques for modeling knowledge. One of these is that Bayesian inference does not have a mechanism to express uncertainty. Hence it also fails in not being able to represent vagueness or ignorance, leading to confusion among these different concepts [65, 77, 79]. In the Bayesian probability framework, ignorance about a fact is modeled by equal prior probabilities, which is equated to choosing the likelihood of occurrence of an event as if a total lack of knowledge existed! Dempster-Shafer theory attempts to overcome this limitation.

6.7 Dempster-Shafer Evidential Theory

Dempster-Shafer evidential theory, a probability-based data fusion classification algorithm, is useful when the sensors (or more generally, the information sources) cannot associate a 100 percent probability of certainty to their output decisions. The algorithm captures and combines whatever certainty exists in the object discrimination capability of the sensors. Knowledge from multiple sensors about events (called propositions) is combined using Dempster's rules to find the intersection or conjunction of the propositions and their associated probabilities. When the intersection of the propositions reported by the sensors is an empty set, Dempster's rule redistributes the conflicting probability to the nonempty set elements. When the conflicting probability becomes large, application of Dempster's rule can lead to counterintuitive conclusions. Several modifications to the original Dempster-Shafer theory have been proposed to accommodate these situations.

6.7.1 Overview of the Process

Figure 6.7 shows the Dempster-Shafer data fusion process as might be configured to identify objects [40]. Each sensor has a set of observable variables corresponding to the phenomena that generate information received about the objects and their surroundings. In this illustration, a sensor operates on the observables with its particular set of classification algorithms (sensor-level fusion). The information gathered by each Sensor k, where $k = 1, \ldots, N$, associates a declaration of object type (referred to in the figure by object o_i where $i = 1, \ldots, n$) with a probability mass or basic probability assignment $m_k(o_i)$ between 0 and 1. The probability mass expresses the certainty of the declaration or hypothesis, that is, the amount of support or belief attributed directly to the declaration by the sensor. Probability masses closer to unity characterize decisions made with more definite knowledge or less uncertainty about the nature of the object. The probability masses for the decisions made by each sensor are then combined using Dempster's rules of combination. The hypothesis favored by the largest accumulation of evidence from all contributing sensors is selected as the most probable outcome of the fusion process. A computer stores the relevant information from each sensor. The converse is also true, that is, objects not supported by evidence from any sensor are not stored.

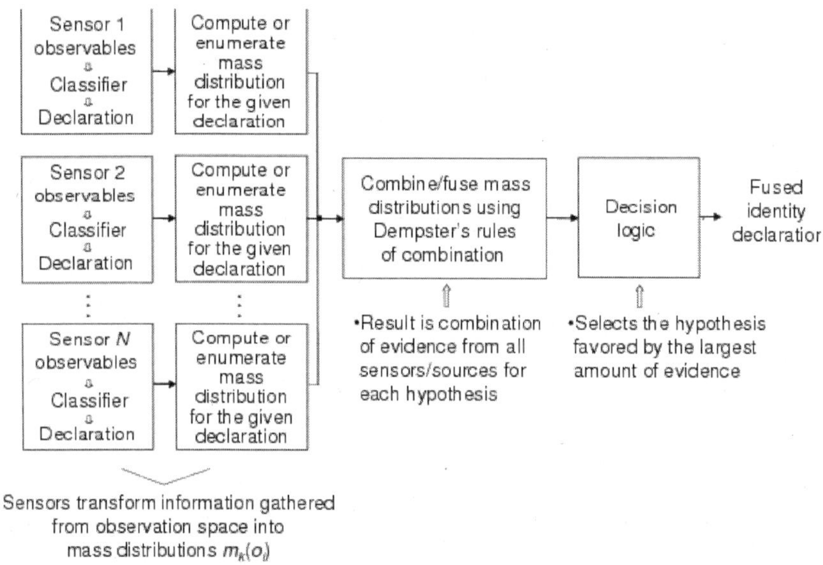

FIGURE 6.7 Dempster-Shafer data fusion process.

In addition to real-time sensor data, other information or rules can be stored in the information base to improve the overall decision or object discrimination capability. Examples of such rules are "Ships detected in known shipping lanes are cargo vessels" and "Objects in previously charted Earth orbits are weather or reconnaissance satellites."

6.7.2 Implementation of the Method

Assume a set of n mutually exclusive and exhaustive propositions exists, for example, an object is of type $a_1, a_2, \ldots$, or a_n. This is the set of all propositions making up the hypothesis space, called the frame of discernment, and is denoted by Θ. A probability mass $m(a_i)$ is assigned to any of the original propositions or to the union of the propositions based on available sensor information. Thus, the union or disjunction that the object is of type a_1 or a_2 (denoted $a_1 \cup a_2$) is assigned probability mass $m(a_1 \cup a_2)$ by a sensor. A proposition is called a focal element if its mass is greater than zero. The number of combinations of propositions that exists (including all possible unions and Θ itself, but excluding the null set) is equal to $2^n - 1$. For example if $n = 3$, there are $2^3 - 1 = 7$ propositions given by a_1, a_2, a_3, $a_1 \cup a_2, a_1 \cup a_3, a_2 \cup a_3$, and $a_1 \cup a_2 \cup a_3$. When the frame of discernment contains n focal elements, the power set consists of 2^n elements including the null set.

In the event that all the probability mass cannot be directly assigned by the sensor to any of the propositions or their unions, the remaining mass is assigned to the frame of discernment Θ (representing uncertainty as to further definitive assignment) as $m(\Theta) = m(a_1, \cup a_2, \cup \ldots, a_n)$ or to the negation of a proposition such as $m(\bar{a}) = m(a_2 \cup a_3, \cup \ldots \cup a_n)$. A raised bar has been used to denote the negation of a proposition. The mass assigned to Θ represents the uncertainty the sensor has concerning the accuracy and interpretation of the evidence [20]. The sum of probability masses over all propositions, uncertainty, and negation equals unity.

To illustrate these concepts, suppose that two sensors observe a scene in which there are three objects. Sensor A identifies the object as belonging to one of the three possible types: a_1, a_2, or a_3. Sensor B declares the object to be of type a_1 with a certainty of 80 percent. The intersection of the data from the two sensors is written as

$$(a_1 \text{ or } a_2 \text{ or } a_3) \text{ and } (a_1) = (a_1), \tag{6.11}$$

or upon rewriting as

$$(a_1 \cup a_2 \cup a_3) \cap (a_1) = (a_1). \tag{6.12}$$

Only a probability of 0.8 can be assigned to the intersection of the sensor data based on the 80 percent confidence associated with the output from Sensor B. The remaining probability of 0.2 is assigned to uncertainty represented by the union (disjunction) of (a_1 or a_2 or a_3) [7].

6.7.3 Support, Plausibility, and Uncertainty Interval

According to Shafer, "an adequate summary of the impact of the evidence on a particular proposition a_i must include at least two items of information: a report on how well a_i is supported and a report on how well its negation $\bar{a}_i$ is supported." [73]. These two items of information are conveyed by the proposition's degree of support and its degree of plausibility.

Support for a given proposition is defined as "The sum of all masses assigned directly by the sensor to that proposition or its subsets." [60, 73]. A subset is called a *focal subset* if it contains elements of Θ with mass greater than zero. Thus, the support for object type a_1, denoted by $S(a_1)$, contributed by a sensor is equal to

$$S(a_1) = m(a_1). \tag{6.13}$$

Support for the proposition that the object is either type a_1, a_2, or a_3 is

$$\begin{aligned} S(a_1 \cup a_2 \cup a_3) = m(a_1) + m(a_2) + m(a_3) + \\ m(a_1 \cup a_2) + m(a_1 \cup a_3) + m(a_2 \cup a_3) + m(a_1 \cup a_2 \cup a_3). \end{aligned} \tag{6.14}$$

Plausibility of a given proposition is defined as "The sum of all mass not assigned to its negation." Consequently, plausibility defines the mass free to move to the support of a proposition. The plausibility of a_i, denoted by $Pl(a_i)$, is written as

$$Pl(a_i) = 1 - S(\bar{a}_i), \tag{6.15}$$

where $S(\bar{a})$ is called the dubiety and represents the degree to which the evidence impugns a proposition, that is, supports the negation of the proposition.

Plausibility can also be computed as the sum of all masses belonging to subsets a_j that have a non-null intersection with a_i. Accordingly,

$$Pl(a_i) = \sum_{a_j \cap a_i \neq 0} m(a_j). \tag{6.16}$$

Thus, when $\Theta = \{a_1, a_2, a_3\}$, the plausibility of a_1 is computed as the sum of all masses compatible with a_1, which includes all unions containing a_1 and Θ such that

$$Pl(a_1) = m(a_1) + m(a_1 \cup a_2) + m(a_1 \cup a_3) + m(a_1 \cup a_2 \cup a_3). \tag{6.17}$$

An uncertainty interval is defined by $[S(a_i), Pl(a_i)]$, where

$$S(a_i) \leqq Pl(a_i). \tag{6.18}$$

The Dempster-Shafer uncertainty interval shown in Figure 6.8 illustrates the concepts just discussed [10, 31]. The lower bound or support for a proposition is equal to the minimal commitment for the proposition based on direct sensor evidence. The upper bound or plausibility is equal to the support plus any potential commitment. Therefore, these bounds show what proportion of evidence is truly in support of a proposition and what proportion results merely from ignorance, or the requirement to normalize the sum of the probability masses to unity. Interpretations of uncertainty intervals are found in Klein [46].

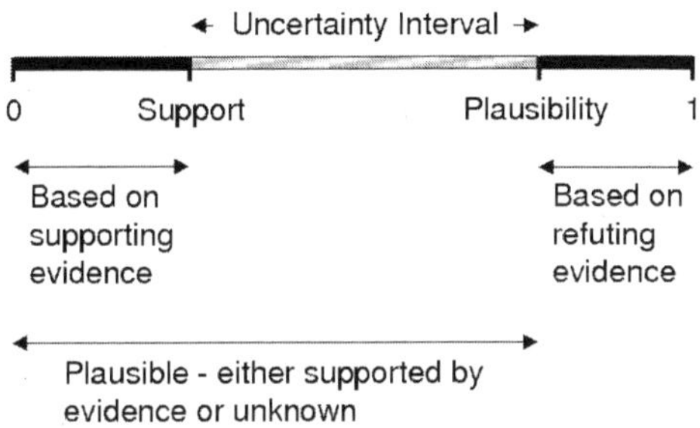

FIGURE 6.8 Dempster-Shafer uncertainty interval for a proposition.

Support and probability mass obtained from a sensor (information source) represent different concepts. Support is calculated as the sum of the probability masses that directly support the proposition and its unions. Probability mass is determined from the sensor's ability to assign some certainty to a proposition based on the evidence.

6.7.4 Dempster's Rule for Combination of Multiple Sensor Data

Dempster's rule supplies the formalism to combine the probability masses provided by multiple sensors or information sources for compatible propositions. The output of the fusion process is given by the intersection of the propositions having the largest probability mass. Propositions are compatible when their intersection exists. Dempster's rule also treats intersections that form a null set, that is, incompatible propositions. In this case the probability masses associated with null intersections are to be zero and the probability masses of the nonempty set intersections are increased by a normalization factor K such that their sum is unity.

The general form of Dempster's rule for the total probability mass committed to an event c defined by the combination of evidence $m_A(a_i)$ and $m_B(b_j)$ from sensors A and B, respectively, is given by

$$m(c) = K \sum_{a_i \cap b_j = c} [m_A(a_i) m_B(b_j)], \tag{6.19}$$

where $m_A(a_i)$ and $m_B(b_j)$ are probability mass assignments on Θ,

$$K^{-1} = 1 - \sum_{a_i \cap b_j = \phi} [m_A(a_i) m_B(b_j)], \tag{6.20}$$

and ϕ is defined as the empty set. If K^{-1} is zero, then m_A and m_B are totally contradictory and the sum defined by Dempster's rule does not exist. The probability mass calculated in Equations (6.19) and (6.20), is termed the orthogonal sum and is denoted as $m_A(a_i) \bigoplus m_B(b_j)$.

Application of Dempster's rule is illustrated with the following four-object, two-sensor example. Suppose that four objects are present:

a_1 = non-threatening type 1 object, a_3 = threatening type 1 object,
a_2 = non-threatening type 2 object, a_4 = threatening type 2 object.

The probability mass matrix for object identification contributed by Sensor A is given by

$$m_A = \begin{pmatrix} m_A(a_1 \cup a_3) = 0.6 \\ m_A(\Theta) = 0.4 \end{pmatrix}, \tag{6.21}$$

where $m_A(\Theta)$ is the uncertainty associated with rules used to determine that the object is of type 1. The probability mass matrix for object identification contributed by Sensor B is given by

$$m_B = \begin{pmatrix} m_B(a_3 \cup a_4) = 0.7 \\ m_B(\Theta) = 0.3 \end{pmatrix}, \tag{6.22}$$

where $m_B(\Theta)$ is the uncertainty associated with the rules used to determine that the object is threatening.

Dempster's rule is implemented by forming a matrix with the probability masses that are to be combined entered along the first column and last row.

TABLE 6.1 Application of Dempster's rule.

First column			
$m_A(\Theta) = 0.4$	$m(a_3 \cup a_4) = 0.28$	$m(\Theta) = 0.12$	
$m_A(a_1 \cup a_3) = 0.6$	$m(a_3) = 0.42$	$m(a_1 \cup a_3) = 0.18$	
	$m_B(a_3 \cup a_4) = 0.7$	$m_B(\Theta) = 0.3$	*Last row*

Inner matrix (row, column) elements are computed as the product of the probability mass in the same row of the first column and the same column of the last row. The proposition corresponding to an inner matrix element is equal to the intersection of the propositions whose masses are multiplied.

Accordingly, matrix element (1, 2) in Table 1.1 represents the proposition formed by the intersection of uncertainty (Θ) from Sensor A and ($a_3 \cup a_4$) from Sensor B, namely that the object is a threatening type 1 or type 2 object. The probability mass $m(a_3 \cup a_4)$ associated with the intersection of these propositions is

$$m(a_3 \cup a_4) = m_A(\Theta)m_B(a_3 \cup a_4) = (0.4)(0.7) = 0.28. \tag{6.23}$$

Matrix element (1, 3) represents the intersection of the uncertainty propositions from Sensor A and Sensor B. The probability mass $m(\Theta)$ associated with the uncertainty intersection is

$$m(\Theta) = m_A(\Theta)m_B(\Theta) = (0.4)(0.3) = 0.12. \tag{6.24}$$

Matrix element (2, 2) represents the proposition formed by the intersection of ($a_1 \cup a_3$) from Sensor A and ($a_3 \cup a_4$) from Sensor B, namely that the object is a threatening type 1 object. The probability mass $m(a_3)$ associated with the intersection of these propositions is

$$m(a_3) = m_A(a_1 \cup a_3)m_B(a_3 \cup a_4) = (0.6)(0.7) = 0.42. \tag{6.25}$$

Matrix element (2, 3) represents the proposition formed by the intersection of ($a_1 \cup a_3$) from Sensor A and (Θ) from Sensor B, namely that the object is of type 1, either non-threatening or threatening. Accordingly, the probability mass associated with this element is

$$m(a_1 \cup a_3) = m_A(a_1 \cup a_3)m_B(\Theta) = (0.6)(0.3) = 0.18. \tag{6.26}$$

The proposition represented by $m(a_3)$ has the highest probability mass in the matrix. Thus, it is typically the one selected as the output to represent the fusion of the evidence from Sensors A and B. Note that the sum of the probability masses associated with the inner matrix elements is unity.

When three or more sensors contribute information, the application of Dempster's rule is repeated using the inner elements calculated from the first application of the rule as the new first column and the probability masses from the next sensor as the entries for the last row (or vice versa).

6.7.5 Dempster's Rule with Empty Set Elements

When the intersection of the propositions that define the inner matrix elements form an empty set, the probability mass of the empty set elements is set equal to zero and the probability mass assigned to the nonempty set elements is increased by the factor K. To illustrate this process, suppose that Sensor B had identified objects 2 and 4, instead of objects 3 and 4, with the probability mass assignments given by $m_{B'}$ as

$$m_{B'} = \begin{pmatrix} m_{B'}(a_2 \cup a_4) = 0.5 \\ m_{B'}(\Theta) = 0.5 \end{pmatrix}. \tag{6.27}$$

Application of Dempster's rule gives the results in Table 1.2, where element (2, 2) now belongs to the empty set. Because mass is assigned to ϕ , we calculate the value K that redistributes this mass to the nonempty set members.

TABLE 6.2 Application of Dempster's rule with an empty set.

First column			
$m_A(\Theta) = 0.4$	$m(a_2 \cup a_4) = 0.20$	$m(\Theta) = 0.20$	
$m_A(a_1 \cup a_3) = 0.6$	$m(\phi) = 0.30$	$m(a_1 \cup a_3) = 0.30$	
	$m_{B'}(a_2 \cup a_4) = 0.5$	$m_{B'}(\Theta) = 0.5$	*Last row*

The value of K^{-1} is computed from Equation (6.20) as unity minus the sum of the products of the probability masses assigned by the information sources to members of the empty set. Thus, K^{-1} is given by

$$K^{-1} = 1 - 0.30 = 0.70, \tag{6.28}$$

and its inverse K by

$$K = 1.429. \tag{6.29}$$

As shown in Table 1.3, the probability mass corresponding to the null set element is set equal to zero and the probability masses of the nonempty set elements are multiplied by K so that their sum is unity. In this example, a type 1 object is declared, but its non-threatening or threatening nature is undetermined.

TABLE 6.3 Probability masses of nonempty set elements increased by K.

First column			
$m_A(\Theta) = 0.4$	$m(a_2 \cup a_4) = 0.286$	$m(\Theta) = 0.286$	
$m_A(a_1 \cup a_3) = 0.6$	0	$m(a_1 \cup a_3) = 0.429$	
	$m_{B'}(a_2 \cup a_4) = 0.5$	$m_{B'}(\Theta) = 0.5$	*Last row*

6.7.6 Comparison with Bayesian Decision Theory

Dempster-Shafer evidential theory accepts an incomplete probabilistic model. Bayesian inference does not. Thus, Dempster-Shafer can be applied when the prior probabilities and likelihood functions or ratios are unknown. The probabilistic information that is available is interpreted as phenomena that impose truth values to various propositions for a certain time period, rather than as likelihood functions. Dempster-Shafer theory estimates how close the evidence is to forcing the truth of a hypothesis, rather than estimating how close the hypothesis is to being true [67].

Dempster-Shafer allows sensor classification error to be represented by a probability assignment directly to an uncertainty class Θ. Furthermore, Dempster-Shafer permits probabilities that express certainty or confidence to be assigned directly to an uncertain event, namely, any of the propositions in the frame of discernment Θ or their unions. Bayesian theory permits probabilities to be assigned only to the original propositions themselves.

Shafer expresses the limitation of Bayesian theory in a more general way: "Bayesian theory cannot distinguish between lack of belief and disbelief. It does not allow one to withhold belief from a proposition without according that belief to the negation of the proposition."

Bayesian theory does not have a convenient representation for ignorance or uncertainty. Prior distributions must be known or assumed with Bayesian. A Bayesian support function [5] ties all of its probability mass to single points in Θ . There is no freedom of motion, that is, no uncertainty interval. The user of a Bayesian support function must somehow divide the support among singleton propositions.

Thus, the difficulty with Bayesian theory lies in representing what we actually know without being forced to over-commit when we are ignorant. With Dempster-Shafer, we use information from the sensors (information sources) to find the support available for each proposition.

Therefore, there is no inherent difficulty in using Bayesian statistics when the required information is available. However, Dempster-Shafer offers an alternative approach when knowledge is not complete, that is, ignorance exists about the prior probabilities associated with the propositions in the frame of discernment. The Dempster-Shafer formulation of a problem collapses into the Bayesian when the uncertainty interval is zero for all propositions and the probability mass assigned to unions of propositions is zero. However, any proposition discrimination information that may have been available from prior probabilities is ignored when Dempster-Shafer in its original formulation is applied. An example showing the equivalence of the Dempster-Shafer and Bayesian solution methods when the uncertainty interval is zero for all propositions and only singleton propositions exist is found in Klein [46]. A discussion of computation times for the two types of algorithms is also found in this reference.

6.7.7 Modifications of Dempster-Shafer Evidential Theory

Criticism of Dempster-Shafer has been expressed concerning the way it reassigns probability mass originally allocated to conflicting propositions and the effect of the redistribution on the proposition selected as the output of the fusion process [87, 88]. This concern is of particular consternation when there is a large amount of conflict that produces counterintuitive results. Several alternatives have been proposed to modify Dempster's rule to better accommodate conflicting beliefs [19, 60, 82]. Among these are the transferable belief model [75, 76, 78], plausibility transformation [14], incorporation of a priori information [28, 55], consensus operator [43], and plausible and paradoxical reasoning [19] as discussed in [46].

In the next section we adopt the Bayesian framework and present an approach for fusing image features from visible and infrared imagery for object tracking using particle filtering.

6.8 Particle Filtering

Sequential Monte Carlo methods (known also as particle filters [1, 3], the condensation [41] and bootstrap filters [34] are some of the most popular Bayesian methods for coping with uncertainties and nonlinearities [68].

The main objective of the particle filter [1, 3, 34, 69] is to keep track of a variable of interest as it evolves over time, describing it with a non-Gaussian and possibly multimodal probability density function (pdf). The method relies on a sample-based construction of the pdf. Multiple particles (samples) of the variables of interest are generated, each one associated with a weight that characterizes the quality of the specific particle. An estimate of the variable of interest is obtained by the weighted sum of the particles. Two major stages can be distinguished: *prediction* and *update*. During prediction each particle is modified according to the state model, including the addition of random noise, in order to simulate its effect on the variable of interest. Then in the *update* stage, each particle's weight is re-evaluated based on the incoming sensor information. A procedure, called *resampling*, deals with the elimination of particles with small weights and replicates the particles with higher weights.

Arulampalam et al. [3, 69] provide one of the most comprehensive introductions to the subject. Del Moral has proposed advanced tools for analyzing these methods [59]. These techniques have blossomed in recent years and are now widespread, particularly for tracking and navigation applications.

Particle filters are generally based on the existence of a Markov model of the physical system:

$$\begin{cases} \boldsymbol{x}_{k+1} = f(\boldsymbol{x}_k, \boldsymbol{v}_k) \\ \boldsymbol{y}_k = g(\boldsymbol{x}_k, \boldsymbol{w}_k) \end{cases} \tag{6.30}$$

where $\boldsymbol{x}_k \in \mathbb{R}^{n_x}$ is the n_x-dimensional state vector; $\boldsymbol{y}_k \in \mathbb{R}^{n_x}$ is n_y the observation vector; $\boldsymbol{v}_k$ and $\boldsymbol{w}_k$ are noises with known probability density functions. Here $f(.)$ and $g(.)$ are the state and measurement functions, respectively. From a probabilistic perspective, Equation (6.30) forms a Markov process characterized by the conditional distributions $p(\boldsymbol{x}_k|\boldsymbol{x}_{k-1})$ and $p(\boldsymbol{y}_k|\boldsymbol{x}_k)$.

Bayes' formula gives the conditional distribution as:

$$p(\boldsymbol{x}_{0:k+1}|\boldsymbol{y}_{1:k+1}) = p(\boldsymbol{x}_{0:k}|\boldsymbol{y}_{1:k}) \times \frac{p(\boldsymbol{y}_{k+1}|\boldsymbol{x}_{k+1})p(\boldsymbol{x}_{k+1}|\boldsymbol{x}_k)}{p(\boldsymbol{y}_{k+1}|\boldsymbol{y}_{1:k})}. \tag{6.31}$$

In general, solving this equation is impossible because the prior probability is unknown analytically, except in the case of linear Gaussian systems, where the recurrence is reduced to the Kalman-Bucy filter [42,44,45]. We must therefore resort to numerical approximation, such as Monte Carlo. The Monte Carlo method uses an approximation for $p(\boldsymbol{x}_{0:k+1}|\boldsymbol{y}_{1:k+1})$.

The aim of sequential Monte Carlo estimation is to evaluate the *posterior* pdf $p(\boldsymbol{x}_k|\boldsymbol{y}_{1:k})$ of the state vector $\boldsymbol{x}_k$, given a set $\boldsymbol{y}_{1:k} = \{\boldsymbol{y}_1, \dots, \boldsymbol{y}_k\}$ of sensor measurements up to time k. The particle filtering approach relies on a sample-based construction to represent the state pdf. Multiple particles (samples) of the state are generated, each one associated with a weight $W_k^{(\ell)}$ that characterizes the quality of a specific particle ℓ, $\ell = 1, 2, \dots, N$.

Within the Bayesian framework, the conditional pdf $p(\boldsymbol{x}_{k+1}|\boldsymbol{y}_{1:k})$ is recursively updated according to the *prediction* step

$$p(\boldsymbol{x}_{k+1}|\boldsymbol{y}_{1:k}) = \int_{\mathbb{R}^{n_x}} p(\boldsymbol{x}_{k+1}|\boldsymbol{x}_k)p(\boldsymbol{x}_k|\boldsymbol{y}_{1:k})d\boldsymbol{x}_k \tag{6.32}$$

and the *update* step

$$p(\boldsymbol{x}_{k+1}|\boldsymbol{y}_{1:k+1}) = \frac{p(\boldsymbol{y}_{k+1}|\boldsymbol{x}_{k+1})p(\boldsymbol{x}_{k+1}|\boldsymbol{y}_{1:k})}{p(\boldsymbol{y}_{k+1}|\boldsymbol{y}_{1:k})}, \tag{6.33}$$

where $p(\boldsymbol{y}_{k+1}|\boldsymbol{y}_{1:k})$ is a normalizing constant. The recursive update of $p(\boldsymbol{x}_{k+1}|\boldsymbol{y}_{1:k+1})$ is proportional to

$$p(\boldsymbol{x}_{k+1}|\boldsymbol{y}_{1:k+1}) \propto p(\boldsymbol{y}_{k+1}|\boldsymbol{x}_{k+1})p(\boldsymbol{x}_{k+1}|\boldsymbol{y}_{1:k}). \tag{6.34}$$

Usually, there is no simple analytical expression for propagating $p(\boldsymbol{x}_{k+1}|\boldsymbol{y}_{1:k+1})$ through Equation (6.34) so numerical methods are used.

The posterior $p(\boldsymbol{x}_{k+1}|\boldsymbol{y}_{1:k+1})$ of the object is approximated by N particles $\boldsymbol{x}_{k+1}^{(\ell)}$ and their normalized importance weights $\widehat{W}_{k+1}^{(\ell)}$:

$$\hat{p}(\boldsymbol{x}_{k+1}|\boldsymbol{y}_{1:k+1}) \approx \sum_{\ell=1}^{N} \widehat{W}_{k+1}^{(\ell)} \delta(\boldsymbol{x}_{k+1} - \boldsymbol{x}_{k+1}^{(\ell)}), \tag{6.35}$$

where $\delta(.)$ is the Dirac delta function. New weights are calculated, putting more emphasis on particles that are important according to the *posterior* pdf (6.35).

It is often impossible to sample directly from the posterior density function $p(\boldsymbol{x}_{k+1}|\boldsymbol{y}_{1:k+1})$. This difficulty is circumvented by utilizing importance sampling from a known *proposal distribution* $p(\boldsymbol{x}_{k+1}|\boldsymbol{x}_k)$. An example of the above procedure is the particle filter with multiple cues given in Table 6.4.

6.9 Object Tracking for Surveillance Systems

A key issue in the development of advanced visual-based surveillance systems is dealing with objects that are usually moving in a highly variable environment [2, 30, 32]. This is typically addressed with sophisticated algorithms for video acquisition, camera calibration, noise filtering, and motion detection that learn and adapt to changing scenes, lighting, and weather. The use of clusters of fixed cameras, typically grouped in areas of interest but covering the whole area, requires coordinated solutions to other issues, such as automatic methods for synchronization of the acquired data (for overlapping and non-overlapping views) and registration of the images or video frames. If tracking is performed on fused data, then suitable methods for data fusion are also necessary. Other issues are related to the sensors (such as their placement, number of sensors and type), models of the moving objects, measurement models, and the function characterizing the similarity between two images or video frames. Multiple sensors of the same type or modality can be used, for example, multiple optical cameras, or sensors providing different modalities, such as optical and infrared cameras, can be employed. The complement of the information supplied by the different sensors can be valuable for solving different kinds of problems. The availability of redundant and complementary data is an advantage, but it also raises many challenges. The amount of data to be processed is often enormous. Furthermore, the image or video sequences need to be registered (aligned) in both time and space in order to combine the information in them.

In image-based tracking, the fusion of data from different sensor modalities and the fusion of different image features can be successfully performed with Bayesian methods. These Bayesian methods are most often used where the problem reduces to reconstruction of the probability density function of object states given the measurements and prior knowledge. They allow data association from multiple targets tracked with multiple sensors by incorporating techniques that address external constraints. Particle filters offer a flexible framework for fusing different image cues derived from image features such as color, edges, texture, and motion in combination or adaptively chosen [11–13, 53, 68]. By assuming the cues are *conditionally independent*, they can be easily combined through a likelihood function that consists of the product of the likelihoods of each cue.

In the remaining part of this chapter, we show that particle filtering techniques are suitable for object tracking in video sequences from different sensors, that is, from separate visible and infrared cameras, and from fused videos.

6.9.1 Multiple Cues

Most video tracking techniques are region based, which means that the object of interest is contained within a region, often of a rectangular or circular

TABLE 6.4 The particle filter with multiple cues.

Initialization

1. $k = 0$, for $\ell = 1, \ldots, N$, generate samples $\{\boldsymbol{x}_0^{(\ell)}\}$ from the prior distribution $p(\boldsymbol{x}_0)$. Set initial weights $W_0^{(\ell)} = 1/N$.

For $k = 1, 2, \ldots,$

Prediction Step

2. For $\ell = 1, \ldots, N$, sample $\boldsymbol{x}_{k+1}^{(\ell)} \sim p(\boldsymbol{x}_{k+1}|\boldsymbol{x}_k^{(\ell)})$ from the dynamic motion model

Measurement Update: evaluate the importance weights

3. Compute the weights

$$W_{k+1}^{(\ell)} \propto W_k^{(\ell)} \mathcal{L}^{fused}(\boldsymbol{y}_{1:k+1}|\boldsymbol{x}_{k+1}^{(\ell)})$$

based on the likelihood $\mathcal{L}^{fused}(\boldsymbol{y}_{1:k+1}|\boldsymbol{x}_{k+1}^{(\ell)})$ from the fused cues from Equation (6.36).

4. Normalize the weights, $\widehat{W}_{k+1}^{(\ell)} = \frac{W_{k+1}^{(\ell)}}{\sum_{\ell=1}^{N} W_{k+1}^{(\ell)}}$.

Output

5. The posterior mean $E[\boldsymbol{x}_{k+1}|\boldsymbol{y}_{1:k+1}]$ is computed using the set of particles

$$\hat{\boldsymbol{x}}_{k+1} = E[\boldsymbol{x}_{k+1}|\boldsymbol{y}_{1:k+1}] = \sum_{\ell=1}^{N} \widehat{W}_{k+1}^{(\ell)} \boldsymbol{x}_{k+1}^{(\ell)}.$$

Resampling Step

Estimate the effective number of particles $N_{eff} = 1/\sum_{\ell=1}^{N}(W_k^{(\ell)})$. If $N_{eff} \leq N_{thresh}$ (N_{thresh} is a given threshold), then perform resampling. Multiply/suppress samples $\boldsymbol{x}_{k+1}^{(\ell)}$ with high/low importance weights $\widehat{W}_{k+1}^{(\ell)}$.

6. For $\ell = 1, \ldots, N$, set $W_{k+1}^{(\ell)} = \widehat{W}_{k+1}^{(\ell)} = 1/N$.

---

shape. This region is then tracked through a sequence of video frames based on certain features (or their histograms), such as color, texture, edges, shape, and their combinations [12, 13, 68, 81]. Next, a distance measure is applied to characterize the distance between the target region and the current region. Distances that have proven good performance are the Bhatacharyya distance (see Section 6.9.2) and the structural similarity measure (see Section 6.9.3 and [53]).

The relationship between different cues is treated differently by different authors. For example, [56] makes the assumption that color and texture are not independent. However, other works [64, 71] assume that color and texture cues are independent. In image classification, the independence assumption between color and texture is applied for feature fusion in a Bayesian framework [74]. There is generally agreement that, in practice, color and texture and color and edges do combine well together.

We assume that the considered cue combinations, for example, color and texture and color and edges are independent. With this assumption, the overall likelihood function $\mathcal{L}^{fused}(\boldsymbol{y}_k|\boldsymbol{x}_k)$ of the particle filter represents a product of the likelihoods of the separate cues [13]

$$\mathcal{L}^{fused}(\boldsymbol{y}_k|\boldsymbol{x}_k) = \prod_{l=1}^{L} \mathcal{L}_l(\boldsymbol{y}_{l,k}|\boldsymbol{x}_k)^{\epsilon_l}. \tag{6.36}$$

The cues are adaptively weighted by coefficients ϵ_l. The measurement vector $\boldsymbol{y}_k$ is composed of the measurement vectors $\boldsymbol{y}_{l,k}$ from the lth cue for $l = 1, \ldots, L$.

6.9.2 The Bhattacharyya Distance Measure

The Bhattacharyya measure [6] was previously used for color cues [15, 63] because it has the important property that $\rho(p, p) = 1$. Here the distributions for each cue are represented by the respective histograms [72],

$$\rho(\boldsymbol{h}_{\text{ref}}, \boldsymbol{h}_{\text{tar}}) = \sum_{i=1}^{B} \sqrt{h_{\text{ref},i} h_{\text{tar},i}}, \tag{6.37}$$

where two normalized histograms $\boldsymbol{h}_{\text{tar}}$ and $\boldsymbol{h}_{\text{ref}}$ describe the cues for a *target region* defined in the current frame and a *reference region* in the first frame, respectively. The measure of the similarity between these two distributions is then given by the Bhattacharyya distance

$$d(\boldsymbol{h}_{\text{ref}}, \boldsymbol{h}_{\text{tar}}) = \sqrt{1 - \rho(\boldsymbol{h}_{\text{ref}}, \boldsymbol{h}_{\text{tar}})}. \tag{6.38}$$

The larger the measure $\rho(\boldsymbol{h}_{\text{ref}}, \boldsymbol{h}_{\text{tar}})$, the more similar the distributions are. Conversely, the smaller the value of d, the more similar are the distributions (histograms). For two identical normalized histograms, we obtain $d = 0$ ($\rho = 1$) indicating a perfect match.

Based on Equation (6.38), a distance D^2 for color can be defined that takes into account all of the RGB color channels

$$D^2(\boldsymbol{h}_{\text{ref}}, \boldsymbol{h}_{\text{tar}}) = \frac{1}{3} \sum_{c \in \{R,G,B\}} d^2(\boldsymbol{h}^c_{\text{ref}}, \boldsymbol{h}^c_{\text{tar}}). \tag{6.39}$$

The distance D^2 for the edges is equal to d^2 because there is only one component. The distance D^2 for texture is

$$D^2(\boldsymbol{h}_{\text{ref}}, \boldsymbol{h}_{\text{tar}}) = \frac{1}{8} \sum_{\omega \in \{1,\ldots,8\}} d^2(\boldsymbol{h}^\omega_{\text{ref}}, \boldsymbol{h}^\omega_{\text{tar}}), \tag{6.40}$$

where ω is the channel in the steerable-pyramid decomposition [13, 29].

The *likelihood function* for the cues can be defined by [68],

$$\mathcal{L}(\boldsymbol{y}|\boldsymbol{x}) \propto \exp\left(-\frac{D^2(\boldsymbol{h}_{\text{ref}}, \boldsymbol{h}_{\boldsymbol{x}})}{2\sigma^2}\right), \tag{6.41}$$

where the standard deviation σ specifies the Gaussian noise in the measurements. Small Bhattacharyya distances correspond to large weights in the particle filter.

6.9.3 Structural Similarity Measure

In contrast to other simple image similarity measures such as the mean square error, mean absolute error, or peak signal-to-noise ratio, the structural similarity measure has the advantage of capturing the perceptual similarity of images or video frames subject to conditions caused by varying luminance, contrast, compression, or noise [86]. Recently, based on the premise that the Hue, Value, Saturation (HVS) space is highly tuned to extract structural information, a new image metric was developed, called the Structural SIMilarity (SSIM) index [86]. The SSIM index between two images I and J is defined as follows:

$$\begin{aligned} S(I,J) &= \left(\frac{2\mu_I\mu_J + C_1}{\mu_I^2 + \mu_J^2 + C_1}\right)\left(\frac{2\sigma_I\sigma_J + C_2}{\sigma_I^2 + \sigma_J^2 + C_2}\right)\left(\frac{\sigma_{IJ} + C_3}{\sigma_I\sigma_J + C_3}\right) \\ &= l(I,J)\, c(I,J)\, s(I,J), \end{aligned} \tag{6.42}$$

where $C_{1,2,3}$ are small positive constants used for the numerical stability purposes, μ denotes the sample mean

$$\mu_I = \frac{1}{L}\sum_{j=1}^{L} I_j, \tag{6.43}$$

σ denotes the sample standard deviation

$$\sigma_I = \sqrt{\frac{1}{L-1}\sum_{j=1}^{L}(I_j - \mu_I)^2}, \tag{6.44}$$

and

$$\sigma_{IJ} = \frac{1}{L-1}\sum_{j=1}^{L}(I_j - \mu_I)(J_j - \mu_J) \tag{6.45}$$

corresponds to the sample covariance. The estimators are defined identically for images I and J, each having L pixels. The image statistics are computed in the way proposed in [86], that is, locally, within a 11×11 normalized circular-symmetric Gaussian window.

For $C_3 = C_2/2$, Equation (6.42) can be simplified

$$S(I, J) = \left(\frac{2\mu_I\mu_J + C_1}{\mu_I^2 + \mu_J^2 + C_1}\right)\left(\frac{2\sigma_{IJ} + C_2}{\sigma_I^2 + \sigma_J^2 + C_2}\right). \tag{6.46}$$

The three components of Equation (6.42), l, c, and s, measure the *luminance*, *contrast*, and *structural similarity* of the two images, respectively. Such a combination of image properties represents a fusion of three independent image cues. The relative independence assumption is based on a claim that a moderate luminance and/or contrast variation does not affect structures of the image objects [85].

6.9.4 Adaptively Weighted Cues

The method presented below utilizes the Bhattacharyya distance (Equation (6.38)) to give significance to the likelihood obtained for each cue based on the current frame. This is different from previous works that use the performance of the cues over the previous frames [68], but do not take into account information from the latest measurements. The proposed approach allows estimation of ϵ_l in Equation (6.36) for each cue. Based on the smallest value of the distance measure $D^2_{l,\min}$, the weight for each cue l is determined as

$$\hat{\epsilon}_l = \frac{1}{D^2_{l,\min}}, \qquad l = 1, \ldots, L. \tag{6.47}$$

The weights are then normalized such that $\sum_{l=1}^{L} \epsilon_l = 1$,

$$\epsilon_l = \frac{\hat{\epsilon}_l}{\sum_{l=1}^{L} \hat{\epsilon}_l}, \qquad l = 1, \ldots, L. \tag{6.48}$$

6.9.5 Video Registration

Due to the temporal and spatial misalignments, the raw videos from the multisensor system cannot be used directly and a spatio-temporal registration is required. When the videos are recorded [48,50], a flash is used (which can be clearly detected by both sensors) as a starting signal. The presence of spatial misalignment is mainly due to the different physical properties of the sensors and the fact that although roughly collocated, they are not at exactly the same position.

Similar to [51], an affine transformation is applied to align the videos. The affine transform parameters are reliably obtained through a least squares estimation using a set of corresponding key points. As the video data is produced by a static multi-sensor system with fixed cameras, we can assume that the local transformations between these two sensors are constant over the recording time. The key points for the registration transformation can be chosen from different frame pairs. This assumption works well only if the distance in the scene between the moving objects and the sensors does not change much. Otherwise a parallax problem has to be solved and the registration transform parameters have to be updated dynamically over time [39].

6.9.6 Image Fusion

Image fusion is the process of combining images from different modalities, for example, visible spectrum and infrared images, in order to construct a composite image that is more informative and more suitable for human visual perception or computer processing. However, when we fuse images, we also fuse the noise and artifacts contained in the input images, and we often may create new artifacts during the fusion process.

Four methods for fusing the visible spectrum and infrared video sequences are considered [47, 48, 50, 58]: 1) simple averaging in the spatial domain (AVE); 2) contrast pyramid (PYR) fusion; 3) (shift-variant version) of Discrete Wavelet Transform (DWT) fusion; and 4) Dual-Tree Complex Wavelet Transform (DT-CWT) fusion. In previous investigations [49,61,62] with still images and short image sequences, we found that DT-CWT significantly outperforms the other fusion methods in terms of (perceptual) fused image quality.

6.10 Results

The experiments presented in this section were performed with real data collected from one visible (VIZ) and one infrared (IR) camera. The testing sequences, taken from the Eden Project Multisensor Dataset of short-range surveillance videos [48] (available at www.scanpaths.org), are very challenging for automatic object tracking due to the following reasons: 1) the color of the moving object is very similar to the background (camouflaged moving targets); 2) the environment contains lush and dense vegetation; 3) frequent obscuration of the targets; and 4) very low signal-to-noise ratio of the visible spectrum videos. The videos were fused using the following methods: Averaging Technique (AVE), Contrast Pyramids (CP), Discrete Wavelet Transforms (DWT), and Dual-Tree Complex Wavelet Transforms (DT-CWT) (see [21] for more details on these techniques).

6.10.1 Performance Evaluation Measure

In order to assess the influence of video fusion on the accuracy of object tracking, video sequences with pre-drawn target maps [21] are used. These are rectangular boxes drawn around the target (a walking human figure) on each frame (Figure 6.9a). The estimate of the target position and corresponding estimated rectangular box are obtained from the tracking algorithm (the small rectangle in Figure 6.9a). The target map (ground truth rectangle) and the estimated rectangle vary in size from frame to frame. In order to measure the performance of the tracking algorithm, the normalized intersection area between the two rectangles is used. The rationale for such a measure is that a larger intersection or overlap between the estimated area (B) and the ground truth rectangle (A), indicates better tracking. This normalized intersection area S is calculated as $S = (A \cap B)/B$. The measure varies between 0 (no intersection between the rectangles) and 1 (estimated rectangle fully within ground truth rectangle).

6.10.2 Experimental Results with Real Data

The particle filter (PF) algorithm with 300 particles was run with the same non-optimized default parameters in all experiments in order to compare object tracking performance with different video sources. The rectangle used to manually initialize the PF algorithm was selected to be significantly smaller than the ground truth rectangle, so that only the core features of the object of interest (the head and the upper part of the body) are included and the influence of the nonstationary background is minimized.

The measure S was calculated for each frame and the fusion methods investigated, as indicated in Figure 6.9b. In order to obtain the global characteristics of the tracking filter performance, the difference between the actual

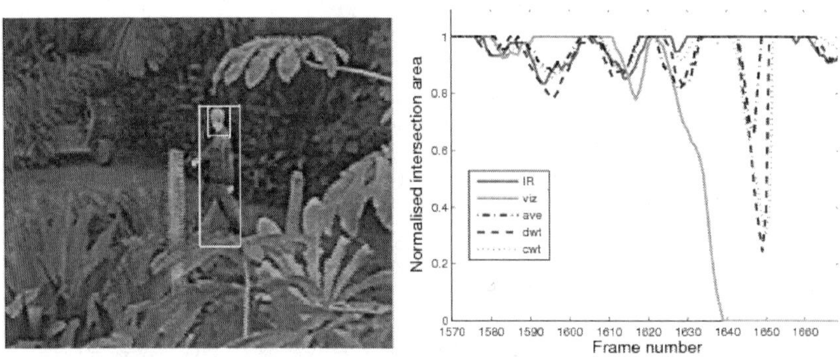

FIGURE 6.9 a) Left: ground truth rectangle (the big box, surrounding the whole person) [21] and PF reference rectangle (the smaller box). b) Right: plot of the normalized intersection area S for the tested tracking algorithms.

and ideal tracking algorithm was characterized: $\sum_k (1 - S_k)$, where k is a frame index. According to this validation, the most accurate tracking results are obtained with the IR video. Other methods (such as fused AVE, DT-CWT, DWT) performed comparably, whereas single modality VIZ suffered from frequent loss of the target.

FIGURE 6.10 A tracking algorithm working with the VIZ video: frames 1570, 1625, and 1670. The object is lost after the full occlusion.

Figure 6.9 shows also that the best overlap between the two rectangles is obtained from the IR video, then the AVE, then DT-CWT, and finally the DWT. The worst result is obtained with VIZ. The color-edge based algorithm failed to track the object in the VIZ videos due to almost the same color as the background. Figures 6.10 through 6.13 show the results from tracking in VIZ, IR, AVE, and DT-CWT videos, respectively, from one of the Tropical Forest sequences used in this study. The results from the CP are not shown because of the many artifacts introduced by the fusion, thus rendering the tracking algorithm unreliable. The results from the DWT are not shown because of their similarity to the DT-CWT.

Figure 6.14 shows results from a video sequence obtained by an optical camera, with PF using color and edges, Figure 6.15 shows results with the structural similarity measure. In the final frame with the color and edges tracking (Figure 6.14), the PF algorithm almost loses the object, whereas the SSIM provides more reliable tracking under the shadow. Both color and edges, and the SSIM show good tracking performance in the IR video sequence (Figure. 6.16 and Figure 6.17).

There is a slight shift in the coordinates of the rectangle when tracking with color and edges both in IR and CWT fused sequences (Figure 6.19). The best results in fused videos are obtained with AVE (Figure 6.12) and DT-CWT (Figure 6.13). The PYR method itself introduces many artifacts (additional information) that have a disturbing effect on the tracking algorithm.

FIGURE 6.11 A tracking algorithm working with the IR video: frames 1570, 1625, and 1670. The object is lost after the full occlusion, but recovered after that.

Figure 6.18 shows results on fused videos with the DT-CWT, where the PF algorithm is working with the Bhattacharyya coefficient, and Figure 6.19 shows results over the same fused video (with the DT-CWT), where the SSIM distance is used.

The algorithms with the best performance are able to keep tracking the object despite severe occlusions. Such an example is shown in Figures 6.10 through 6.13 where the person is completely hidden by the tree. Because the head features are well distinguishable from the surrounding environment, the algorithm succeeded in recovering the person. The tracking algorithm may still lose the object in the presence of long full occlusions (5-10 frames or more) or when the target region does not correspond completely to the initially chosen target region. This could be solved by adding a detection scheme to the tracking technique and updating the target region.

The obtained results are similar to findings of a psycho-visual study with human eye tracking, conducted independently of these experiments [21], over the same sequence of data. In this case, the AVE also produced the best results.

FIGURE 6.12 A tracking algorithm working with AVE fused video sequences: frames 1570, 1625, and 1670. Reliable performance is observed.

FIGURE 6.13 A tracking algorithm working with DT-CWT fused video sequences: frames 1570, 1625, and 1670. Reliable performance is observed.

FIGURE 6.14 Frames 1, 35, and 100 from the tracking algorithm working with color and edges over a video sequences, from an optical camera.

FIGURE 6.15 Frames 1, 35, and 100 from the tracking algorithm working with the SSIM, from an optical camera.

FIGURE 6.16 Frames 1, 35, and 100 of a tracking algorithm working with color and edges over IR video sequences.

FIGURE 6.17 Frames 1, 35, and 100 of a tracking algorithm working with the SSIM over an infrared sequence.

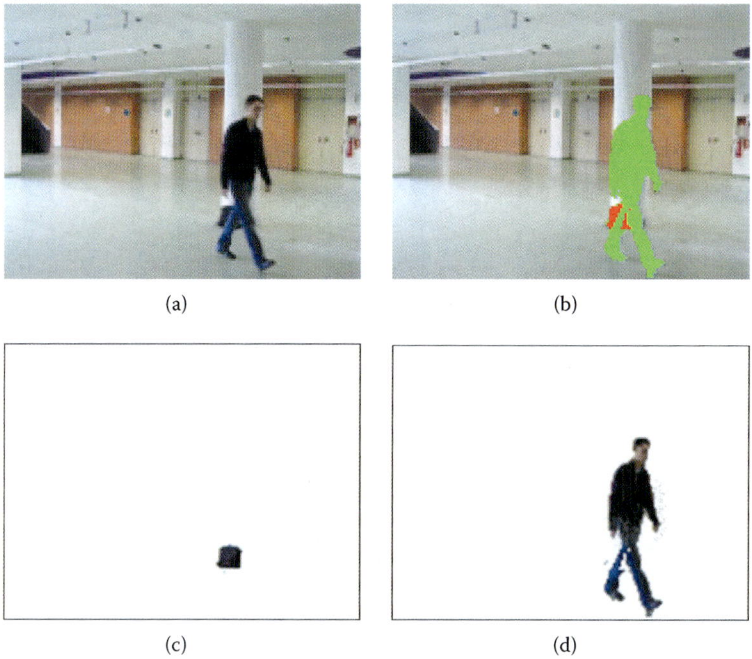

FIGURE 3.4 Stopped Foreground Subtraction for sequence *Msa*: (a) original frame; (b) original frame with moving (green) and stopped (red) foreground objects; (c) moving foreground model; (d) stopped foreground model.

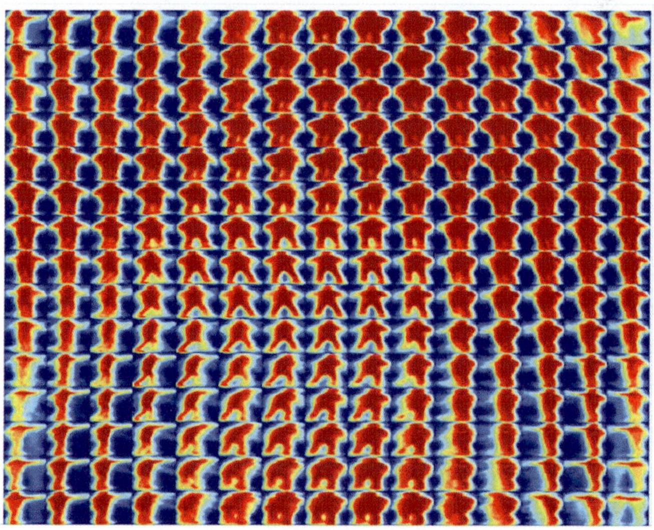

FIGURE 7.2 Codebook of the trained SOM for action punch.

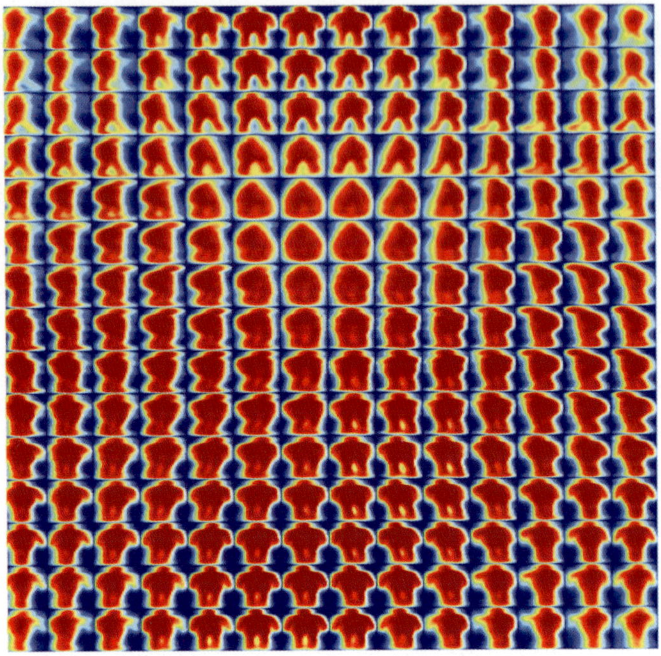

FIGURE 7.3 Codebook of the trained SOM for all actions.

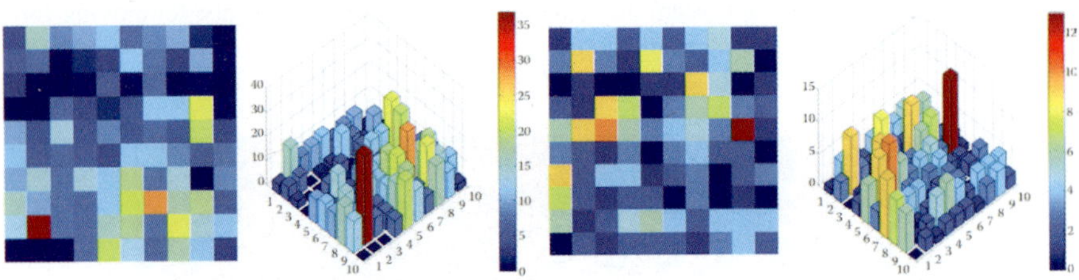

FIGURE 7.5 (Left) Histogram of winner-neurons for k-SOM when action was ω_k, (Right) Histogram of winner-neurons for k-SOM when action was ω_i with $i \neq k$.

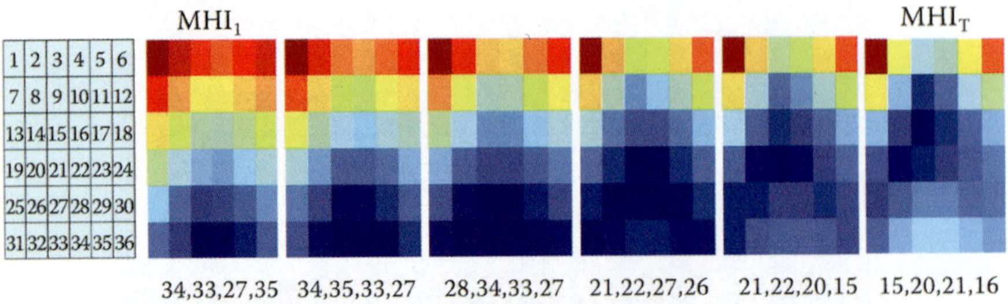

FIGURE 7.7 Q-error maps for a sequence of temporal MHIs. Color represents the Euclidean distance from the pattern to the SOM. Hot colors imply high distance. The winner neuron is the darkest blue cell.

(a) Frame no. 900

(b) Frame no. 2340

FIGURE 8.1 In this example, component-based clustering is done in components *object's position, object's size, object's color correlogram, and object's time of presence.* Then, the cluster algebraic operation is applied to these component clusters to get the composed clusters. Each color indicates a composed cluster of the object properties as depicted in the image by plotting the centroid of the objects across time. Thus, the trajectories of the objects are discovered by composition of clusters from different component spaces, including time. It can be seen in (b) that the composed cluster of yellow color is not continuous although the cluster corresponds to the same object across frames in some of which the object was either occluded or not detected. This cluster shows that despite occlusion/nondetection, continuity and similarity are maintained by composing these component clusters.

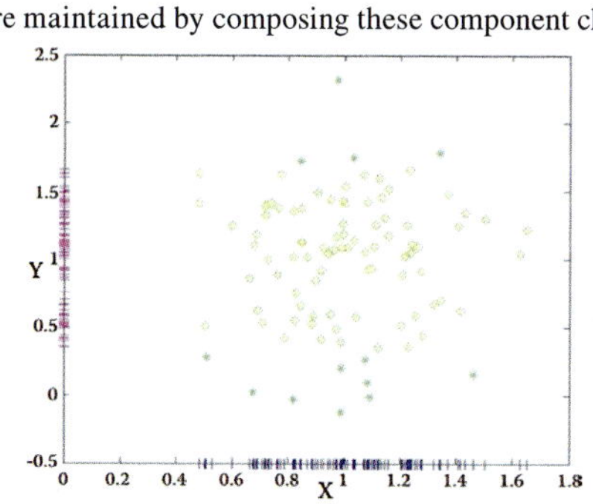

FIGURE 8.2 The blue points belong to the cluster in the X-component and the magenta points belong to the cluster in the Y-component. The green points belong to the cluster formed by directly clustering in the 2D space $C_{(x^*,y^*)}$ and the yellow points belong to the composed cluster $C_{x^* \otimes y^*}$ as well as to $C_{(x^*,y^*)}$. The degree of approximation is 0.0022, showing that the composed cluster is a good approximation of the 2D cluster.

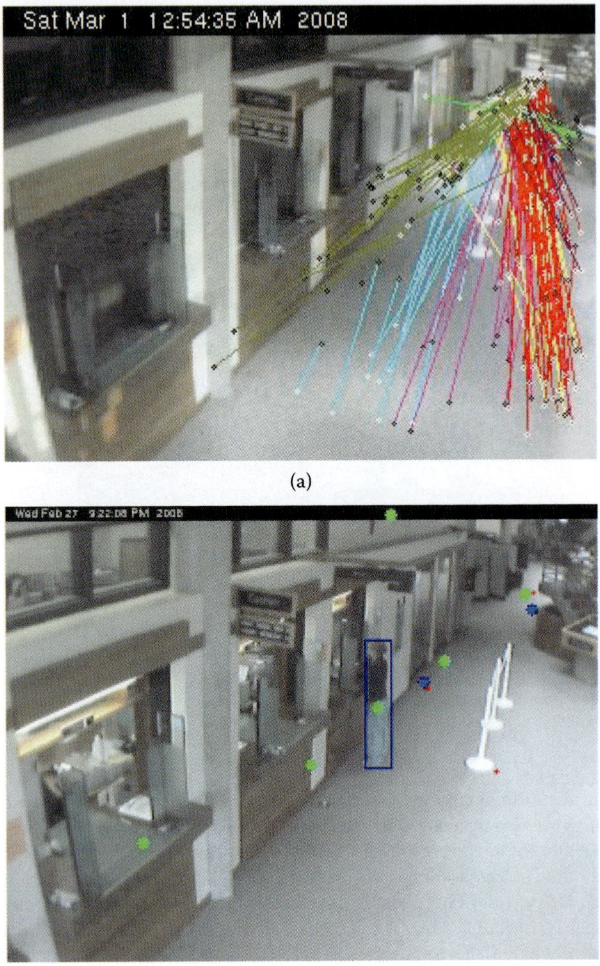

(a)

(b)

FIGURE 8.3 The usual paths in the scene are discovered by clustering the trajectories of the objects in the scene, as shown in (a). The white point indicates the starting point and the black point indicates the ending location of the trajectory. Each trajectory cluster is represented by a separate cluster. As the rate of arrival of frames via the Internet from the camera at [25] is very low, some objects are detected for the first time in the middle of the scene while some vanish from the scene suddenly from non-exit locations. In spite of this, our system discovers the true entry (blue), exit (red), and wait (green) locations, as shown in (b). We allow the black strip at the top of the image to remain and be detected as a foreground object because the time information is necessary. However, it results in it getting detected as a wait location, where objects tend to remain static for long periods of time.

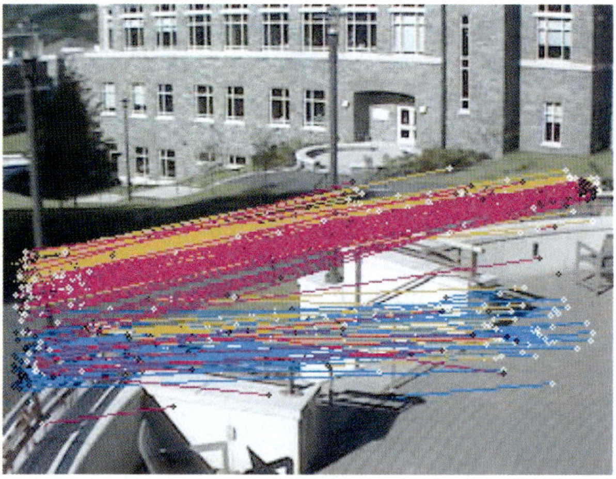

FIGURE 8.7 The usual paths in the scene are shown as clusters of trajectories, where each color represents a separate cluster. Each line is a trajectory in the scene with the white dot indicating the start location and the black dot indicating the end position.

FIGURE 10.2 Pixel clustering during MSER formation. Clustered pixels are colored using the region's average color. Non-assigned pixels are shown in blue. Results represent clusters after iterations 2, 5, and 35 (left to right).

FIGURE 10.3 Left: An example of the feed-forward process. Dark gray pixels are preserved, light gray pixels are re-clustered. Center: MSERs are modeled and displayed using ellipses and average color values. Right: An example of MSER image segmentation. Regions are filled with their average color, detected lines are shown in green, and the path of the tracked hand is represented as a red line.

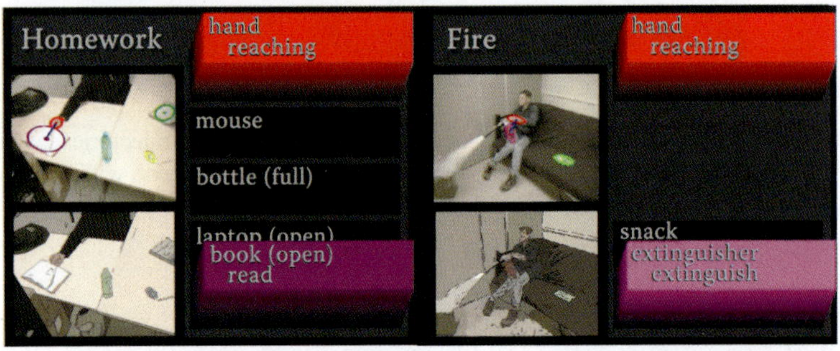

FIGURE 10.7 An example of activity recognition using our system. Each object is associated by a color bar at the right of the image. The apparent height of the bar corresponds to the computed probability that the person's hand is interacting with that object. In the scenario shown on the left, a person engaged in typical homework-type behaviors, including typing on a laptop, turning pages in a book, moving a mouse, and drinking from a bottle. In the scenario on the right, a person reached into a bag of chips multiple times, and extinguished a trash fire with a fire extinguisher.

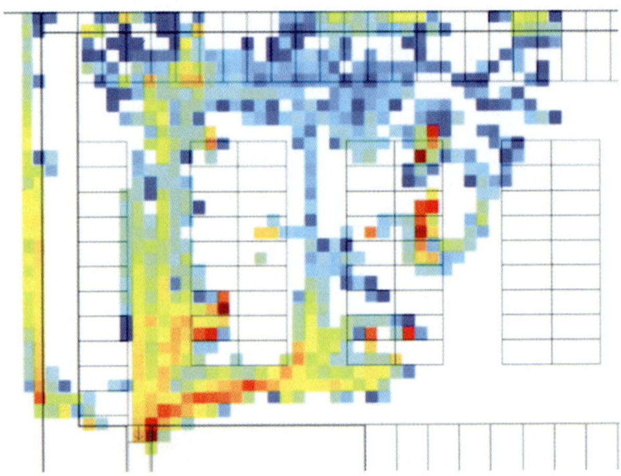

FIGURE 11.4 A relevance map defined by audio sensors, superimposed on a map of the observed environment of a parking lot.

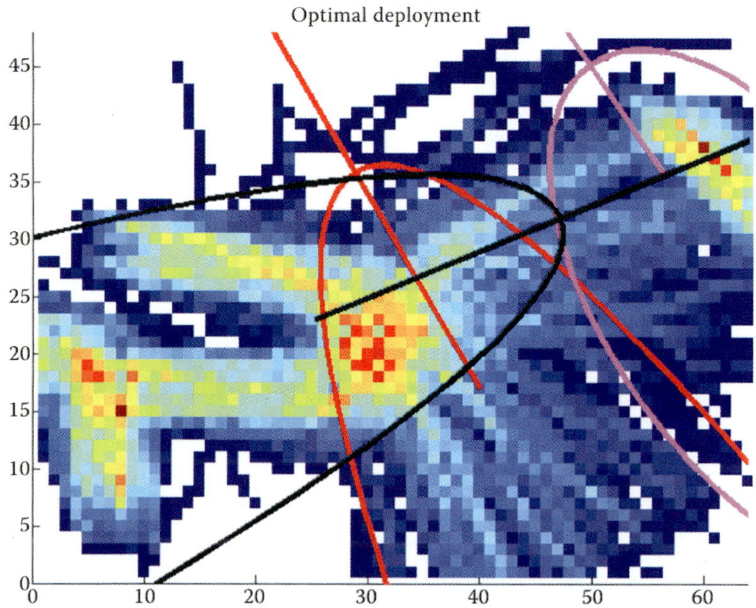

FIGURE 11.15 Optimal configuration for a camera network composed of three video sensors.

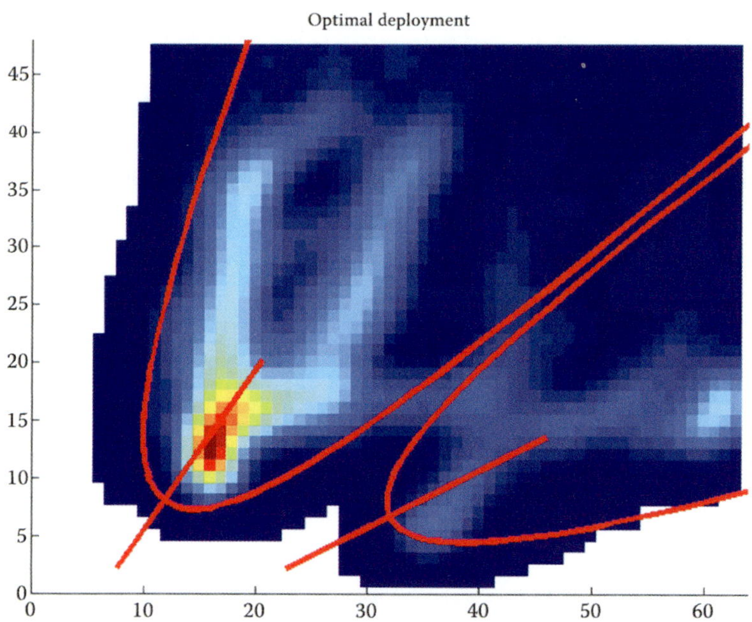

FIGURE 11.16 A real-world camera reconfiguration example. The relevance map is built according to audio data acquired during a 4-hour monitoring session. The ellipses show the areas observed by two PTZ cameras after reconfiguration.

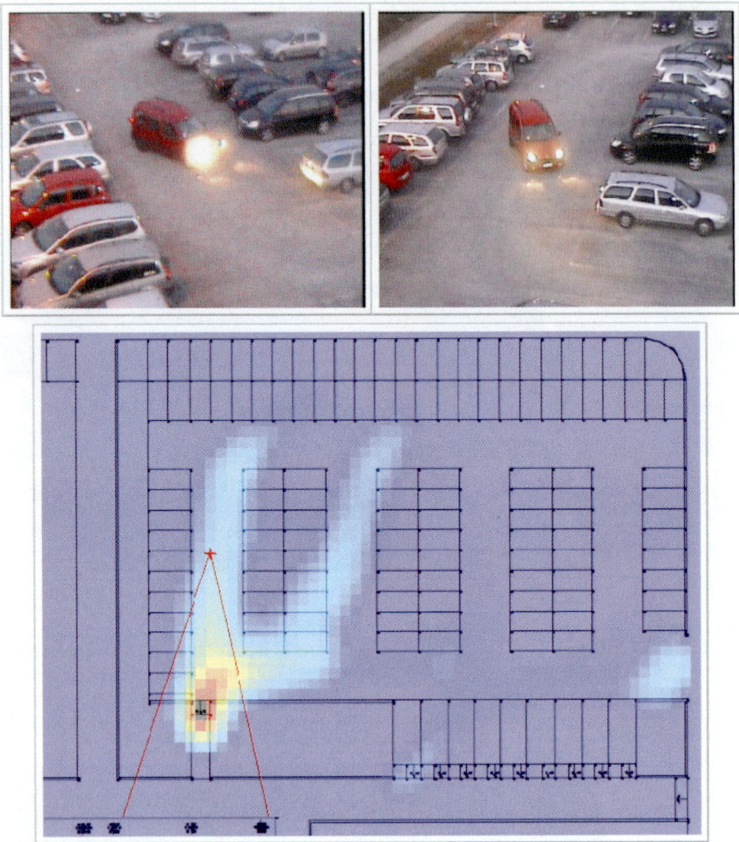

FIGURE 11.17 Two PTZ cameras are automatically focused on an audio source detected by the audio subsystem.

FIGURE 6.18 Frames 1, 35, and 100 of a tracking algorithm working with color and edges over fused video sequences, with DT-CWT.

FIGURE 6.19 Frames 1, 35, and 100 of a tracking algorithm working with the SSIM over fused video sequences, with DT-CWT.

6.11 Conclusions

This chapter described a six-level sensor and data fusion model; sensor-level, centralized and hybrid architectures; data fusion algorithm taxonomies; and challenges. Dempster-Shafer and Bayesian inference data fusion algorithms were considered. A particular example discussed the impact of the image and video fusion for enhancing the tracking process in the presence of changeable illumination, shadows, and other ambiguous conditions.

Extension of the developed methods for video systems to more complicated environments that take into account intensity changes and multiple object tracking is envisaged. The problem of multiple people tracking is still unresolved due to various challenges such as occlusions and illumination changes. Person tracking has become increasingly important for a series of military and civilian applications such as security, surveillance, smart-environments, and medicine. Detection and tracking of a moving and/or speaking person in a cluttered environment typically requires microphones or cameras that respond to spatio-temporal changes in the data they collect. Open avenues exist in exploiting multiple modalities such as video and audio data. This research envisages new and robust systems able to provide accurate multimodal location estimates in real-time. This assumes development of new technologies that can detect the presence of speech, identify prespecified objects of interest and improve the image quality.

Behavior recognition is another emerging area where the main challenge is in the recognition of different activities in video, for example, the normal from abnormal ones. A possible realistic scenario is tracking a person as used to detect a specific area of interest in a public space. For example, people can be tracked in a shop or parking areas. Multiple-view tracking is another challenge where the best camera streams must be chosen.

Acknowledgments

The authors acknowledge the support from the UK DIF DTC Centre and the EU COST Action 0702.

References

1. A. Doucet and N. Freitas and N. Gordon, Eds. *Sequential Monte Carlo Methods in Practice.* New York: Springer-Verlag, 2001.
2. H. Aghajan and A. Cavallaro. *Multi-Camera Networks: Principles and Applications.* New York: Academic Press, 2009.
3. M. Arulampalam, S. Maskell, N. Gordon, and T. Clapp. A tutorial on particle filters for online nonlinear/non-Gaussian Bayesian tracking. *IEEE Trans. on Signal Processing*, 50(2):174–188, 2002.
4. Y. Bar-Shalom and W. Blair. *Multitarget-Multisensor Tracking: Applications and Advances,* Ch. 6 by S.S. Blackman, Association and fusion of multiple sensor data, volume III. Boston: Artech House, 2000.
5. J.A. Barnett. Computational methods for a mathematical theory of evidence. In *Seventh International Joint Conference on Artificial Intelligence II*, IJCAI'81, pages 868–875, 1981.
6. A. Bhattacharayya. On a measure of divergence between two statistical populations defined by their probability distributions. *Bulletin Calcutta Math. Society*, 35:99–110, 1943.
7. S. S. Blackman. *Multiple Target Tracking with Radar Applications.* Boston: Artech House, 1986.
8. E.P. Blasch and S. Plano. JDL level 5 fusion model 'user refinement' issues and application in group tracking. In *Proceedings of SPIE 4729, Aerosense*, pages 270–279, 2002.
9. H.A.P. Bloom and Y. Bar-Shalom. The interacting multiple-model algorithm for systems with switching coefficients. *IEEE Transactions on Automatic Control*, 33:780–783, 1988.
10. P.L. Bogler. Shafer-Dempster reasoning with applications to multisensor target identification systems. *IEEE Transactions on Systems, Man, and Cybernetics*, 17(6):968–977, 1987.
11. P. Brasnett, L. Mihaylova, N. Canagarajah, and D. Bull. Improved proposal distribution with gradient measures for tracking. In *IEEE International Workshop on Machine Learning for Signal Processing*, pages 105–110. Mystic, CT, USA, Sept. 28–30 2005.

12. P. Brasnett, L. Mihaylova, N. Canagarajah, and D. Bull. Particle filtering with multiple cues for object tracking in video sequences. In *Proceedings of SPIE's 17th Annual Symposium on Electronic Imaging, Science and Technology*, 5685:430–441, 2005.
13. P. Brasnett, L. Mihaylova, N. Canagarajah, and D. Bull. Sequential Monte Carlo tracking by fusing multiple cues in video sequences. *Image and Vision Computing*, 25(8):1217–1227, Aug. 2007.
14. B.R. Cobb and P.P. Shenoy. A comparison of methods for transforming belief function models to probability models. In *LNCS in Artificial Intelligence. Symbolic and Quantitative Approaches to Reasoning with Uncertainty,* Eds. T.D. Nielsen and N.L. Zhang, pages 255–266. Berlin: Springer-Verlag, 2003.
15. D. Comaniciu, V. Ramesh, and P. Meer. Kernel-based object tracking. *IEEE Trans. Pattern Analysis Machine Intelligence*, 25(5):564–575, 2003.
16. V.G. Comparato. Fusion - The key to tactical mission success. Sensor fusion. In *Proc. SPIE 931*, pages 2–7, 1988.
17. Data Fusion Development Strategy Panel. Functional description of the data fusion process. Technical report, Office of Naval Technology, 1991.
18. Data Fusion Subpanel. Data fusion lexicon. Technical report, Joint Directors of Laboratories Technical Panel for C3, 1991.
19. J. Dezert. Foundations for a new theory of plausible and paradoxical reasoning. *Information and Security*, 9:1–45, 2002.
20. R.A. Dillard. Computing confidences in tactical rule-based systems by using Dempster-Shafer theory. Technical Report NOSC TD 649 (AD A 137274), San Diego: Naval Ocean Systems Center, 1983.
21. T.D. Dixon, J. Li, J.M. Noyes, T. Troscianko, S.G. Nikolov, J. Lewis, E.-F. Canga, D.R. Bull, and C.N. Canagarajah. Scanpath analysis of fused multi-sensor images with luminance change: A pilot study. In *Proceedings of the 9th International Conference on Information Fusion*, Italy, July 2006.
22. O.E. Drummond and S.S. Blackman. Challenges of developing algorithms for multiple sensor, multiple target tracking. In *Proc. SPIE 1096*, pages 244–255, 1989.
23. D. Dubois, H. Prade, and H. Hguyen. Possibility theory, probability and fuzzy sets: Misunderstandings, bridges and gaps. In D. Dubois and H. Prade, Editors, *Fundamentals of Fuzzy Sets. The Handbook of Fuzzy Sets.* Kluwer Academic Pulications, Dordrecht, 1999.
24. N.-E. El Faouzi. Data fusion: Concepts and methods. Technical Report No. 01.01, Licit, INRETS-ENTPE, 2001.
25. N.-E. El Faouzi, H. Leung, and A. Kurian. Data fusion in intelligent transportation systems: Progress and challenges information fusion. *Information Fusion*, 12(1):4–10, 2011.
26. M. Fauvel, J. Chanussot, and J.A. Benediktsson. Decision fusion for the classification of urban remote sensing images. *IEEE Transactions on Geoscience and Remote Sensing*, 44(10):2828–2838, 2006.

27. F.E. White, Jr. Joint directors of laboratories data fusion subpanel report: Sigint session. In *Tech. Proc. Joint Service Data Fusion Symposium I*, DFS'90, pages 469–484, 1990.
28. D. Fixsen and R.P.S. Mahler. The modified Dempster-Shafer approach to classification. *IEEE Transactions on Systems, Man, and Cybernetics. Part A: Systems and Humans*, 27(1):96–104, 1997.
29. W. Freeman and E.H. Adelson. The design and use of steerable filters. *IEEE Trans. Pattern Analysis & Machine Intelligence*, 13:891–906, 1991.
30. T. Gandhi and M. Trivedi. Pedestrian protection systems: Issues, survey and chllenges. *IEEE Transactions on Intelligent Transportation Systems*, 8(3):413–430, 2007.
31. T.D. Garvey, J.D. Lowrance, and M.A. Fischler. An inference technique for integrating knowledge from disparate sources. In *Proc. Seventh International Joint Conference on Artificial Intelligence I*, IJCAI'81, pages 319–325, 1981.
32. D. Gerónimo, A. M. López, Angel D. Sappa, and T. Graf. Survey of pedestrian detection for advanced driver assistance systems. *IEEE Transactions on Pattern Analysis and Machine Intelligence*, 32:1239–1258, 2010.
33. M. Goldstein and W.R. Dillon. *Discrete Discriminant Analysis.* Wiley, New York, 1978.
34. N. Gordon, D. Salmond, and A. Smith. A novel approach to nonlinear / non-Gaussian Bayesian state estimation. In *IEE Proceedings-F*, 140:107–113, April 1993.
35. D. L. Hall and J. Llinas, Editors. *Handbook of Multisensor Data Fusion. Second Edition.* Boca Raton, FL, USA, CRC Press, 2008.
36. D.L. Hall. *Mathematical Techniques in Multisensor Data Fusion.* Norwood: Artech House, 1992.
37. D.L. Hall and R.J. Linn. Algorithm selection for data fusion systems. In *Tri-Service Data Fusion Symposium Technical Proceedings I*, DFS'87, Warminster: Naval Air Development Center, 1987.
38. D.L. Hall and R.J. Linn. A taxonomy of algorithms for multi-sensor data fusion. In *Tech. Proc. Joint Service Data Fusion Symposium I*, DFS'90, pages 594–610, Warminster: Naval Air Development Center, 1990.
39. J. Heather and M. Smith. Multimodel image registration with applications to image fusion. In *Proceedings of the 8th International Conference on Information Fusion.* USA, July 2005.
40. S.A. Hovanessian. *Introduction to Sensor Systems.* Chapter 7 in S.S. Blackman, Multiple sensor tracking and data fusion. Boston: Artech House, 1988.
41. M. Isard and A. Blake. Condensation – Conditional density propagation for visual tracking. *International Journal of Computer Vision*, 28(1):5–28, 1998.
42. J. Jazwinski. *Stochastic Processes and Filtering Theory.* New York: Academic Press, 1970.
43. A. Josang. The consensus operator for combining beliefs. *Artificial Intelli-*

gence Journal, 14(1-2):157–170, 2002.

44. R. Kalman. A new approach to linear filtering and prediction problems. *Trans ASME, Serie D: J. Basic Eng.*, 82:35–45, 1960.
45. R. Kalman and R. Bucy. New results in linear filtering and prediction theory. *Trans ASME, Serie D: J. Basic Eng.*, 83:95–108, 1961.
46. L.A. Klein. *Sensor and Data Fusion: A Tool for Information Assessment and Decision Making. Monograph 138 (3rd printing).* Bellingham: SPIE Press, 2010.
47. J. Lewis, S. Nikolov, N. Canagarajah, and D. Bull. Region-based image fusion. Technical Report UOB-DIF-DTC-PROJ201-TR02, Data and Information Fusion, Defence Technology Centre, University of Bristol, 2004.
48. J. Lewis, S. Nikolov, A. Łoza, E.-F. Canga, N. Cvejic, J. Li, A. Cardinali, N. Canagarajah, and D. Bull. Multi-sensor data gathering. Technical Report UOB-DIF-DTC-PROJ201-TR11, Univ. of Bristol, Bristol, UK, Oct. 2005.
49. J. J. Lewis, R. J. O'Callaghan, S. G. Nikolov, D. R. Bull, and C. N. Canagarajah. Pixel- and region-based image fusion using complex wavelets. *Information Fusion, Special Issue on Image Fusion: Advances in the State of Art*, 8(2):119–130, 2007.
50. J.J. Lewis, S.G. Nikolov, A. Loza, E. Fernandez-Canga, N. Cvejic, L. Li, A. Cardinali, C.N. Canagarajah, D.R. Bull, T. Riley, D. Hickman, and M.I. Smith. The Eden project multi-sensor data set. Technical report, University of Bristol, TR-UoB-WS-Eden-Project-Data-Set (ImageFusion.org), 2006.
51. J. Li, C. Benton, S. Nikolov, and N. Scott-Emanuel. Multi-sensor motion computation, analysis and fusion. Technical Report UOB-DIF-DTC-PROJ201-TR14, Department of Electrical and Electronic Engineering, University of Bristol, Bristol, UK, February 2006.
52. J. Liu. *Monte Carlo Strategies in Scientific Computing.* Springer Verlag, 2001.
53. A. Loza, L. Mihaylova, D. Bull, and N. Canagarajah. Structural similarity-based object tracking in multimodality surveillance videos. *Machine Vision and Applications*, 20(2):71–83, February 2009.
54. R.C. Luo and M.G. Kay. Multisensor integration and fusion: Issues and approaches. sensor fusion. In *Proc. SPIE 931*, pages 42–49, 1988.
55. R.P.S. Mahler. Combining ambiguous evidence with respect to ambiguous a priori knowledge, I: Boolean logic. *IEEE Transactions on Systems, Man, and Cybernetics-Part A: Systems and Humans*, 26(1):27–41, 1996.
56. R. Manduchi. Bayesian fusion of colour and texture segmentations. In *IEEE International Conference on Computer Vision*, ICCV'99, September 1999.
57. M.E. Liggins II, C.-Y. Chong, I. Kadar, M.G. Alford, V. Vannicola, and S. Thomopoulos. Distributed fusion architectures and algorithms for target tracking. *Proceedings of IEEE*, 85(1):95–107, 1997.

58. L. Mihaylova, A. Łoza, S. G. Nikolov, J. J. Lewis, E.-F. Canga, J. Li, C. N. Canagarajah, and D. R. Bull. Object tracking in multi-sensor video. Technical Report UOB-DIF-DTC-PROJ202-TR10, University of Bristol, UK, 2006.
59. P. Del Moral, A. Doucet, and A. Jasra. Sequential Monte Carlo for Bayesian computation. In J. M. Bernardo, M. J. Bayarri, J. O. Berger, A. P. Dawid, D. Hecherman, A. F. M. Smith, and M. West, Editors, *Bayesian Statistics 8*, pages 115–148. Oxford University Press, 2007.
60. C.K. Murphy. Combining belief functions when evidence conflicts. *Decision Support Systems*, pages 1–9, 2000.
61. S. G. Nikolov, P. Hill, D. Bull, and N. Canagarajah. Wavelets for image fusion. In A. Petrosian and F. Meyer, Editors, *Wavelets in Signal and Image Analysis*, pages 213–244. Kluwer Academic Publishers, 2001.
62. S. G. Nikolov, J. J. Lewis, E. F. Canga, A. Łoza, C. N. Canagarajah, and D. R. Bull. *Fusion of Visible and Infrared Image Sequences Using Wavelets.* Progress report UOB-DIF-DTC-PROJ201-TR06, DTC-UoB, 2004.
63. K. Nummiaro, E. B. Koller-Meier, and L. Van Gool. An adaptive color-based particle filter. *Image and Vision Computing*, 21(1):99–110, 2003.
64. E. Ozyildiz, N. Krahnstöver, and R. Sharma. Adaptive texture and color segmentation for tracking moving objects. *Pattern Recognition*, 35:2013–2029, 2002.
65. P. Smets (Ed.). *Quantified Representation of Uncertainty and Imprecisions. Handbook of Defeasible Reasoning of Uncertainty Management Systems*, volume I. Kluwer Academic Publ., Dordrecht, The Netherlands, 1998.
66. C.R. Parikh, M.J. Pont, N.B. Jones, and F.S. Schlindwein. Improving the performance of CMFD applications using multiple classifiers and a fusion framework. *Transactions of Institute of Measurement and Control*, 25(2), 2003.
67. J. Pearl. *Probabilistic Reasoning in Intelligent Systems: Networks of Plausible Inference.* San Mateo, CA. Morgan Kaufmann, 1988.
68. P. Pérez, J. Vermaak, and A. Blake. Data fusion for tracking with particles. *Proceedings of the IEEE*, 92(3):495–513, 2004.
69. *Proceedings of the IEEE.* Special issue. sequential state estimation: From Kalman filters to particle filters. 92(4), 2004.
70. G.S. Robinson and A.O. Aboutalib. Trade-off analysis of multisensor fusion levels. In *Proc. of the 2nd National Symp. on Sensors and Sensor Fusion II*, GACIAC PR 89-01, pages 21–34. Chicago: IIT Research Institute, 1990.
71. E. Saber and A. M. Tekalp. Integration of color, edge, shape and texture features for automatic region-based image annotation and retrieval. *Journal of Electronic Imaging (Special Issue)*, 7(3):684–700, 1998.
72. D. W. Scott. *Multivariate Density Estimation: Theory, Practice and Visualization.* Wiley Series in Probability and Mathematical Statistics. New York: John Wiley & Sons, 1992.

73. G. Shafer. *A Mathematical Theory of Evidence.* Princeton University Press, 1976.
74. X. Shi and R. Manduchi. A study on Bayes feature fusion for image classification. *IEEE Workshop on Statistical Analysis in Computer Vision*, 8:95, 2003.
75. P. Smets. Constructing the pignistic probability function in a context of uncertainty. *Uncertainty Artificial Inteligence*, 5:29–39, 1990.
76. P. Smets. The transferable belief model and random sets. *International Journal of Intelligent Systems*, 7:37–46, 1992.
77. P. Smets. Imperfect information: Imprecision - uncertainty. In A. Motro and P. Smets, Editors, *Uncertainty Management in Information Systems: From Needs to Solutions*, pages 225–254. Kluwer Academic Publishers, 1997.
78. P. Smets and R. Kennes. The transferable belief model. *Artificial Intelligence*, 66:191–234, 1994.
79. M. Smithson. *Ignorance and Uncertainty: Emerging Pradigms.* Spinger-Verlag, New York, 1989.
80. A.N. Steinberg, C.L. Bowman, and F.E. White. Revisions to the JDL data fusion model. In *Procceedings of SPIE 3719*, pages 430–441, 1999.
81. J. Triesch and C. von der Malsburg. Democratic integration: Self-organized integration of adaptive cues. *Neural Computation*, 13(9):2049–2074, 2001.
82. L. Valet, G. Mauris, and Ph. Bolon. A statistical overview of recent literature in information fusion. *IEEE Aerospace and Electronic Systems Magazine*, pages 7–14, 2001.
83. R. Viswanathan and P.K. Varshney. Distributed detection with multiple sensors: Part I - fundamentals. *Proc. IEEE*, 85(1):54–63, 1997.
84. E. Waltz and J. Llinas. *Multisensor Data Fusion.* Boston, MA: Artech House, 1990.
85. Z. Wang, A. C. Bovik, and E. P. Simoncelli. Structural approaches to image quality assessment. In A. Bovik, Editor, *Handbook of Image and Video Processing, 2nd Edition*, chapter 8.3. Academic Press, 2005.
86. Z. Wang, A.C. Bovik, H.R. Sheikh, and E.P. Simoncelli. Image quality assessment: from error visibility to structural similarity. *IEEE Transactions on Image Processing*, 13(4):600–612, 2004.
87. L.A. Zadeh. A theory of approximate reasoning. In J. Hayes, D. Michie, and L. Mikulich, Editors, *Machine Intelligence 9*, pages 149–194. 1979.
88. L.A. Zadeh. Review of Shafer's mathematical theory of evidence. *Artificial Intelligence Magazine*, 5(3):81–83, 1984.

7

Independent Viewpoint Silhouette-Based Human Action Modeling and Recognition

Carlos Orrite
University of Zaragoza, Zaragoza, Spain

Francisco Martínez-Contreras
University of Zaragoza, Zaragoza, Spain

Elías Herrero
University of Zaragoza, Zaragoza, Spain

Hossein Ragheb
Kingston University, London, United Kingdom

Sergio A. Velastin
Kingston University, London, United Kingdom

7.1 Introduction

With the ubiquitous presence of video data and the increasing importance of real-world applications such as visual surveillance, it is becoming increasingly necessary to automatically analyze and understand human motions from large amounts of video data. Machine learning for vision-based motion analysis is the research field that tries to bring together aspects from motion analysis such as detection, tracking, and object identification with statistical machine learning techniques. In this chapter we address the problem of silhouette-based human action modeling and recognition independently of the camera point of view. The ability to do so is crucial for deployment in practical CCTV systems where it would be impossible to train each camera to recognize actions for its particular point of view and where many cameras are of the moving pan-tilt-zoom (PTZ) type, that is, can be moved under either computer or human control. Basically, there are two main approaches for human pose modeling: model based top-down and model-free bottom-up strategies. Model-based approaches presuppose the use of an explicit model of a person and basically match a projection of the human body with the image observation. Bottom-up methods do not use such an explicit representation, but directly infer human pose/action from the image features previously extracted. Essentially, an example-based method or a learning-based approach is followed from a dataset of exemplars.

This chapter focuses on bottom-up methods where action recognition is achieved by comparing a 2D motion template, built from observations, with learned models of the same type captured from a wide range of viewpoints. 2D templates can be directly observable in the image, which constitutes a significant advantage over using 3D models. In addition, humans are able to recognize actions from a single viewpoint. Johansson [11] showed that the trajectories of the 2D joints provide sufficient information to interpret the performed action. However, the main disadvantage of 2D templates is their dependence on viewpoint. Recently, many authors have proposed a common approach consisting of discretizing the space considering a series of view-based 2D models [16,32]. This approach gives some preliminary goods results, but there are two main problems to be addressed: spatial discontinuities due to viewpoint discretization and temporal discontinuities. The approach presented here tries to integrate spatial and temporal templates into a common framework, simultaneously combining motion state and viewpoint changes. The novelty is that a 3D reconstruction is not required at any stage. Instead, learned 2D motion templates, captured from different viewpoints, are used to produce a temporal-viewpoint map where 2D observations are projected for recognition.

Human action recognition and pose recovery have been studied extensively in recent years; see [7] and [27] for a survey. Because human actions can be characterized as motion of a sequence of human silhouettes over time, silhouette-based methods have received significant attention lately [1,3,4,9, 19,20,28,30,31].

Motion Energy Images (MEI) and Motion History Images (MHI) were introduced by Davis and Bobick [3] to capture motion information in images. This approach works effectively under the assumption that the viewpoint is relatively fixed (usually a frontal or a lateral view), possibly with a small variance. The lack of a view-invariant action representation limits the applications of such 2D-based approaches. To overcome this limitation, the authors propose to use multiple cameras. Weinland et al. [31] extend the motion history image concept, introducing the motion history volumes as a free-viewpoint representation for human actions in the case of multiple calibrated video cameras. The practicality of their method is limited because it requires multiple test cameras as well as multiple training cameras. More recently [30], the same authors propose a new framework to model actions using three-dimensional occupancy grids, built from multiple viewpoints, in an exemplar-based HMM. Lv and Nevatia [19] exploit a similar idea as that proposed by this chapter, where each action is modeled as a series of synthetic 2D human poses rendered from a wide range of viewpoints, instead of using an explicit 3D representation. The constraints on transition of the synthetic poses are represented by a graph model called Action Net. Given the input, silhouette matching between the input frames and the key poses is performed first using an enhanced Pyramid Match Kernel algorithm. The best matched sequence of actions is then tracked using the Viterbi algorithm. As their proposal is based on synthetic 2D human poses and given the type of shape context descriptor used, their approach does not seem to be robust under noisy conditions, which is one of the objectives of our work, as will be shown in the experimental section.

Recently, Meng and Pears [20] proposed a new feature called the Motion History Histogram (MHH), which extends previous work on temporal template (MHI related) representations by additionally storing frequency information as the number of times motion is detected at every pixel, further categorized into the length of each motion. In essence, maintaining the number of contiguous motion frames removes a significant limitation of MHI, which only encodes the time from the last observed motion at every pixel. Ahad and Ishikawa [1] developed an optical flow-based directional motion history approach that can solve the motion overwriting problem due to its self-occlusion in the MHI.

The idea of dimensionality reduction to analyze human actions in a low-dimensional subspace rather than the ambient space has been proposed before [28]. Space-time points are projected into a low-dimensional space exploiting locality preserving projections (LPP). Action classification is achieved in a nearest-neighbor framework using median Hausdorff distance or normalized spatio-temporal correlation for similarity measures. Shechtman and Irani [25] introduced a behavior-based similarity measure to identify whether two different space-time intensity patterns have resulted from a similar motion field. Because it requires no prior modeling or learning of activities, it is not restricted to a small set of predefined activities. However, their method is not invariant to large geometric deformations of the video template, such as large variations in the point of view of the camera as we do, allowing only some small changes in scale and orientation. Basically, the main difference of these works in relation to the approach presented in this chapter is that the information encoded by the specific-action SOM explicitly models camera viewpoint and temporal action, while the others use parameters to encode such information.

Most dimensionality reduction techniques cannot handle large variations within a dataset, such as an action performed by different people. As a result, they tend to capture the intrinsic structure of each manifold separately without generalization. Consequently, the common embedded space shows separate and highly distorted manifolds. To deal with this fundamental issue, Lewandowski et al. [17] use the TLE algorithm, which shows excellent generalization properties.

The use of SOM networks for human activity recognition has not received much attention in the past. It has mainly focused on describing movement in terms of a sequence of flow vectors. Johnson et al. [12] describe object movement as positions and velocities in the image plane. A statistical model of object trajectories is formed with two competitive learning networks. Hu et al. [8] improve this work by introducing a hierarchical self-organizing approach for learning the patterns of motion trajectories that has smaller scale and faster learning speed. Owens et al. [21] apply a SOM feature map to find flow vector distribution patterns. These patterns are used to determine whether a point on a trajectory is normal or abnormal. These works are based on some flow parameters, and the use of the SOM is restricted to learning trajectories, rather than modeling human motion from different point of views and dynamic figure evolution.

Recently, HMMs have been successfully applied to the task of human action recognition. Work related to ours includes that of Kellokumpu et al. [14], which describes actions as a continuous sequence of discrete postures derived from an invariant descriptor. They extract the contour from human silhouettes and calculate affine invariant Fourier descriptors from this contour. In order to classify posture, these descriptors are used as the inputs of a radial basis SVM. They use HMM to model different activities and calculate the probabilities of the activities based on the posture sequence from SVM. Wang and Suter, in [29], propose the following method. First, features are extracted from image sequences, with two alternatives: a silhouette and a distance transformed silhouette. Second, a low-dimensional feature representation is found embedded in high-dimensional image data. Finally, motions are analyzed and recognized using a continuous HMM. Kale et al., in [13], propose a continuous HMM-based approach to represent and recognize gait. Their approach involves deriving a low-dimensional observation sequence (the width of the silhouette of the body) during a gait cycle. Learning is achieved by training an HMM for each person over several gait cycles. View-invariance is a crucial issue for real-world applications. In [5] the author proposed a three-layer model for view-invariant action and identity recognition. In this approach, HMMs with a fixed number of states are used to cluster each sequence into a fixed number of poses to generate the observation data for an asymmetric bilinear model. Ahmad and Lee [2] take into account multiple viewpoints and use a multidimensional HMM to deal with the different observations. Instead of modeling viewpoint, Lu and Little [18] use a hybrid HMM where one process denotes the closest shape-motion template while the other models position, velocity, and scale of the person in the image.

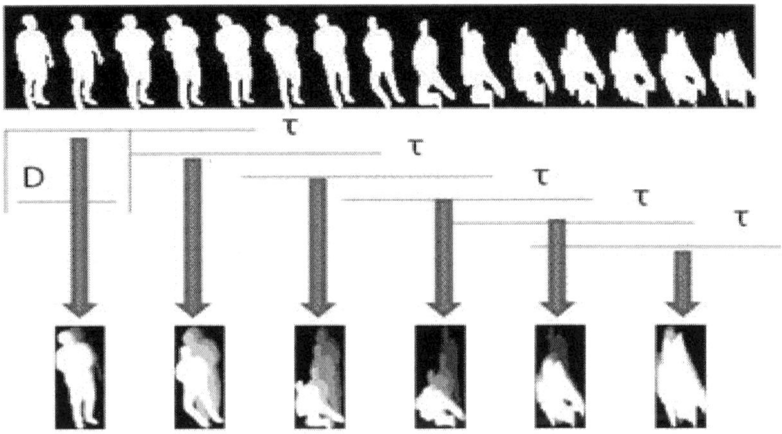

FIGURE 7.1 Overlapping MHI.

7.2 Human Action Modeling Based on Motion Templates

Motion templates are based on the observation that in video sequences, a human action generates a space-time shape in the space-time volume. MHIs were introduced to capture motion information in images, encoding how recently motion occurred at a pixel [3]. These images contain both the spatial information about the pose of the human figure at any time (location and orientation of the torso and limbs, aspect ratio of different body parts), as well as the dynamic information (global body motion and motion of the limbs relative to the body). Each MHI can be seen as a "piece" of an action and the link of successive pieces will constitute the proper action:

$$\mathrm{MHI}_t(x,y) = \begin{cases} \tau & \text{if D(x,y,t)} = 1 \\ max(0, \mathrm{MHI}_{t-1}(x,y) - 1) & \text{otherwise,} \end{cases} \tag{7.1}$$

where τ represents the number of images used for generating the motion template.

Given an image sequence of a particular action, a window of length τ and displacement D is used to calculate the sequence of MHI, as depicted in Figure 7.1 with the action "Collapse."

As the goal is to recognize human body activities from different persons who are free to move in space, with different orientations and sizes, some normalization step must be carried out. Location and scale dependencies are removed by centering with respect to the center of mass, and scale normalizing with respect to a predefined mask. To account for point rotation, or camera point of view, it might be possible to generate a 3D model of the body as [31]. However, this approach, using many calibrated cameras, is not feasible for many applications. In this chapter we introduce a novel approach where 3D reconstruction is not required at any stage. Instead, learned 2D motion templates, captured in different viewpoints, are used to produce 2D image information that is compared to the observations.

The main problem with the previously described temporal representation is that it involves dealing with a very high dimensional space. An alternative to alleviate this problem consists of mapping temporal features into a lower dimensional space. The simplest way to reduce dimensionality is via Principal Component Analysis (PCA), which assumes that the data lies on a linear subspace. Except in very special cases, data does not lie on a linear subspace; thus methods are required that can learn the intrinsic geometry of the manifold from a large number of samples [27]. Nonlinear dimensionality reduction techniques allow for representation of data points based on their proximity to each other on nonlinear manifolds, for example, using the Kohonen Self-Organizing feature Map (SOM) [15]. All the 2D motion templates, encompassing temporal movement, are projected into a new subspace by means of the SOM, which provides the grouping of viewpoint (spatial) and movement (temporal) in a principal manifold.

A self-organizing map (SOM) is an artificial neural network that is trained using unsupervised learning to produce a low-dimensional (typically two-dimensional) discretized representation of the input space of the training samples. Self-organizing maps are different from other artificial neural networks in the sense that they use a neighborhood function to preserve the topological properties of the input space. An SOM map consists of components called neurons. Associated with each neuron is a weight vector of the same dimension as the input data vectors and a position in the map space. The usual arrangement of neurons is regular spacing in a hexagonal or rectangular grid. The SOM describes a mapping from a higher dimensional input space to a lower dimensional map space. The procedure for placing a vector from data space onto the map is to find the neuron with the closest weight vector to the vector taken from data space and then to assign the map coordinates of this neuron to our vector. The weights of the neurons are initialized either to small random values or sampled evenly from the subspace spanned by the two largest principal component eigenvectors. With the latter alternative, learning is much faster because the initial weights already give good approximations of SOM weights.

Training uses competitive learning. The network must be fed with a large number of example vectors that represent, as closely as possible, the kinds of vectors expected during mapping. When a training example is fed to the network, its Euclidean distance is computed to all weight vectors. The neuron with the weight vector most similar to the input is called the best matching unit (BMU). The weights of the BMU and neurons close to it in the SOM lattice are adjusted toward the input vector. The magnitude of the modification

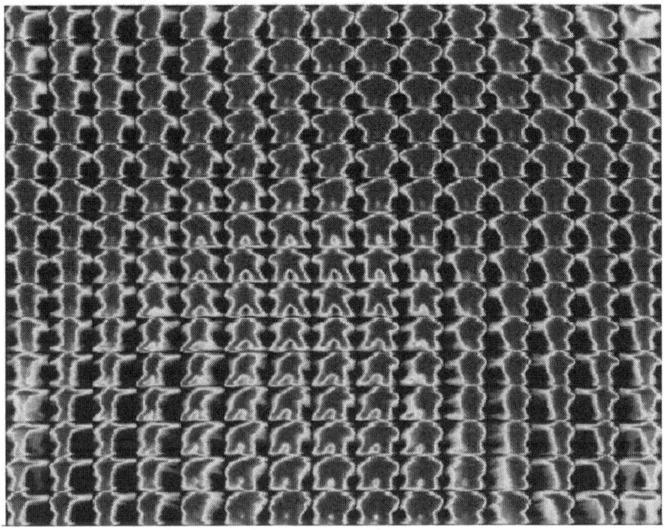

FIGURE 7.2 Codebook of the trained SOM for action punch. (See color insert.)

decreases with time and with distance from the BMU. The update formula for a neuron with weight vector $W_i(t)$ is given by Equation (7.2):

$$W_i(t+1) = W_i(t) + \Theta(i,t)\alpha(t)(D(t) - W_i(t)), \tag{7.2}$$

where $\alpha(t)$ is a monotonically decreasing learning coefficient and D(t) is the input vector. The neighborhood function $\Theta(i,t)$ depends on the lattice distance between the BMU and the i-th neuron. In its simplest form, it has a value of one for all neurons close enough to BMU and zero for others, but a Gaussian function is a common choice, too. Regardless of the actual functional form, the neighborhood function decreases with time. At the beginning, when the neighborhood is broad, self-organization takes place on the global scale. When the neighborhood has shrunk to adjust a couple of neurons, the weights are converging to local estimates. This process is repeated for each input vector for a large number of cycles.

During mapping, there will be one single winning neuron: the neuron whose weight vector lies closest to the input vector. This can be simply determined by calculating the Euclidean distance between an input vector and the weight vectors, obtaining the Quantization Error Distribution (Q-error).

In this approach, the inputs to the SOM algorithm are the MHIs reshaped as vectors, with size 1×2250, the topology of the map is a rectangular grid with a toroid shape. The neighborhood function used is a Gaussian function, and the learning coefficient is a linear decreasing function. The size of the weight vector of each neuron is the same as that of the input vector (1×2250).

Figure 7.2 shows the representation of the codebook of the trained SOM

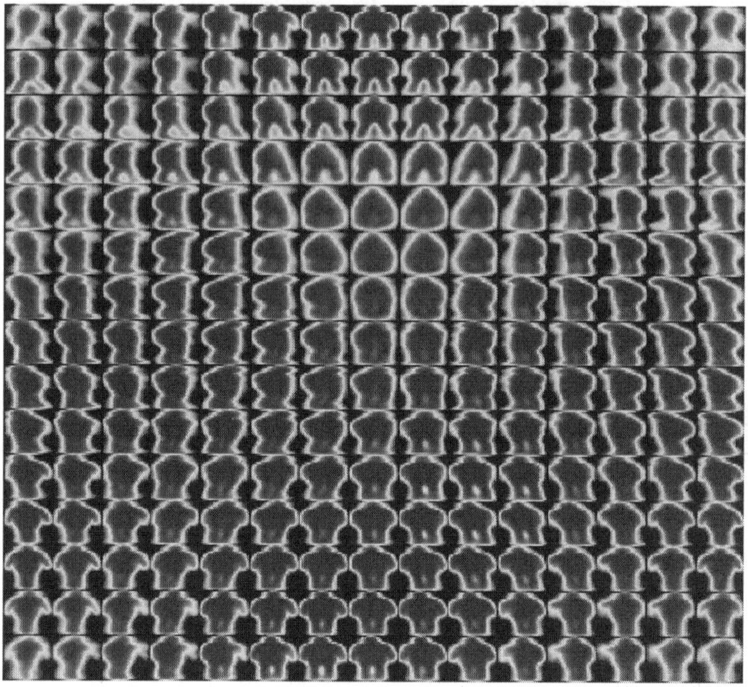

FIGURE 7.3 Codebook of the trained SOM for all actions. (See color insert.)

for action "punch," and Figure 7.3 shows a representation of the codebook of the trained SOM of size 15 $\times$ 15 for a set of different actions. Each neuron is represented in the map by the pattern learned during the training. So, the SOM integrates spatial and temporal models into a 2D map, combining simultaneously motion and viewpoint changes. It establishes motion correspondences between viewpoints without having to use a mapping to a complex 3D model.

7.3 Learning the Viewpoint Manifold for Action Recognition

First, let us assume that there are enough motion templates to train different activity-specific SOMs, one per each different action. Given a pre-segmented video sequence, corresponding to a particular action, a set of MHIs are obtained from it, constituting the temporal inputs to all action-specific SOMs.

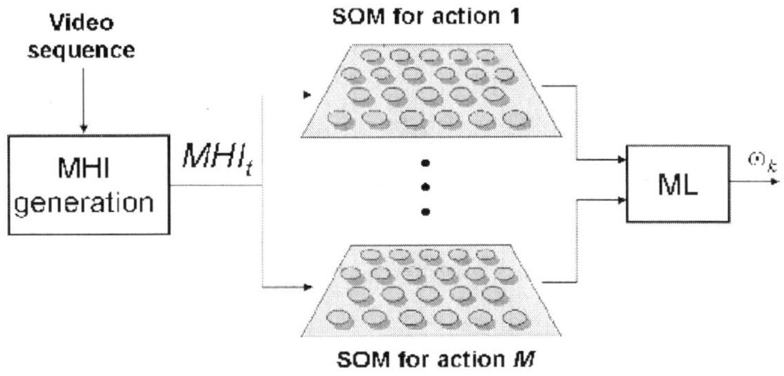

FIGURE 7.4 Maximum likelihood classifier for action recognition based on action-specific SOMs.

From now on we will use k-SOM to refer to a particular k-action-specific SOM. Action recognition is accomplished using the outputs of these SOMs through a Maximum Likelihood (ML) classifier. Figure 7.4 shows the proposed architecture for action recognition.

For every MHI_t at time t, the probability that this pattern belongs to class-action ω_k, that is, $P(\omega_k|MHI_t)$, is found. This probability can be expressed as

$$p(\omega_k|MHI_t) = p(wSOM = k|MHI_t) \cdot p(\omega_k|wSOM = k), \qquad (7.3)$$

where $wSOM = k$ denotes that the k-SOM was the winner.

In the previous expression, the index of the winner-neuron (i.e., $BMU = j$) corresponding to k-SOM, is not taken into account to calculate the posterior. Bearing in mind that all actions may start from a common pose, for example, standing up, all action-specific SOMs include some clusters representing these poses. Taking into account the information about the index of the winner-neuron, the classification may become more robust. The new expression becomes

$$\begin{aligned} &p(\omega_k|MHI_t) = \\ &p(wSOM = k, BMU = j|MHI_t) \cdot p(\omega_k|wSOM = k, BMU = j). \end{aligned} \qquad (7.4)$$

To calculate the first conditional probability in the previous expression, we use the following equation:

$$p(BMU_k|MHI_t) \propto e^{-\lambda \cdot Qe_k(j)^2}, \qquad (7.5)$$

where $Qe_k(j)$ is the Q-error of the neuron j for the specific k-action SOM, and λ is a smooth parameter chosen experimentally. So, a distance is transformed into a probability.

The term $p(\omega_k|wSOM = k, BMU = j)$ is calculated by the Bayes rule:

$$p(\omega_k|wSOM = k, BMU = j) = \frac{p(\omega_k) \cdot p(wSOM = k, BMU = j|\omega_k)}{p(wSOM = k, BMU = j)}. \quad (7.6)$$

From the chain rule of conditional probabilities

$$p(wSOM = k, BMU = j|\omega_k) = p(BMU = j|\omega_k, wSOM = k) \cdot p(wSOM = k|\omega_k). \quad (7.7)$$

The denominator can be expressed as a conditional probability:

$$p(wSOM = k, BMU = j) = p(BMU = j|wSOM = k) \cdot p(wSOM = k), \quad (7.8)$$

giving as a result

$$p(\omega_k|wSOM = k, BMU = j) = \frac{p(\omega_k) \cdot p(BMU = j|\omega_k, wSOM = k) \cdot p(wSOM = k|\omega_k)}{p(BMU = j|wSOM = k) \cdot p(wSOM = k)}. \quad (7.9)$$

The previous probabilities can be obtained using the dataset used for the training process for every map.

In this regard, it is worth mentioning that a problem can arise if the number of data for training is low. Let us consider the example depicted in Figure 7.5, taken from the results given in Section 7.5.1. These figures represent the test for a new actor in relation to a particular action. Figure 7.5(left) represents the histogram for $p(BMU = j|\omega_k, wSOM = k)$ when k-SOM was the winner and the action was ω_k. Figure 7.5(right) represents the histogram when k-SOM was the winner but the action was ω_i, with $i \neq k$. For the sake of clarity, both histograms are represented in the same grid as the SOM to associate cells with neurons. There are some particular neurons that are more representative of the specific action, and on the other hand, there are some neurons where misclassification happens more often. Moreover, there are some neurons that have never been the winner. This can be due to the small number of examples for training and constitutes a serious problem indeed, as for a particular test pattern it could happen that the wining neuron was one of these and then the probability taken from the previous histogram would be null.

7.3.1 Nonparametric Probabilities Modeling

To solve this problem, a nonparametric technique, such as Kernel Density Estimation (KDE) [10], can be used to learn the unknown probability distribution, constituting one of the novel contributions presented in this work. KDE does

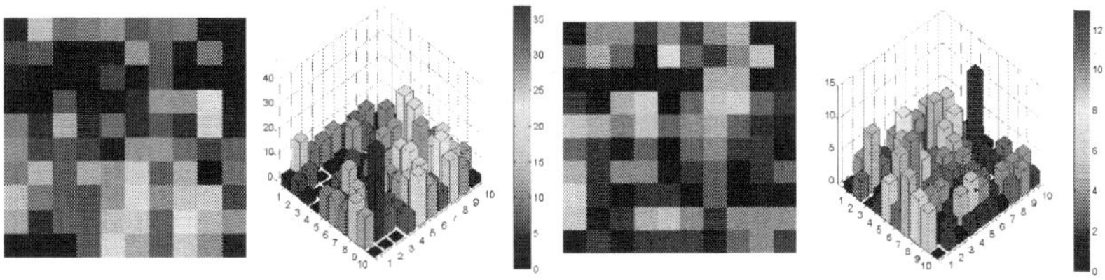

FIGURE 7.5 (Left) Histogram of winner-neurons for k-SOM when action was ω_k, (Right) Histogram of winner-neurons for k-SOM when action was ω_i with $i \neq k$. (See color insert.)

not assume any specific underlying distribution (for example, Gaussian) and, theoretically, the estimate can converge to any density shape with enough samples. In this technique, the underlying probability density function, for a nonspecific variable x, is estimated as

$$p(x) = \frac{1}{N}\sum_{i=1}^{N} K(x - x_i), \tag{7.10}$$

where N is the number of examples, and K is a kernel function (typically a Gaussian) centered at the data points x_i, $i = 1, \ldots, N$.

To calculate, for example $p(BMU = j|wSOM = k, w_k)$, depicted in Figure 7.5(left) as a histogram, we will use

$$p(BMU = j|wSOM = k, w_k) \propto \frac{1}{N}\sum_{i=1}^{N} e^{-D^2(i,j)/2\sigma^2}. \tag{7.11}$$

In the previous expression, $D^2(i, j)$ denotes the weight between neurons i and j assigned by the codebook corresponding to the trained k-SOM, that is, the Euclidean distance between prototype nodes, σ being a smooth parameter chosen experimentally.

7.3.2 Combining Temporal SOM Activations

Once all MHI_t templates corresponding to the same action are fed to the specific-action SOM, see Figure 7.4, individual likelihood outputs can be combined to obtain a consensus decision. Different features can offer complementary information about the movement, and combining all this information obtains a better result. Different combination alternatives are evaluated in this work: sum (7.12), product (7.13), and maximum (7.14) rules, given by the equations below:

$$\sum_{t=1}^{T} P(\omega_j|MHI_t) = max_{k=1}^{M} \sum_{t=1}^{T} P(\omega_k|MHI_t), \tag{7.12}$$

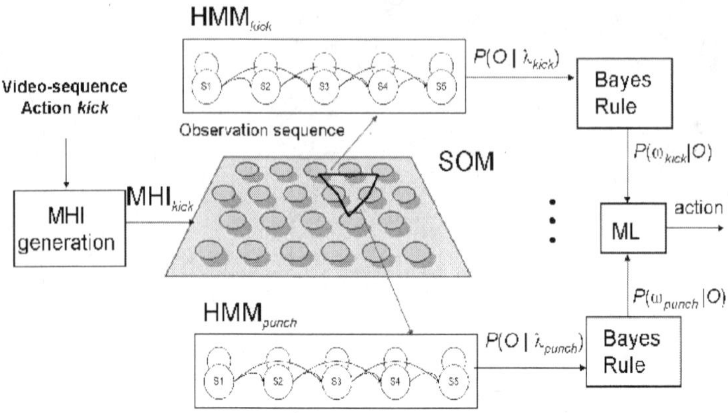

FIGURE 7.6 Maximum likelihood classifier for a single SOM with action recognition specific HMM.

$$\prod_{t=1}^{T} P(\omega_j|MHI_t) = max_{k=1}^{M} \prod_{t=1}^{T} P(\omega_k|MHI_t), \tag{7.13}$$

$$P(\omega_j|MHI_t) = max_{k=1}^{M} P(\omega_k|MHI_t), \tag{7.14}$$

where T is the number of MHIs in an action sequence, and M the number of classes. The sequence is classified as the action corresponding to the class model giving the highest probability.

7.4 HMM-Based Human Action Recognition

In order to train action-specific SOM, many samples are needed to reach stable and accurate performance. When this is not possible, due to the scarcity of training samples, a single SOM can be trained with all the temporal templates. This implies an additional procedure to track the map behavior on temporal templates using, for example, different action-specific HMMs. Any HMM would have as observations the indices of the winner-neurons of the SOM. Figure 7.6 shows the structure of the classifier, where a sequence of MHIs produces a temporal activation in the SOM. The lines represent the activation trajectory for this particular temporal sequence of patterns. Different HMMs receive this observation sequence as input for estimating the probability of class membership. Following Bayes rule, the posterior is given by

$$p(\omega_i|O) = \frac{p(\omega_i) \cdot p(O|\omega_i)}{p(O)}. \tag{7.15}$$

In Equation (7.15), $p(\omega_i)$ is the priori likelihood, where ω_i is the class i. The conditional term, $p(O|\omega_i) \approx p(O|\lambda_i)$, with λ_i the parameters of the HMM$_i$, can be calculated by the forward-backward algorithm and the probability of the observation $p(O)$ is the same for all classes so it is not taken into account to calculate the maximum a posteriori.

Finally, temporal pattern recognition is accomplished by a Maximum Likelihood (ML) classifier.

7.4.1 Recognition Using HMM

In this section we give a brief review of HMMs in order to define the notation used later. Formally, the parameters of the Markov model, notated as $\lambda = \{A, B, \pi\}$, are specified in our particular problem as follows [22]:

1. $\boldsymbol{N}$ states, that is, $S = S_1, \ldots S_N$
2. The initial state probability distribution, $\pi = \{\pi_i\}$, where $\pi_i = Pr(q_1 = S_i)$, $1 \leq i \leq N$
3. State transition matrix, $A = a_{ij}$, where the transition probability a_{ij} represents the frequency of transiting from state i to state j calculated as $a_{ij} = Pr(q_{t+1} = S_j \mid q_t = S_i)$
4. Observation probability distribution $B = b_j(o_t)$, where $b_j(o_t) = Pr(o_t \mid q_t = S_j)$ and o_t the observation vector

The problem of estimating the parameters λ of an HMM, given a sequence of observations $O = \{o_1, \ldots o_T\}$, can be approached as an ML problem. To solve this we have the Baum-Welch algorithm, which is an EM procedure, estimating the parameters of an HMM in an iterative procedure [22].

7.4.2 Discrete Observations

Given a trained SOM map of M neurons, we can consider the index k, ($k = 1, \ldots, M$) of the winner-neuron in the map as a discrete observation, that is, $o_t = k$. This is a simple approach that exhibits a significant limitation when the number of training patterns is low in relation to the number of neurons in the map. HMM training also requires a significative amount of data. Taking into account that there are neighbor neurons around the winner neuron with low distance as well, we can take more observations for the same input pattern. In this regard, we can apply sampling techniques [6].

Among sampling methods, the Monte Carlo methods approximate the target density by numbers of samples that are distributed according to the target density.

The naive approach consists of considering that every map is independent of the previous ones. So, we have

$$p(O) = p(o_1, \ldots, o_T) = \prod_{t=1}^{T} p(o_t). \tag{7.16}$$

TABLE 7.1 Several sequences of observations given by the resampling of the Q-error maps depicted in Fig. 7.7.

o_1	o_2	o_3	o_4	o_5	o_T
34	34	28	21	21	15
34	35	28	21	22	15
33	34	28	22	22	16
33	35	28	21	22	16

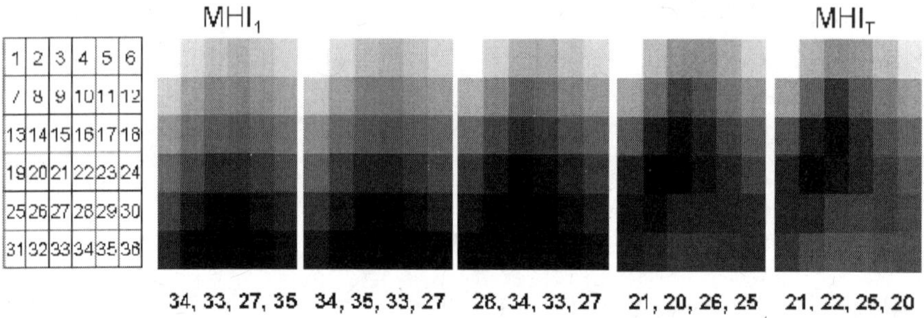

FIGURE 7.7 Q-error maps for a sequence of temporal MHIs. Gray level represents the Euclidean distance from the pattern to the SOM. The winner neuron is the darkest cell. (See color insert.)

The goal is to obtain an index for any Q-error map, that is, $o_t = k$.

$$p(o_t = k) \propto e^{-\lambda \cdot Qe_t(k)^2}, \tag{7.17}$$

where $Qe_t(k)$ is the Q-error of the neuron k at time t and λ is a smooth parameter chosen experimentally. In this way, a distance is transformed into a probability. If the quantization error is null, then the probability is 1. On the other hand, if the distance tents to infinity, the probability is 0.

Following this approach, given a temporal sequence of MHIs, we are able to obtain several observation sequences for training a discrete HMM, as shown in Table 7.1. These sequences have been obtained by the naive example given in Figure 7.7, where some Q-error maps corresponding to the test MHI sequence are depicted. Gray level represents the Euclidean distance from the MHI_t pattern to the SOM. The winner neuron is the darkest cell. So, for this particular test sequence we have only one observation sequence given by the indexes of the BMUs, that is, $O_T = \{34, 34, 28, 21, 21, 15\}$. However, by means of the SIR approach introduced here, we are able to obtain more observation sequences.

7.5 Examples

Different experiments, using three different datasets, are conducted to test the previous approaches. One dataset consists of virtual actors (ViHASi) and many camera views and the others, called Inria Xmas Motion Acquisition

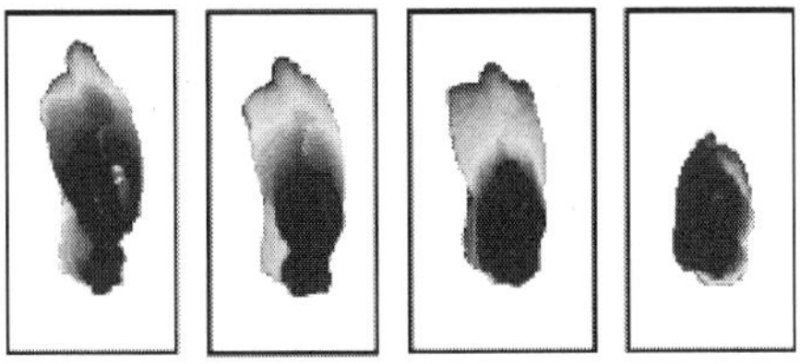

FIGURE 7.8 Some HMIs for action sit-down corresponding to the IXMAS dataset.

Sequences (IXMAS) and the Multicamera Human Action Video (MuHaVi) Manually Annotated Silhouette data (MuHaVi-MAS), have real actors.

7.5.1 Experiments with the IXMAS Dataset

The Inria Xmas Motion Acquisition Sequences (IXMAS) [24], contains eleven actions, each performed three times by ten actors. Action (1) corresponds to check watch, action (2) cross arms, action (3) scratch head, action (4) sit down, action (5) get up, action (6) turn around, action (7) walk, action (8) wave, action (9) punch, action (10) kick, and finally, action (11) pick up. Acquisition was achieved by five cameras and the actors can freely change their orientation for each acquisition in order to demonstrate view-invariance. In the experiment reported here, the zenithal-viewpoint camera is not taken into consideration.

Action-specific SOMs

The method followed for pre-segmented sequences is depicted in Figure 7.4, using different combination rules for the Maximum Likelihood module.

The original dataset was split into two: the training and the evaluation sets. The first was used for training the SOMs and obtaining the posterior probabilities used in the Bayes rule, while the second was dedicated exclusively for testing. In this approach the kernel used for estimating the probability density function in Equation (7.11) was a Gaussian with σ equal to 1.

A test was carried out by the leave-one-person out (LOO) cross-validation. At each iteration, three samples corresponding to one single action class performed by one single actor and captured from four camera views were tested. Testing was done for each of the four cameras in isolation. It is worth mentioning that for each iteration of LOO, new PCAs and SOMs are learned for every action.

Every sequence generates ten MHIs; some of them can be seen in Figure 7.8 for action sit-down. The pixels of the motion history image are hotter where the movement is more recently, and cooler otherwise.

To combine the output of every action-specific PCA/SOM for these MHIs, we follow different combination rules. Tables 7.2 and 7.3 show the number of misclassifications (out of 120 samples per action class), and average recogni-

TABLE 7.2 Number of misclassifications (out of 120 per action) and average recognition (%) for the sum rule applied to the ten outputs given by the action-specific PCA.

1	2	3	4	5	6	7	8	9	10	11
35	49	87	47	93	91	62	67	107	118	108
70.8	59.1	27.5	60.8	22.5	24.1	48.3	44.1	10.8	1.6	10.0

TABLE 7.3 Number of misclassifications (out of 120 per action) and average recognition (%) for the sum rule applied to the ten outputs given by the action-specific SOM.

1	2	3	4	5	6	7	8	9	10	11
42	27	47	5	11	23	0	48	38	16	18
65.0	77.5	60.8	95.8	90.8	80.8	100.0	60.0	68.3	86.6	85.0

tion rate (%) per action applying the sum rule over the outputs given by the PCA or SOM, respectively. Similar results have been obtained when applying the product and the maximum rules.

The confusion matrix for all actions provided by the SOM is shown in Table 7.4. Each column of the matrix represents the instances in a predicted class, while each row represents the instances in an actual class. As can be noticed, actions 1, 3, and 8 are similar to some point. The same happens with action 2, in relation to actions 1 or 3. On the other hand, actions 4 and 11 are quite similar in the initial instant of movement, so there are some misclassifications between both.

In relation to other works using the same dataset, this paper's approach achieves an overall recognition rate of 79.2% using SOM, and 34.5% over ten actors using PCA. In [31], Weinland et al. report a higher action classification rate (93.3%) over ten actors on the same dataset. They use all five cameras to build visual hulls and classify actions based on a 3D action representation. As mentioned before, the practicality of their method is limited because it requires multiple test cameras as well as training cameras. In [19], Lv et al. report an overall action recognition rate of 80.6% over ten actors, quite similar to our approach.

TABLE 7.4 Confusion Matrix using SOM (120 samples per action).

1	2	3	4	5	6	7	8	9	10	11
78	10	6	0	2	6	2	5	10	0	1
5	93	5	1	1	4	2	2	6	1	0
8	7	73	0	1	3	3	17	3	1	0
0	0	0	115	0	0	0	0	0	0	5
0	0	0	4	109	0	0	0	1	1	5
1	1	1	0	4	97	5	2	6	3	0
0	0	0	0	0	0	120	0	0	0	0
12	4	15	1	2	9	1	72	2	1	1
7	5	1	0	2	7	1	3	82	12	0
2	0	0	0	3	1	1	0	7	104	2
0	0	0	9	0	2	3	0	0	4	102

TABLE 7.5 Average recognition rates for a five-state HMM.

1	2	3	4	5	6	7	8	9	10	11
32.6	25.0	29.8	59.0	52.7	56.2	65.9	39.5	40.2	40.2	47.2

A single SOM and several action-specific HMMs

To evaluate the performance of the HMM as neuron action tracker, several left-to-right HMMs, with different numbers of hidden states (from 2 to 10) were tested. The results for the five-state HMM approach are shown in Table 7.5. As one can see, these results are quite poor in relation to those given in Table 7.2. The explanation for this bad achievement can be found in the different action point of view, given as a result of several trajectories difficult to model by an HMM.

7.5.2 Experiments with the ViHASi Dataset

Virtual Human Action Silhouette (ViHASi) data [23] provides a large body of synthetic video data generated for the purpose of evaluating different algorithms on human action recognition based on silhouettes. This dataset exhibits an interesting property, as in the different clothes worn by the actors. The data consist of twenty action classes, nine actors, and up to forty synchronized perspective camera views split into two sets of twenty cameras. The actions were (C1) HangOnBar, (C2) JumpGetOnBar, (C3) JumpOverObject, (C4) JumpFromObject, (C5) RunPullObject, (C6) RunPushObject, (C7) Run-Turn90Left, (C8) RunTurn90Right, (C9) HeroSmash, (C10) HeroDoorSlam, (C11) KnockoutSpin, (C12) Knockout, (C13) Granade, (C14) Collapse, (C15) StandLook, (C16) Punch, (C17) JumpKick, (C18) Walk, (C19) WalkTurn-Back, and (C20) Run; see Figure 7.9. The cameras are located around two circles in a surround configuration, with 18° tilt angle between neighbor cameras, where camera numbers are assigned in the anti-clockwise direction starting with V1 for one 45° slant set and V21 for 27° slant set.

Figure 7.10 shows some MHIs corresponding to action RunTurn90Right.

Action-specific SOMs

The first test was carried out by the leave-one-out cross-validation over all twenty actions (C1 to C20), all actors (A1 to A9). At each iteration, twelve samples corresponding to one single action class performed by one single actor and captured from twelve camera views were tested. Different overlapping window sizes were evaluated. Tables 7.6 and 7.7 show the number of errors and the average recognition rate per action applying the sum rule for a window size given as a result of the division of the action sequence into 5 MHIs, using action-specific PCAs and SOMs, respectively. Similar results were reported by applying the product and the maximum rules. The average recognition rate is 92.0% using PCA, with 211 misclassifications, and 98.5%, that is, 40

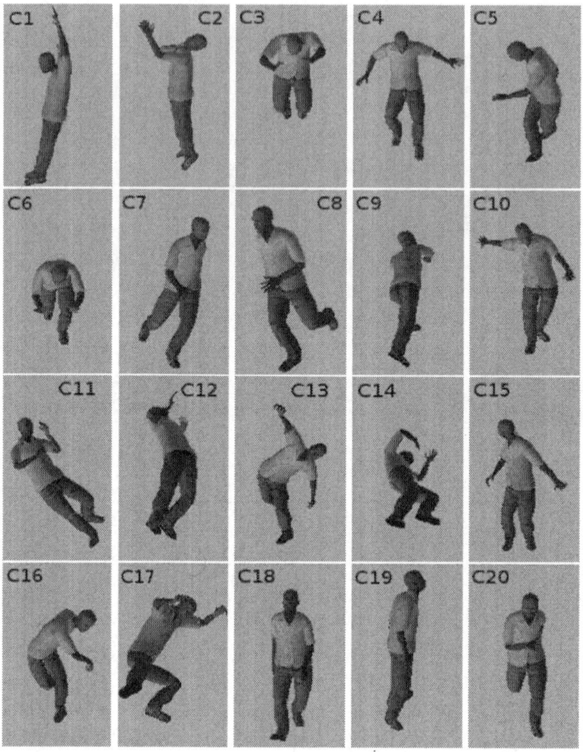

FIGURE 7.9 Actions of the ViHaSi dataset [23].

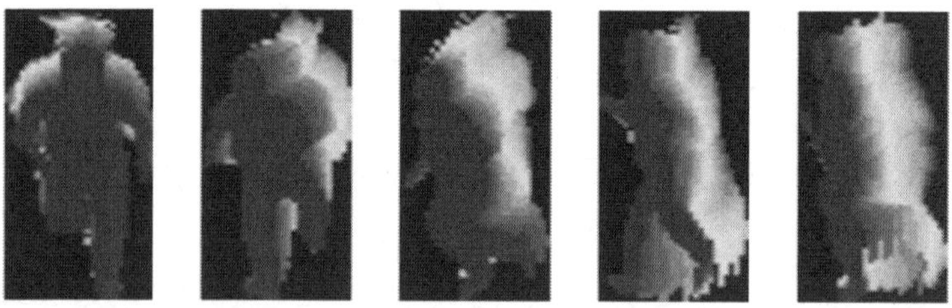

FIGURE 7.10 Some MHIs for action RunTurn90Right corresponding to the ViHaSi dataset.

TABLE 7.6 Number of misclassifications (out of 240 per actor) and average recognition (%) against test actors using PCA.

A1	A1b	A2	A2b	A3	A4	A5	A6	A7	A8	A9
6	1	4	22	59	32	6	0	19	61	1
97.5	99.5	98.3	90.8	75.4	86.6	97.5	100	92.1	74.5	99.5

TABLE 7.7 Number of misclassifications (out of 240 per actor) and average recognition (%) against test actors using SOM.

A1	A1b	A2	A2b	A3	A4	A5	A6	A7	A8	A9
1	1	0	2	1	2	6	0	2	2	23
99.5	99.5	100	99.1	99.5	99.1	97.5	100	99.1	99.1	90.4

misclassifications out of 2,640 test samples, using SOM.

It can be seen in Table 7.7 that most of the misclassification corresponds to actor A9, which is in fact a virtual child and no other child actor exists in the training data. Size normalization carried out in every silhouette was not enough here, as the child morphology is different from the rest of the actors. Tables 7.8 and 7.9 show the number of errors and the average recognition rate per action using PCA/SOM. Action C13 gives the worst score.

The next experiment was carried out to test the robustness of the proposal in the presence of noise. For this purpose, silhouettes were corrupted by "salt and pepper" noise. A parameter (α) defines the percentage of the pixels in each silhouette image that are swapped from "1" to "0". In order to evaluate the system under unlearning conditions, the actors and the camera views used for training were different from those used for testing, for all twenty actions. The actors corresponding to the training data were A1 and A2, while those corresponding to the test data were A3, A4, A5, A6, A7, A8, and A9. The actions performed by A1 and A2 were captured from eight camera views (V2, V5, V7, V10, V12, V15, V17, and V20) corresponding to the first camera set, and eight second camera views (V22, V25, V27, V30, V32, V35, V37, and V40) corresponding to the second camera set. Test data contains twenty-four novel camera views. Actors (A3, A4, A5, A9) were captured from the first set of twelve camera views (V1, V3, V4, V6, V8, V9, V11, V13, V14, V16, V18, and V19) while those performed by the other three actors (A6, A7, A8) were captured from the second set of twelve camera views (V21, V23, V24, V26, V28, V29, V31, V33, V34, V36, V38, V39). The test data then contains 1,680 actions, 240 per test actor. This experiment is more challenging compared to the previous one because the actors and the camera views in the test data are

TABLE 7.8 Number of misclassifications (out of 72 samples per action class) and average recognition (%) for each action class, using PCA.

C1	C2	C3	C4	C5	C6	C7	C8	C9	C10	C11	C12	C13	C14	C15	C16	C17	C18	C19	C20
10	8	13	10	0	0	3	4	5	4	1	5	20	11	19	12	19	15	37	15
92.4	93.9	90.1	92.4	100	100	97.7	96.9	96.2	96.9	99.2	96.2	84.8	91.6	85.6	90.9	85.6	88.6	71.9	88.6

TABLE 7.9 Number of misclassifications (out of 72 samples per action class) and average recognition (%) for each action class, using SOM.

C1	C2	C3	C4	C5	C6	C7	C8	C9	C10	C11	C12	C13	C14	C15	C16	C17	C18	C19	C20
0	0	0	6	1	0	0	0	2	0	0	4	11	0	6	1	6	1	2	0
100	100	100	95.4	99.2	100	100	100	98.4	100	100	96.9	91.6	100	95.4	99.2	95.4	99.2	98.4	100

TABLE 7.10 Average recognition rates (%) using the VNBA08 and ours adding α percent "salt and pepper" noise to the test data.

Method	α=5%	α=10%	α=20%	α=25%	α=50%
VNBA08	89.0	86.8	81.7	–	25.0
ours	86.0	85.5	–	84.8	75.7

novel.

Table 7.10 lists the average recognition rate using different values of α. The average recognition rate without noise is 85.8%, that is, 238 misclassifications out of 1,680 test samples. This lower percentage, in relation to the one given by the previous experiment, is mainly due to the low number of examples used for the SOM training stage, (only two actors, eight view points for one slant set, and eight view points for the other slant set, per actor). In spite of this, results are quite impressive, as the recognition rate hardly goes down even for a noise level of about 25.0%. As it can be noticed, our approach is more robust to noise than the VNBA08. This positive achievement is due to the kind of motion template features used, as well as the good behavior of the neural approach to deal with corrupted patterns.

A single SOM and several action-specific HMMs

The final test was carried out by the "leave-one-out" cross-validation over all twenty actions, all actors, and twelve camera views. Each action sequence, over 2,640, is formed by 5 MHIs, and only one SOM, of size 29 × 20 is trained for the whole set of actions. Different numbers of states of HMM are tested. The average recognition rates per action are shown in Table 7.11. The overall recognition rate is 99.92%.

7.5.3 Experiments with the MuHaVi-MAS Dataset

The Multicamera Human Action Video (MuHaVi) Manually Annotated Silhouette data (MuHaVi-MAS) [26] is composed of five action classes (WalkTurnBack, RunStop, Punch, Kick, ShotgunCollapse), and the authors manually annotated the image frames to generate the corresponding silhouettes

TABLE 7.11 Successful rate, in percent, per action.

C1	C2	C3	C4	C5	C6	C7	C8	C9	C10	C11	C12	C13	C14	C15	C16	C17	C18	C19	C20
100	100	100	100	100	100	99.24	100	100	100	100	100	99.24	100	100	100	100	100	100	100

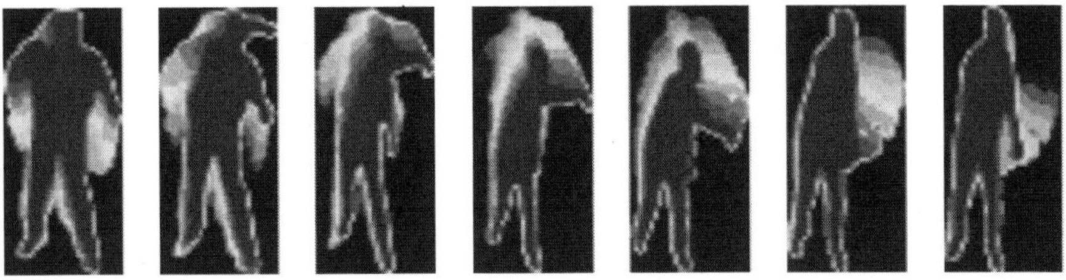

FIGURE 7.11 Some MHIs for action Punch Right corresponding to the MuHaVi-MAS dataset.

TABLE 7.12 Successful rate, in percent, against action.

C1	C2	C3	C4	C5	C6	C7	C8
100	100	100	100	93.75	93.75	100	100

of the actors. There are only two actors and two camera views for these five actions. From these action combinations, fourteen simpler actions (collapse right, walk right to left, walk left to right, turn back right...) can be obtained, which may also reorganized into eight classes (Collapse, Stand up, Kick Right, Punch Right, Guard, Run, Walk, Turn back) where similar actions make a single class. This dataset, only has 136 sequences. Figure 7.11 represents some MHIs corresponding to action Punch Right.

The parameters used in order to obtain the motion templates of the MHIs, are $\tau = 10$ frames, and the displacement of the overlapping windows is 1 frame. Due to the small number of sequences per action, which makes it impossible to have good training of separate SOMs for each action, we have decided to train only a single SOM, of size 15×15, for the whole set of actions, and apply the "leave-one-out" technique over this training. That is, we train with 135 sequences, and test with the remaining one.

A single SOM and several action-specific HMMs

To evaluate the performance of the HMM as a neuron tracker, several HMMs, with different numbers of hidden states (from 2 to 11) have been tested, using a different number of neurons as discrete input. The best results are shown in the Table 7.12. The overall recognition rate is 98.43%. The confusion matrix using the "leave-one-out" cross-validation approach showing the number of test sequences classified en each class is shown in Table 7.13. Each column of the matrix represents the instances in a predicted class, while each row represents the instances in an actual class. Figure 7.12 shows the evolution of the overall rate from a 2-states HMM, depending on the number of resampling neurons. It can be clearly observed that the resampling technique improves the results for a given HMM.

TABLE 7.13 Confusion matrix.

1	2	3	4	5	6	7	8
16	0	0	0	0	0	0	0
0	12	0	0	0	0	0	0
0	0	16	0	0	0	0	0
0	0	0	16	0	0	0	0
0	0	0	0	30	0	2	0
0	0	1	0	0	15	0	0
0	0	0	0	0	0	16	0
0	0	0	0	0	0	0	12

7.6 Conclusions

In this chapter some methods for human action modeling and recognition from video sequences of human silhouettes were explored. MHIs were used to capture both the spatial information about the pose of the human figure at any time (location and orientation of the torso and limbs, aspect ratio of different body parts), as well as the dynamic information (global body motion and motion of the limbs relative to the body). These kinds of motion features have reported good results in situations where human silhouettes were corrupted by noise.

The modeling step relies on a Kohonen self-organized feature map, trained from 2D motion templates recorded in different viewpoints and velocities. The SOM approach enables the integration of spatial and temporal templates into a common framework, combining simultaneously motion state and viewpoint changes. It enables one to reduce the high dimensionality of the temporal templates, improving the results obtained using a linear approach such as PCA. The main novelty of this work is that a 3D reconstruction is not required at any stage. It is true that better results have been reported using a 3D model. However, working with many cameras, each one fully calibrated, is not feasible for many applications.

To train action-specific SOM, many samples are needed to reach stable and accurate performance. When this is not possible, due to the scarcity of training samples, a single SOM can be trained with all the temporal templates. This implies an additional procedure to track the map behavior on temporal templates using, for example, different discrete HMMs. Any HMM would have as observations the indices of the winner neurons of the SOM. However, HMM training also requires a significant amount of data. To solve this problem, we have used a resampling technique such as the SIR algorithm, widely used for particle filters, to increase the number of training sequences, useful when the number of action samples is not enough. The results shown in Figure 7.12 prove that the use of this algorithm significantly improves recognition rates through the use of resampling.

Test average recognition rates given in two real datasets, as well as a virtual one, are very promising, considering the complexity of the test, using a single camera, and the small amount of data used for training.

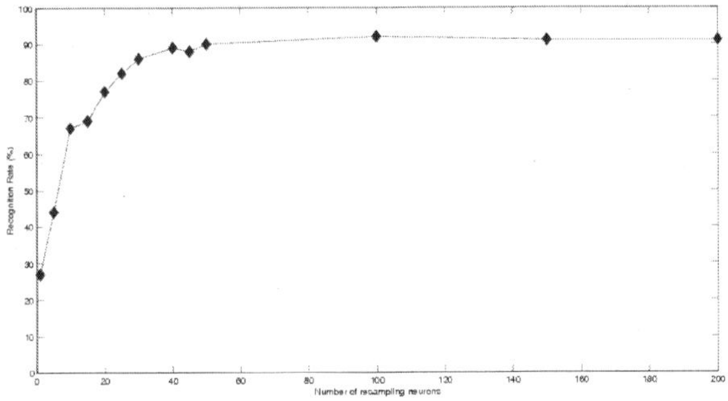

FIGURE 7.12 Successful rate as a function of the number of resampling neurons.

References

1. M.A.R. Ahad, J.K. Tan, H.S. Kim, and S. Ishikawa. Action recognition by employing combined directional motion history and energy images. In *CVCGI10*, pages 73–78, 2010.
2. M. Ahmad and S.-W. Lee. Human action recognition using shape and CLG-motion flow from multi-view image sequences. *Pattern Recogn.*, 41(7):2237–2252, 2008.
3. A.F. Bobick and J.W. Davis. The recognition of human movement using temporal templates. *IEEE Transactions on Pattern Analysis and Machine Intelligence*, 23(3):257–267, 2001.
4. P.-C. Chung and C.-D. Liu. A daily behavior enabled hidden Markov model for human behavior understanding. *Pattern Recogn.*, 41(5):1589–1597, 2008.
5. F. Cuzzolin. Using bilinear models for view-invariant action and identity recognition. In *IEEE Computer Society Conference on Computer Vision and Pattern Recognition*, 2:1701–1708, 2006.
6. A. Doucet, N. de Freitas, and N. Gordon. *Sequential Monte Carlo Methods in Practice*. Springer-Verlag, 2001.
7. W. Hu, T. Tan, L. Wang, and S. Maybank. A survey on visual surveillance of object motion and behaviors. *IEEE Trans. on Systems, Man, and Cybernetics-Part C: Applications and Reviewers*, 34(3):334–352, 2004.
8. W.M. Hu, D. Xie, and T.N. Tan. A hierarchical self-organizing approach for learning the patterns of motion trajectories. *Chin. J.Comput.*, 26(4):417–426, 2003.
9. Y. Hu, L. Cao, F. Lv, S. Yan, Y. Gong, and T.S. Huang. Action detection in complex scenes with spatial and temporal ambiguities. In *IEEE International Conference on Computer Vision*, 2009.
10. J. Hwang, S. Lay, and A. Lippman. Nonparametric multivariate density estimation: A comparative study. *IEEE Trans. Signal Processing*, 42:2795–

2810, 1994.

11. G. Johanson. Visual interpretation of biological motion and a model for its analysis. *Percept. Psychophys.*, 73(2):201–211, 1973.
12. N. Johnson and D. Hogg. Learning the distribution of object trajectories for event recognition. *Image Vis. Comput.*, 14(8):609–615, 1996.
13. A.A. Kale, N. Cuntoor, and V. Krüger. Gait-based recognition of humans using continuous HMMs. In *FGR '02: Proceedings of the Fifth IEEE International Conference on Automatic Face and Gesture Recognition*, page 336. IEEE Computer Society, 2002.
14. V. Kellokumpu, M. Pietikäinen, and J. Heikkilä. Human activity recognition using sequences of postures. In *Proceedings of the IAPR Conference on Machine Vision Applications (IAPR MVA 2005)*, pages 570–573, 2005.
15. T. Kohonen. Self-organization and associative memory. *Springer Series in Information Sciences, Berlin, Springer-Verlag*, 1988.
16. C.S. Lee and A.M. Elgammal. Simultaneous inference of view and body pose using torus manifolds. In *Proc. ICPR*, pages 489–494, 2006.
17. M. Lewandowski, D. Makris, and J.C. Nebel. View and style-independent action manifolds for human activity recognition. In *ECCV10*, pages VI: 547–560, 2010.
18. W.L. Lu and J.J. Little. Simultaneous tracking and action recognition using the PCA-HOG descriptor. In *The Third Canadian Conference on Computer and Robot Vision*, CRV'06, Quebec, Canada, June 2006.
19. F. Lv and R. Nevatia. Single view human action recognition using key pose matching and viterbi path searching. In *IEEE CVPR*, pages 1–8, 2007.
20. H. Meng and N. Pears. Descriptive temporal template features for visual motion recognition. *Pattern Recogn. Lett.*, 30(12):1049–1058, 2009.
21. J. Owens and A. Hunter. Application of the self-organizing map to trajectory classification. In *Proc. IEEE Int. Workshop Visual Surveillance*, pages 77–83, 2000.
22. L. R. Rabiner. A tutorial on hidden Markov models and selected applications in speech recognition. *Proceedings of the IEEE*, 77(2):257–286, 1989.
23. H. Ragheb, S.A. Velastin, P. Remagnino, and T. Ellis. ViHASi: virtual human action silhouette data for the performance evaluation of silhouette-based action recognition methods. In *Proceeding of the 1st ACM Workshop on Vision Networks for Behavior Analysis*, VNBA '08, pages 77–84, Vancouver, British Columbia, Canada, 2008. ACM.
24. INRIA Rhne-Alpes'. The inria IXMAS motion acquisition sequences. *https://charibdis.inrialpes.fr*, 2006.
25. E. Shechtman and M. Irani. Space-time behavior based correlation or how to tell if two underlying motion fields are similar without computing them? *IEEE Transactions on Pattern Analysis and Machine Intelligence (PAMI)*, 29(11):2045–2056, November 2007.
26. S. Singh, S.A. Velastin, and H. Ragheb. MuHAVi: A multicamera human action video dataset for the evaluation of action recognition methods. In

2nd Workshop on Activity Monitoring by Multi-Camera Surveillance Systems (AMMCSS), Boston, MA, USA, Aug. 29, 2010.

27. P.K. Turaga, R. Chellappa, V.S. Subrahmanian, and O. Udrea. Machine recognition of human activities: A survey. *IEEE Trans. Circuits Syst. Video Techn.*, 18(11):1473–1488, 2008.
28. L. Wang and D. Suter. Learning and matching of dynamic shape manifolds for human action recognition. *IEEE Trans. on Image Processing*, 16(6):1646–1661, 2007.
29. L. Wang and D. Suter. Visual learning and recognition of sequential data manifolds with applications to human movement analysis. *Comput. Vis. Image Underst.*, 110(2):153–172, 2008.
30. D. Weinland, E. Boyer, and R. Ronfard. Action recognition from arbitrary views using 3D exemplars. *Proc. IEEE ICCV*, pages 1–7, 2007.
31. D. Weinland, R. Ronfard, and E. Boyer. Free viewpoint action recognition using motion history volumes. *Computer Vision and Image Understanding*, 104(2):249–257, 2006.
32. J. Zhang, R. Collings, and Y. Liu. Bayesian body localization using mixture of non-linear shape models. *Proc. IEEE ICCV*, pages 725–732, 2005.

8

Clustering for Multi-Perspective Video Analytics: A Soft Computing-Based Approach

Ayesha Choudhary
Indian Institute of Technology, Delhi, India

Santanu Chaudhury
Indian Institute of Technology, Delhi, India

Subhashis Banerjee
Indian Institute of Technology, Delhi, India

8.1 Introduction

Among the various modes of surveillance that exist today, visual surveillance is gaining importance primarily due to its ability to record everything in an area under observation. This recorded data offers a human being complete information of the area, including the scene layout, the weather conditions during that time period, and the various objects that come into the scene and their actions and activities. Therefore, this recorded data can not only be used for security and real-time monitoring, but also for a wide range of analysis of the area on the basis of object behaviors, patterns of movements, and types of objects that come into the scene. This analysis, in turn, can be used for applications such as security monitoring, traffic planning and management, customer behavior in a retail shop, egress planning in large stores and offices, and layout designs of retail shops and stores.

The ability to capture every detail is a strength of a visual surveillance system. To be able to leverage the vast amount of information, captured by even a single camera, one has to view the complete video and analyze the information. Manual analysis of the huge volume of data captured by a camera is impractical and prone to errors because of the limitations of the human brain to process multiple signals over a large period of time as well as the limited span of attention of human beings. Therefore, with the increase in the requirement of surveillance in many areas both for security and planning, there is also a growing need for automated analysis of the surveillance videos.

Automated analysis of videos requires detection of objects of interest, tracking these objects and analyzing their movement and/or behavior patterns, and then summarizing the information in the video for offline mining and analysis.

We develop an automated analysis and summarization framework that takes into consideration that video data consists of information at various levels of granularity (e.g., object, frame, group of frames, etc.) and in various attributes within each layer (such as time, size, shape, position of objects, number of objects within a time interval, etc.). We believe that the attributes from the various layers of granularity by themselves and in conjunction with other attributes are important for the analysis and summarization of the video data. The framework gives the user the ability to choose these attributes to analyze the video data based on the context and requirements of the analysis. These attributes also play an important role in automated analysis of surveillance videos for learning the usual events and detecting unusual events. A particular attribute, such as time or type of object, can help in labeling an event as usual or unusual. For example, in a bank, entering the building is a usual event; but if this event occurs after office hours, then it should be detected as an unusual one. However, if security personnel enter the building everyday at a particular time after office hours, then it should not be flagged as a breach in the security. Therefore, a multi-perspective analysis of the same data needs to be carried out for learning the usual patterns, which can then be used for various purposes.

Analysis of the video data in multiple attributes from the various layers requires application of unsupervised learning techniques. This is because an event can be characterized by various attributes and it is not practical to predefine all possible events that can occur in an area under observation. In the past, the systems such as [1,18,51] were only capable of recognizing predefined events. More recently, methods have been developed for discovering various activity classes in the video sequences such as in [11, 19]. These methods are based on discovery of events based on a specific set of features and do not provide the flexibility of mining video data information from multiple perspectives using combinations of summaries of the attributes. To learn the patterns in each of the attribute spaces, we develop the concept of component-based clustering. We represent the video data as a tuple of the various components and cluster in each of the component spaces independent of the others. These clusters can be viewed as a summary of the information in each attribute space. Because the co-occurrence of the attribute values from different attribute spaces also plays an important role in characterizing the events that occur in the area under observation, we develop a cluster algebra that enables combining clusters from the same as well as different attribute spaces. These combined clusters give a summary of the video data and allow the user to mine various information by choosing the clusters and the algebraic operations on these clusters. Our component-based clustering method is independent of the clustering technique used and allows for the use different clustering techniques in each of the component spaces.

In the literature, Bayesian methods are popular for analysis of events, starting from recognition of simple actions [7, 8, 28, 49] to complex events involving multiple entities [21, 22, 33, 39, 42, 43]. Another popular technique used for recognizing events is the hidden Markov models (HMMs) [2, 5, 6, 13, 17, 31, 56, 66]. HMMs and their variants such as the parameterized-HMM, coupled-HMM, hidden semi-Markov models (HSMM), switching semi-Markov models, and coupled hidden semi-Markov models(CHSMM) have been applied for recognition of complex events and activities [6, 16, 23, 38, 62]. Context-free grammars and stochastic parsers have also been applied for recognition of events and identification of known patterns of activities [29, 36, 37, 52].

In the recent past, topic discovery models such as probabilistic Latent Semantic Analysis (pLSA) [20], Latent Dirichlet Allocation (LDA) [4], Hierarchical Dirichlet Process (HDP) [57], and their variants such as the Absolute position pLSA (ABS-pLSA), Translation and Scale Invariant pLSA (TSI-pLSA), and pLSA with Implicit Shape Model (pLSA-ISM) have gained popularity for recognition of actions and activities in videos [32, 40, 53, 61, 63, 69].

Most of the methods for analysis of events and activities are based on discovering and learning the usual events and/or recognizing them. In these systems, the events that cannot be recognized are labeled as unusual events. Our framework is capable of discovering, learning, and recognizing usual events, and can detect unusual ones based on a *usualness* measure that we define for clusters. Along with this, our framework provides a flexible tool to explore the space of clusters from various components, which may result in learning of important information leading to better characterization of the same data. It also provides a summary of the information in the video and gives the user a powerful and generic tool for multi-perspective analysis and mining of the summarized data.

We view the clusters in each of the component spaces and the clusters formed by algebraic operations on the clusters from various component spaces as a summary of the information in the video. In the literature, a video summary or a video synopsis implies making a shorter video of the original video such hat the content and context of the original video are well-communicated to the viewer [9, 30, 47, 48, 54]. This summary is usually formed by removing the temporal or spatio-temporal redundancies, but leads to loss of chronological consistency. Various other forms of video summary includes pictorial summaries [60], video skimming [12, 55, 71], and adaptive fast-forward [3, 45]. Recently, in [46], clustered summaries were developed for faster browsing of surveillance videos by showing together multiple objects that have similar motion.

Offline mining of video data is especially important for surveillance and forensic applications. It has been discussed in [14, 15] that video mining systems should be computationally simple and based on unsupervised learning so that they can be applied to various types of videos. But many of the mining methods [34, 35, 41, 50, 59, 65, 67, 70, 71] developed in the past required supervised learning techniques along with various pattern recognition methods and were meant for mining specific types of videos such as news [67], medical videos [70, 72], sports [35, 64], and traffic [10]. Some methods of video mining [73–75] explore the video data based on the associations present in video units, that is, based on video association mining.

Rough set theory was first introduced by Z. Pawalak [44] to characterize insufficient and incomplete information in a system. A rough set $\mathcal{A}$ is characterized as a set with approximate or uncertain boundary, and is represented by two sets, $\underline{\mathcal{A}}$, its lower approximation, and $\overline{\mathcal{A}}$, its upper approximation. The set $\underline{\mathcal{A}}$ contains elements that only belong to $\mathcal{A}$ while $\overline{\mathcal{A}}$ contains elements that may or may not belong to $\mathcal{A}$. The boundary region of a rough set is non-empty, unlike that of a crisp set, and contains elements that cannot be uniquely classified to a set or its complement based on the available knowledge. Therefore, a rough set captures the approximation present in the data. On the other hand, fuzzy set theory was introduced by L. A. Zadeh [68] to characterize sets that contain elements with a certain degree of belonging. An element of a set is said to belong completely to that set if its membership value is 1, otherwise it belongs to the set with the degree of membership between $(0, 1)$. If the fuzzy membership of an element in a fuzzy set is 0, it does not belong to that set. Therefore, a fuzzy set captures the vagueness present in the data with respect to the concept that the data represents. In the fuzzy rough set theory, both the vagueness and the approximate knowledge present in the system are captured.

Activity analysis from surveillance videos requires classification of events based on various parameters, in spite of incomplete and approximate information present in the system. This is because many a times the features may not always be correctly and crisply extracted, thus leading to the presence of incomplete and approximate information in the system. It also requires classification of events, in spite of vagueness, associated with the events themselves, for example, as discussed earlier, an event can be usual in some context while unusual in another context. Therefore, in automated analysis of surveillance videos, a fuzzy-rough interpretation of the data is helpful in dealing with the approximate, incomplete, and vague characteristics of the video data. However, in the literature fuzzy sets, rough sets, and fuzzy-rough sets have not been employed for the task of automated analysis of videos specially in the context of analysis of events from surveillance videos. In our framework, we show that the concept of component-based clustering and the application of cluster algebra on these component clusters result in rough clustering of the data in high-dimensional spaces. We also show how fuzzy membership functions can be applied in the presence of labeled training data for classification of test events as usual and unusual.

8.2 Video Representation Scheme

A video consists of information at various levels of abstraction and in various attributes that can be efficiently represented as a tuple V whose components consist of the analytics data of the video. Let

$$V =< v_1, v_2, \ldots, v_k >, \qquad k \geq 2, \tag{8.1}$$

where each component $v_i, 1 \leq i \leq k$, of the tuple represents an attribute of the video data that can belong to any of the various levels of abstractions that exist in a video. For example,

$$\begin{aligned} V =<&\text{Number of objects, color correlogram of objects,} \\ &\text{position of objects, size of objects,} \\ &\text{trajectories of objects, time of presence of objects} > \end{aligned} \tag{8.2}$$

is a video data tuple in which the first four components represent attributes from the level of a frame while the last two components represent attributes from the level of a group of frames.

If the video data tuple is considered a vector, it forms a high-dimensional heterogeneous data vector, as each component can be multidimensional and can be of a different data type. Each component of the video data tuple defines an n-dimensional space, $n \geq 1$, such that n is less than the dimension of the video data tuple taken as a monolithic vector. Representing the video data as a tuple is a simple and flexible scheme for unified representation of the attributes from the multilayered video data, allowing each component to be analyzed independently as well in conjunction with the other components. It gives the user the ability to choose, add, or delete attributes of the video data based on the context and final goal of the analysis. It is in tune with the fact that each component as well as a combination of components define important characteristics of the data that is being analyzed.

8.3 Component-Based Clustering Framework

Rough set theoretic representation of clusters is a well-studied concept. A conventional or crisp cluster consists of elements that are very similar to each other based on a well-defined similarity measure and belong to only one crisp cluster. Therefore, elements from two different crisp clusters are dissimilar based on the similarity measure used for clustering the data. On the other hand, a rough cluster is similar to the concept of a rough set. A rough cluster $\mathcal{C}$ is represented by its lower, $\underline{\mathcal{C}}$, and its upper, $\overline{\mathcal{C}}$, approximations. The lower approximation $\underline{\mathcal{C}}$ consists of elements that belong only to cluster $\mathcal{C}$. The upper approximation $\overline{\mathcal{C}}$ consists of elements that can belong to other rough clusters also. The boundary region of a rough cluster, given by $\overline{\mathcal{C}} - \underline{\mathcal{C}}$, is non-empty.

We introduce the concept of component-based clustering of tupular data and develop a cluster algebra. We show later that by clustering in the component spaces, we get crisp clusters; however, the cluster algebraic operations on these component clusters may result in rough clusters in the high-dimensional spaces. We first discuss the concept of component-based clustering, its necessity and advantages - and then discuss cluster algebra. Later, we show the rough set-based interpretation of the higher dimensional clusters that are formed using cluster algebraic operations on the component clusters.

It is easily seen that the tupular representation scheme gives a flexible and simple representation of the multilayered information in a video. The question that arises at this point is "how do we analyze the data usefully and from multiple perspectives to be able to answer different questions pertaining to the same data?" Clustering the video tuple as a monolithic vector is a difficult task, primarily because of the heterogeneity of the video data. In such a case, defining a single similarity measure would require converting all components into numerical data, which may lead to loss of important information. Moreover, clustering the video tuple as a monolithic vector would lead to loss of information of the patterns in each component and the ability to analyze various combinations of these attribute/components.

Therefore, to be able to analyze the data of each component as well as their combinations, we have proposed the concept of component-based clustering. Component-based clustering implies clustering the data of each component of the tuple independent of the other components. Clusters of each component space depict the summary and distribution of data in that component. A major advantage of the tupular representation scheme is that it enables application of a different similarity measure in each component, based on the characteristics of the data of that component. This in turn allows clustering in each of the component spaces, based on the similarity measures relevant for that component. Another advantage is that different clustering schemes can be used on different component spaces, based on the requirement and characteristics of the data of that component.

8.3.1 Similarity Measure For Tuples

Let $\mathfrak{T}$ be the space of all tuples from a data stream. Then, let $T_i = < t_1^i, t_2^i, \ldots, t_m^i¿ \in \mathfrak{T}$, where each component t_k^i represents semantic or numeric data. Let S_k denote the similarity measure for the kth component $t_k^i \in T_i$ such that $S_k(t_k^i) = t$ iff t and t_k^i are similar to each other with respect to the similarity measure S_k. Then, $S = (S_1, S_2, \ldots, S_m)$ defines the similarity measure between two tuples $T_i \in \mathfrak{T}$ and $T_j \in \mathfrak{T}$ such that $S(T_i) = T_j$ iff $S_1(t_1^i) = t_1^j, S_2(t_2^i) = t_2^j, \ldots, S_m(t_m^i) = t_m^j$. This similarity measure S for tuples defines an equivalence relation on all the space of all tuples $t \in \mathfrak{T}$.

8.3.2 Component-Based Clustering

Let $T_i =< t_1^i, t_2^i, \ldots, t_m^i >\in \mathrm{T}$ be a tuple, $1 \leq i \leq N$ and S be the similarity measure for these tuples. Then, component-based clustering on T leads to $\Omega =< C_1, C_2, \ldots, C_m >$, where Ω is the space of all clusters from all the component spaces of the tuples in T and $C_i = \{c_1^i, c_2^i, \ldots, c_k^i\}$ is the set of all clusters from the i-th component space. Moreover, $c_j^i, 1 \leq j \leq k$ is the j-th cluster of C_i containing tuples $T_n = < t_1^n, t_2^n, \ldots, t_m^n > \in \mathrm{T}$, and $T_p = < t_1^p, t_2^p, \ldots, t_m^p > \in \mathrm{T}$ such that $S_i(t_i^n) = T_i^p$.

8.3.3 *Usualness* Measure Associated with a Cluster

Clusters impart information about which values are similar, and the number of elements in a cluster represents how often a particular value occurs. But, based on the attribute space and the values in that attribute space, different cluster sizes would represent how common a value is. Therefore, we defined a *usualness* measure for clusters to be able to distinguish between the usual and unusual values of the data being clustered. This measure is independent of the data type and the clustering technique used for clustering the data, and therefore it can be used in any attribute space to quantify the usualness of an attribute value.

Let Ω be the set of all clusters and $C \in \Omega$ be a cluster of size m, where the size of a cluster is the number of elements in the cluster. Then, the *usualness* measure of a cluster C is defined as the function

$$p(C) = \begin{cases} -1 & m < t_1 \\ e^{-(m-t_2)^2/(2(\frac{t_2-t_1}{3})^2)} & t_1 \leq m \leq t_2 \\ 1 & m > t_2, \end{cases} \tag{8.3}$$

where t_1 and t_2 are predefined thresholds.

If the *usualness* measure is -1, the cluster is treated as noise. This is because for video data there is a time period associated with every occurrence of an attribute value, whether usual or not. Based on the *usualness* measure, we need to decide whether or not a cluster represents a usual value. We assume that *usualness* is a set, $\mathfrak{U}$, such that a cluster is an element of this set only if it represents a usual value. Because the *usualness* measure of a cluster is not a binary measure, *usualness* is a fuzzy set, with each cluster having a membership in this fuzzy set. We define a fuzzy membership function, $f(\cdot)$, of a cluster C based on the *usualness* measure such that

$$f(C) = \begin{cases} 0 & p(C) \leq T \\ p(C) & T < p(C) \leq 1, \end{cases} \tag{8.4}$$

where T is a prespecified threshold such that $T \geq t_1$.

If the fuzzy membership of a cluster is 0, it represents an unusual cluster and if $f(C) \in [T, 1]$, the cluster belongs to $\mathfrak{U}$ with the degree of membership $f(C)$. This helps in deciding whether or not a cluster represents a usual value. It also eliminates clusters that are formed by the occurrence of noise in the dataset by setting the membership function of such a cluster to zero. The value of the membership function of a cluster in $\mathfrak{U}$ gives an indication of the usualness of the cluster.

8.3.4 Properties of *Usualness* Measure

These are the properties that the *usualness* measure of clusters satisfy:

- $0 \leq p(A) \leq 1$
- $p(\phi) = 0$
- $p(A \cup B) = max\{p(A), p(B)\}$
- $p(A \cap B) \leq min\{p(A), p(B)\}$

where ϕ is an empty cluster, and A and B are clusters either from the same component space or different component spaces of clusters. It is easy to see that the union of two clusters is well-defined only if both clusters belong to the same component space. It is the *OR* operation and defines a commutative monoid on the space of clusters from a component space.

8.3.5 Algebraic Operations on Clusters

Combining clusters from various component spaces is essential for a multi-perspective analysis of the video data. It is necessary to discover co-occurrences of values from various component spaces, because it is quite possible that the combination of component values is an unusual one, even if the values in each component are usual. Because component-based clustering clusters the data in each of the attribute space independently, we develop a cluster algebra that can be applied to combine clusters from both the spatial and temporal domains. These algebraic operations can be applied on the component clusters when the complete tuple is stored while clustering in each of the component spaces.

Let (X, Y) be a tuple; then $X \times Y$ is a two-dimensional space. Let $S_x(x) = x^*$ imply that x and x^* are *close* (or similar) based on the distance measure used on the X-component. Then, $C_{x^*} = \{(x, y)|S_x(x) = x^*\}$ is the cluster for the value $x^* \in X$ and $C_{y^*} = \{(x, y)|S_y(y) = y^*\}$ be the cluster for the value $y^* \in Y$ for all $(x, y) \in X \times Y$. The algebraic operations on C_{x^*} and C_{y^*} are defined as

1. Composition:

$$\begin{aligned} C_{x^*\otimes y^*} = \{(x, y)|S_x(x) = x^*, (x, y) \in C_{y^*} \\ \text{and } S_y(y) = y^*, (x, y) \in C_{x^*}\} \end{aligned} \tag{8.5}$$

2. Intersection:

$$\begin{aligned} C_{x^*\cap y^*} = \{(x, y)|S_x(x) = x^*, (x, y) \in C_{x^*} \\ \text{and } S_y(y) = y^*, (x, y) \in C_{y^*}\} \end{aligned} \tag{8.6}$$

3. Union: Union of clusters is defined for clusters from the same component space. Let C_{x^*} and $C_{x'}$ be two clusters from the X-component space. Then,

$$\begin{aligned} C_{x^*\cup x'} = \{(x, y)|S_x(x) = x^*, (x, y) \in C_{x^*} \\ \text{or } S_x(x) = x', (x, y) \in C_{x'}\} \end{aligned} \tag{8.7}$$

4. Temporal composition: Let $C^R_{t^*} = \{(x, t)|S^R_t(t) = t^*\}$ be a cluster in the temporal domain, where, R is any one of the temporal similarity measures: equality, follow, containment or overlap. Then, if all points in C_{x^*} occur during the time interval t^*, the clusters C_{x^*} and $C_{x'}$ are temporally composed as:

$$\begin{aligned} C_{(x^*\cup x')\otimes t^*} = \{(x, t)|S_x(x) = x', (x, t) \in C^R_{t^*} \\ \text{and } S_x(x) = x^*, (x, t) \in C^{Equality}_{t^*}\} \end{aligned} \tag{8.8}$$

Remarks

1. In relational databases, a join operation combines records from two or more tables. The intersection of clusters can be seen as a join operation between clusters, instead of records.

2. Equation 8.9 gives the bound on the *usualness* measure of the set $C_{x^*\cap y^*}$,

$$p(C_{x^*\otimes y^*}) \leq min\{p(C_{x^*}), p(C_{y^*})\}. \tag{8.9}$$

   This can be used for predicting the possibly unusual events that can occur in the area under observation.

In the PETS2001 video [26], clustering is done in each of the components: *(s)ize, (c)olor correlogram, (p)osition, and (t)ime of presence of the objects*. Then, the composed cluster is given as

$$\begin{aligned} &C_{s\otimes p\otimes c\otimes t} = \{(s_i, p_i, c_i, t_i)| \\ &S_s(s_i) = s, (s_i, p_i, c_i, t_i) \in C_{p\otimes c\otimes t}, \\ &\text{and } S_p(p_i) = p, (s_i, p_i, c_i, t_i) \in C_{s\otimes c\otimes t}, \\ &\text{and } S_c(c_i) = c, (s_i, p_i, c_i, t_i) \in C_{s\otimes p\otimes t}, \\ &\text{and } S_t^{Follow}(t_i) = t, (s_i, p_i, c_i, t_i) \in C_{s\otimes p\otimes c}^{Follow}\}. \end{aligned} \tag{8.10}$$

Each composed cluster depicts the trajectory of an object. Each color represents a composed cluster in Figure 8.1. To depict a cluster, we plot the positions of objects across time. It can be seen that the yellow cluster is broken into various contiguous pieces. This indicates that the object was either occluded or not detected in some frames but despite occlusion and nondetection, the continuity and similarity are maintained.

8.3.6 Rough Set-Based Interpretation of Cluster Algebraic Operations

Component-based clustering and composition of clusters from the component spaces may not result in the same clusters that would have formed had the data been clustered as a monolithic vector. We show, using Rough set theory [44], that compositions of component clusters are good approximations of the high-dimensional clusters. A rough set is a set without a crisp boundary and therefore best suited for working with imperfect data, attribute dependencies, and for representing incomplete information.

Suppose $C_{(x^*,y^*)}$ denotes the cluster formed by directly clustering in $X \times Y$ space, and (x^*, y^*) is its cluster representative. Also, suppose that $C_{x^*\otimes y^*}$ denotes the composition of clusters C_{x^*} and C_{y^*}, from X and Y component spaces, respectively. In general, it is not necessary that $C_{x^*\otimes y^*}$ will be the same as $C_{(x^*,y^*)}$. We call high-dimensional clusters as those formed in a high-dimensional space by clustering the tuple as a monolithic vector. The clusters obtained after applying the algebraic operations on the clusters from the different component spaces are approximations of the high-dimensional clusters, assuming that the components taken into consideration are the same in both cases. This is true when the similarity measure used in the high-dimensional space is a monotonicity-preserving function of the similarity measures used in each of the component spaces.

Let $\overline{C}_{(x^*,y^*)}$ be the upper approximation of $C_{(x^*,y^*)}$, and $\underline{C}_{(x^*,y^*)}$ be the

(a) Frame no. 900

(b) Frame no. 2340

FIGURE 8.1 In this example, component-based clustering is done in components *object's position*, *object's size*, *object's color correlogram*, and *object's time of presence*. Then, the cluster algebraic operation is applied to these component clusters to get the composed clusters. Each color indicates a composed cluster of the object properties as depicted in the image by plotting the centroid of the objects across time. Thus, the trajectories of the objects are discovered by composition of clusters from different component spaces, including time. It can be seen in (b) that one of the composed clusters is not continuous although the cluster corresponds to the same object across frames in some of which the object was either occluded or not detected. This cluster shows that despite occlusion/nondetection, continuity and similarity are maintained by composing these component clusters. (See color insert.)

lower approximation of the cluster $C_{(x^*,y^*)}$. Then,

$$\begin{aligned}&\overline{C}_{(x^*,y^*)} = \{(x,y)|0 \le S_x(x,x^*) \le 1, (x,y) \in C_{y^*},\\&\text{and } 0 \le S_y(y,y^*) \le 1; (x,y) \in C_{x^*} \text{ such that}\\&\text{at least one of } S_x(x,x^*) \text{ and } S_y(y,y^*) \text{ is non-zero.}\}\\&\implies \overline{C}_{(x^*,y^*)} = C_{x^* \cap y^*},\end{aligned} \tag{8.11}$$

and

$$\begin{aligned}&\underline{C}_{(x^*,y^*)} = \{(x,y)|S_x(x,x^*) = 1, (x,y) \in C_{y^*},\\&\text{and } S_y(y,y^*) = 1, (x,y) \in C_{x^*}\}\\&\implies \underline{C}(x^*,y^*) = C_{x^* \otimes y^*},\end{aligned} \tag{8.12}$$

where $0 \le S_x(x,x^*) \le 1$ is the extent of similarity of x with x^*. It is defined as

$$S_x(x,x^*) = \begin{cases}1 & x \in C_{x^*}\\ e^{-d(x,x')} & x \notin C_{x^*},\end{cases} \tag{8.13}$$

where d is a well-defined distance measure, m is the size of the cluster C_{x^*}, and $x' \in C_{x^*}$ such that

$$d(x,x') = \min_{1 \le i \le m} \{d(x,x_i)|x_i \in C_{x^*}\} \tag{8.14}$$

By definition, $\underline{C}_{(x^*,y^*)} \subseteq \overline{C}_{(x^*,y^*)}$.

We define the rough membership of elements of a 2D cluster as

$$R(x,y) = \frac{S_x(x,x^*) + S_y(y,y^*)}{2}. \tag{8.15}$$

Thus, $0 \le R(x,y) \le 1$. The rough membership of all elements in the cluster $\underline{C}_{(x^*,y^*)} = 1$ and that of the elements in $\overline{C}_{(x^*,y^*)}$ is in the interval $(0,1]$. Therefore, the rough membership defines the degree of overlap between the cluster $C_{(x^*,y^*)}$ and its upper and lower approximations. Therefore, all elements of the composed cluster $C_{x^* \otimes y^*}$ also belong to the high-dimensional cluster $C_{(x^*,y^*)}$. If the rough membership of an element of $C_{x^* \cap y^*}$ is 1, then that element also belongs to the high-dimensional cluster $C_{(x^*,y^*)}$. Therefore, using the rough membership, we can reconstruct the high-dimensional cluster, as required. Although here we have defined the extent of similarity for a 1D cluster and rough membership for a 2D cluster, they are extendible to higher dimensions. Using the rough membership function in Equation(8.15), we define the degree of approximation of a cluster formed by the algebraic operations when compared with the high-dimensional cluster $C_{(x^*,y^*)}$.

We define the degree of approximation as

$$DoA_{(x^*,y^*)} = 1 - \frac{\sum_{i=1}^{N} R(x_i,y_i)}{N}, \tag{8.16}$$

where, (x_i, y_i) are elements of the 2D cluster and N is the size of that cluster. If the degree of approximation is 0 then $C_{(x^*,y^*)} = C_{x^* \otimes y^*}$. In this case, $\overline{C}_{(x^*,y^*)} = \underline{C}_{(x^*,y^*)}$.

In Figure 8.2, the points on the X-axis are the cluster C_{x^*} in X-space, while the points on the Y-axis belong to the cluster C_{y^*} in Y-space. The dark points belong to the cluster $C_{(x^*,y^*)}$, while the lighter points belong to both $C_{(x^*,y^*)}$ and $C_{x^* \otimes y^*}$. Here, the degree of approximation, $DoA_{(x^*,y^*)}$ is 0.0022.

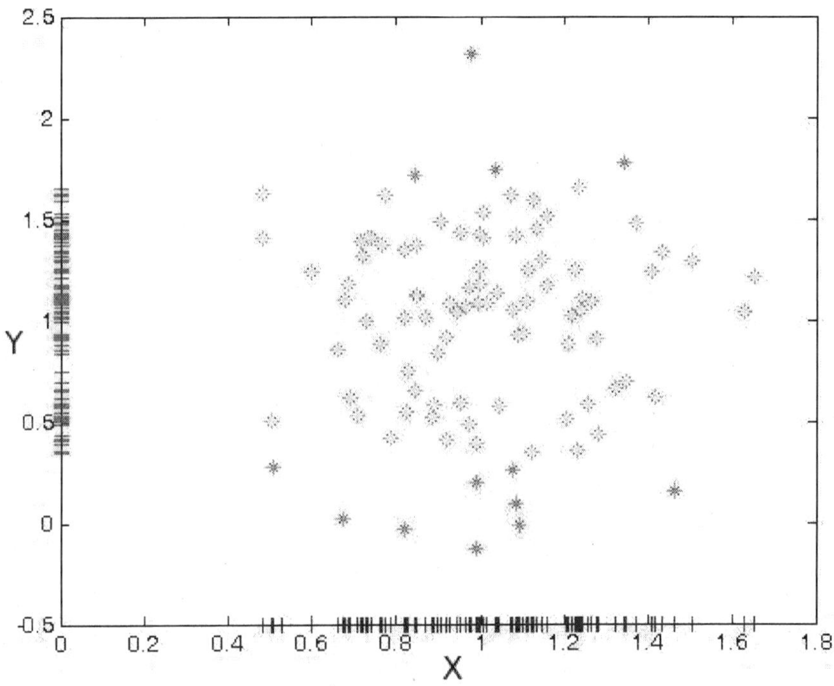

FIGURE 8.2 The points on the X-axis belong to the cluster in the X-component and the points on the Y-axis belong to the cluster in the Y-component. All the points belong to the cluster formed by directly clustering in the 2D space $C_{(x^*,y^*)}$ but the lighter points at the core belong to the composed cluster $C_{x^* \otimes y^*}$ as well as to $C_{(x^*,y^*)}$. The degree of approximation is 0.0022, showing that the composed cluster is a good approximation of the 2D cluster. (See color insert.)

In the PETS2001 video [26], we cluster the trajectories by clustering in the high-dimensional space as well as clustering in the component spaces and then composing the clusters. For the largest cluster, the degree of approximation is 0.0318, showing that the composed cluster is a good approximation of the high-dimensional cluster.

(a)

(b)

FIGURE 8.3 The usual paths in the scene are discovered by clustering the trajectories of the objects in the scene, as shown in (a). The white point indicates the starting point and the black point indicates the ending location of the trajectory. Each trajectory cluster is represented by a separate cluster. As the rate of arrival of frames via the Internet from the camera at [25] is very low, some objects are detected for the first time in the middle of the scene while some vanish from the scene suddenly from non-exit locations. In spite of this, our system discovers the true entry (black), exit (dark), and wait (gray) locations, as shown in (b). We allow the black strip at the top of the image to remain and be detected as a foreground object because the time information is necessary. However, it results in it getting detected as a wait location, where objects tend to remain static for long periods of time. (See color insert.)

8.4 Unusual Event Classification in Case of Supervised Learning

Many times, an application requires recognizing predefined events in the area under observation. Assume that there exists a large amount of labeled training data, based on the application requirements. The first step is to form clusters of these labeled data, simply on the basis of their labels. It is not always possible to crisply classify the test data, based on the features used for labeling the test data, in an automated manner. This is primarily because of the vagueness associated with the components in the video tuple or the features used for labeling the test data itself. We show that using a fuzzy membership function, it is not only possible to classify the test data, but also to discover unusual or unexpected events that occur in the area under observation. In this case, we assume that the clusters denoting the events are composed clusters formed by composition of clusters from the various component spaces, which were used for initially labeling the test data. However, the concept of using the fuzzy measure for classification of data to labeled clusters as well as discovering the unusual cluster is a general one and can be applied in various disciplines, independent of the clustering technique used.

Assume that the labeled training data is crisply clustered based on the labels on the data points. Then, it is possible to compute a cluster representative for each of these clusters such that it represents the cluster in the best possible manner. This cluster representative is used for clustering the test data into one of the labeled clusters. For each test data point, we compute the distance of the data point from each of the cluster representatives and define a fuzzy membership of that data point in each of the labeled clusters, based on this distance. Let q be the test data point and q' be a cluster representative of a cluster $\mathcal{C}$. Then, Equation(8.17) gives the distance decay measure for a data point q from the cluster representative q', where d is a well-defined distance measure for that component. For example, for numerical data, d can be taken to be the Euclidean distance measure,

$$m(q,q') = e^{-d(q,q')}. \tag{8.17}$$

Then, the fuzzy membership of the data point in the cluster $\mathcal{C}$ is given by

$$f(q) = \begin{cases} 0 & m(q,q') < z_1 \\ m(q,q') & m(q,q') \geq z_1, \end{cases} \tag{8.18}$$

where $0 \leq z_1 \leq 1$ is a predefined threshold on the distance decay function m.

Using this fuzzy membership function, the fuzzy membership of the test data point q is computed for each of the labeled datasets and the data is classified to the labeled cluster for which the membership is maximum among all the clusters. In case the membership of the data point is zero in all the clusters, we cluster the test data into a cluster labeled as *unusual* cluster.

Therefore, not only are we able to cluster the test data into one of the usual or labeled data points using the fuzzy measure, which helps overcome the vagueness present in the data, but we are also able to discover the unusual events that occur in the test data.

We illustrate this through a small example. In a parking video, labeled data denoting correct parking is provided for the components *object size* and *object position* across time. The composition of clusters from these two component spaces results in the composed cluster $C_{labeled}$. This cluster has labeled elements depicting correct parking. The test data in Figure 8.4(a) has fuzzy membership 0.9 in $C_{labeled}$, while the one in Figure 8.4(b) has fuzzy membership 0.85 in $C_{labeled}$. However, the test data in Figure 8.4(c) has a fuzzy membership 0 in $C_{unlabeled}$ because the similarity with the cluster representative is 0.1 and the threshold z_1 is taken as 0.35, which has been empirically determined. This leads to the formulation of a new cluster $C_{unlabeled}$ that in this context depicts a deviant parking style.

8.5 Experimental Results

For our experiments, we worked with frames from the Internet from AXIS [27] network cameras, whose geographical locations in the world are unknown to us. We have used adaptive background subtraction [58] for segmenting the foreground objects from the background in the scene. We have considered simple features such as size, color correlogram, and position of objects, which are the segmented foreground blobs. Our framework is generic and although we have used simple features for our experiments, our framework can also incorporate complex features.

The video representation used for object level is

$$V =< size, color\ correlogram, position, time > . \tag{8.19}$$

Temporal composition of clusters from the component spaces of the tuple in Equation(8.19) gives the trajectories of these objects. The trajectories are represented as:

$$V =< start\ location, end\ location, start\ time, end\ time > . \tag{8.20}$$

Trajectory clusters are formed by composition of clusters from each of the component spaces of tuple in Equation(8.20).

8.5.1 Online Video 1

The first experiment is carried out using 24 hours of video from the camera at [25]. Composition of clusters from each of the component spaces in Equation(8.19) gives the object-level summarization. The trajectory clusters are obtained as explained above. The fuzzy membership of these clusters in the

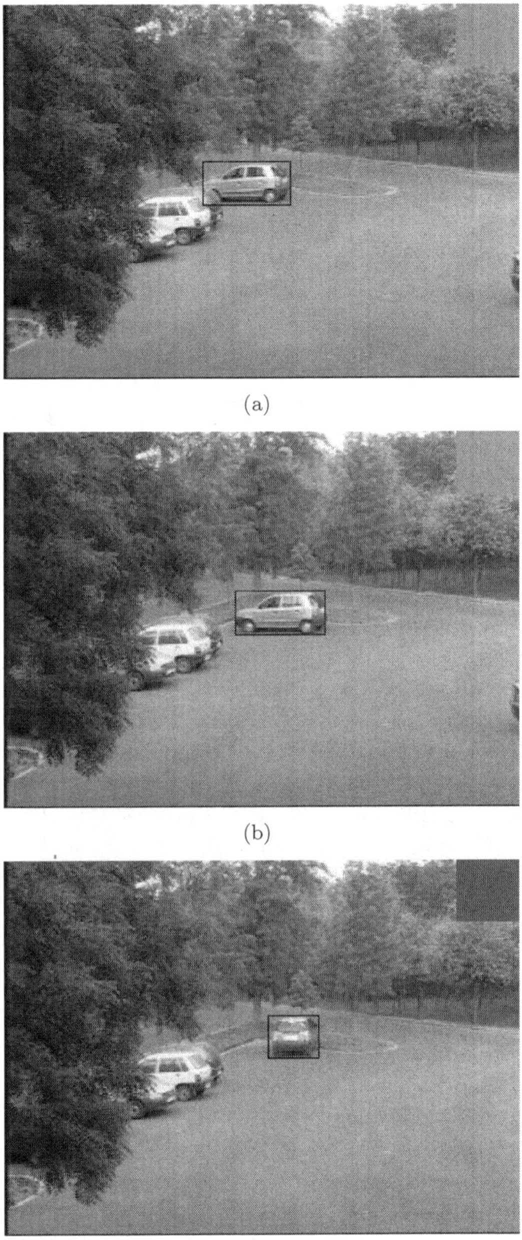

(a)

(b)

(c)

FIGURE 8.4 In a parking experiment, labeled data is provided for components $object's\ size$ and $object's\ position$ and the composition of these clusters from these two component spaces gives the composed cluster $C_{labeled}$, which contains examples of correct parking. The test data in (a) has fuzzy membership 0.9 in $C_{labeled}$ while that in (b) has fuzzy membership 0.85 in $C_{labeled}$. However, the test data in (c) has fuzzy membership 0 in $C_{labeled}$. Hence, it belongs to a new label $C_{unlabeled}$. In this context, it depicts a deviant parking style while (a) and (b) depict the correct parking style.

set $\mathfrak{U}$ determines the usual paths in the scene. The thresholds for usualness measure are $t_1 = 30$ and $t_2 = 50$, while the threshold for the fuzzy membership is set to $T = 35$.

Figure 8.3(a) shows the usual paths in the scene with the white dot representing the starting point of the trajectory while the black dot represents the ending point. Because the data is collected from the Internet where the rate of arrival of the data is 1 frame in 2 to 3 seconds. Many times, the first occurrence of an object is in the middle of the frame or the objects suddenly vanish from the scene from a non-edit location. This leads to incorrect detection of the entry-exit points. However, in spite of the noise, our system is able to locate the actual entry and exit locations because their fuzzy membership in $\mathfrak{U}$ is 11. In Figure 8.3(b), the blue dots show the entry locations, the red points depicts the exit locations, while the locations where objects tend to be static are shown by green dots. The visualization of these clusters at any time t depicts the summary of the events that have occurred until that point in time.

8.5.2 Online Video 2

From the camera at [24], we have analyzed the data of 2 days. We obtain the trajectories of the object as composition of component clusters from the tuple from Equation(8.19). The objects are also clustered on the basis of their trajectories, their time of presence, and the size of these objects. Because mostly only people walk on these paths in the scene, most objects are of similar sizes. However, in a few instances we find two instances that show a cat prowling in the area during two different time periods, as shown in Figures 8.5 and 8.6. Although the cat moves on a usual path, it is detected as a deviant object because of its size. Figure 8.7 shows the usual paths in the scene.

8.6 Conclusion

Multi-perspective video surveillance is important for learning the various patterns present in the video data, especially for unusual activity detection. Representing the video data as a tuple of attributes is a simple and flexible method for taking into consideration the fact that video data consists of multiple attributes at various levels of abstraction. These attributes, individually as well as together, describe various properties of the area under observation and therefore need to be analyzed both independently and in conjunction with each other for a multi-perspective analysis of the video data. We have proposed a component-based clustering scheme that allows clustering the components of the tuple independently and gives the data patterns of each component. These clusters can be viewed as a summarization of the video data and can be leveraged upon for offline mining of video data. We have also developed a cluster

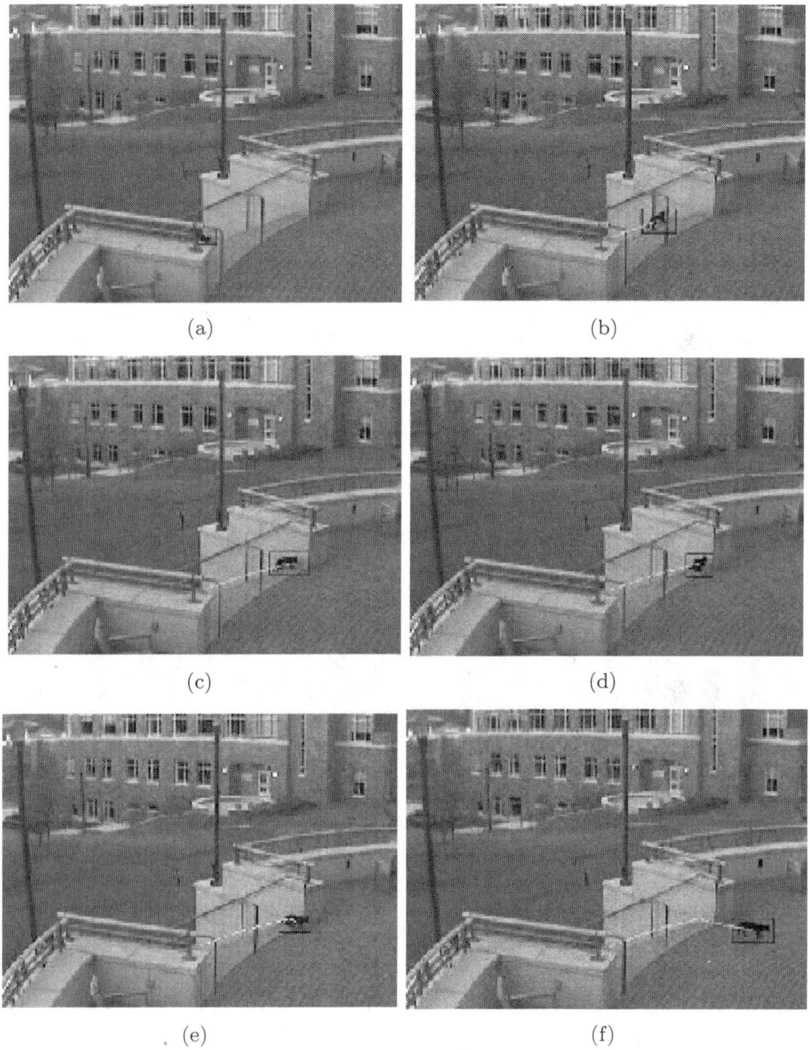

(a) (b) (c) (d) (e) (f)

FIGURE 8.5 A cat is seen prowling in the scene, and this event is detected as an unusual event. This is primarily because of the size of the object.

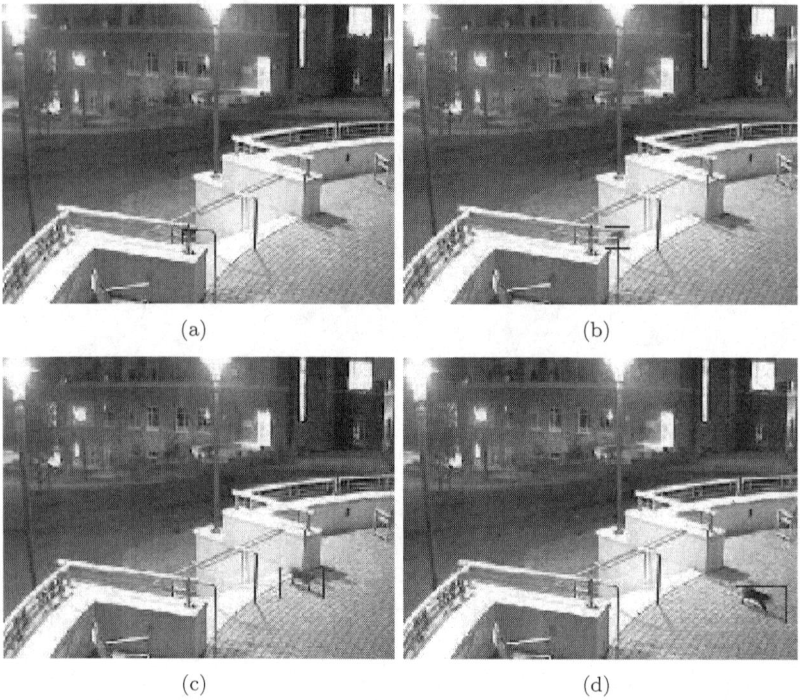

FIGURE 8.6 The cat is again seen in the area under observation, at a different time period. This event is also detected as an unusual event because of the size of the object and time of occurrence of the event.

FIGURE 8.7 The usual paths in the scene are shown as different clusters of trajectories. Each line is a trajectory in the scene with the white dot indicating the start location and the black dot indicating the end position. (See color insert.)

algebra that allows analyzing the co-occurrences of values from two or more clusters, thereby learning the co-occurrences of patterns in the video data components. These algebraic operations can be used on clusters from various component spaces having heterogeneous data types. Therefore, component-based clustering along with the cluster algebra give a complete and in-depth analysis of the video data while enabling the user to combine and analyze clusters from various attribute spaces. Therefore, the system proposed by us gives the user a powerful, flexible, and easy-to-use tool for multi-perspective analysis of video data for learning the usual event patterns as well as detecting unusual events.

References

1. D. Ayers and M. Shah. Monitoring human behavior from video taken in an office environment. *Image and Vision Computing*, 19(2):833–846, 2001.
2. F. Bashir, A. Khokhar, and D. Schonfeld. Object trajectory-based activity classification and recognition using hidden Markov models. *IEEE Transactions on Image Processing*, 16:1912–1919, 2007.
3. E. Bennett and L. McMillan. Computational time-lapse video. In *Proceedings of the ACM SIGGRAPH*, 2007.
4. D. M. Blei, A. Y. Ng, M. I. Jordan, and J. Lafferty. Latent Dirichlet allocation. *Journal of Machine Learning Research*, 3:993–1022, 2003.

5. M. Brand and V. Kettnaker. Discovery and segmentation of activities in video. *IEEE Transactions on Pattern Analysis and Machine Intelligence*, 22(8):844–851, 2000.
6. M. Brand, N. Oliver, and A. Pentland. Coupled hidden Markov models for complex action recognition. In *Proceedings of the IEEE Conference on Computer Vision and Pattern Recognition (CVPR)*, 1997.
7. H. Buxton and S. Gong. Advanced visual surveillance using Bayesian networks. In *Proceedings of the International Conference on Computer Vision (ICCV)*, pages 111–123, 1995.
8. H. Buxton and S. Gong. Visual surveillance in a dynamic and uncertain world. *Artificial Intelligence*, 78(2):431–459, 1995.
9. F. Chen, M. Cooper, and J. Adcock. Video summarization preserving dynamic content. In *Proceedings of the International Workshop on TRECVID Video Summarization*, pages 40–44, 2007.
10. S. Chen, M. Shyu, C. Zhang, and J. Strickrott. Multimedia data mining for traffic video sequence. In *International Workshop on Multimedia Data Mining (MDM/KDD)*, 2001.
11. A. Choudhary, M. Pal, S. Banerjee, and S. Chaudhury. Unusual activity analysis using video epitomes and pLSA. In *Proceedings of ICVGIP*, 2008.
12. M. G. Christel, A. G. Hauptmann, A.S. Warmack, and S.A. Crosby. Adjustable filmstrips and skims as abstractions for a digital video library. *IEEE Advances in Digital Libraries Conference*, 1999.
13. N. P. Cuntoor, B. Yegnanarayana, and R. Chellapa. Activity modeling using event probability sequences. *IEEE Transactions on Image Processing*, 17(4):594–607, 2008.
14. A. Divakaran, K. Miyahara, K.A. Peker, R. Radhakrishnan, and Z. Xiong. Video mining using combinations of unsupervised and supervised learning techniques. In *Proceedings of the SPIE Conference on Storage and Retrieval for Multimedia Databases*, 5307:235–243, 2004.
15. A. Divakaran, K. A. Peker, S. Chang, R. Radhakrishnan, and L. Xie. Video mining: Pattern discovery versus pattern recognition. In *Proceedings of the International Conference on Image Processing*,4:2379–2382, 2004.
16. T. V. Duong, H. H. Bui, D. Q. Phung, and S. Venkatesh. Activity recognition and abnormality detection with the switching hidden semi-Markov model. In *Proceedings of the IEEE Conference on Computer Vision and Pattern Recognition (CVPR)*, pages 838–845, 2005.
17. A. Galata, N. Johnson, and D. Hogg. Learning variable length Markov models of behavior. *Computer Vision and Image Understanding*, 81(3):398–413, 2001.
18. A. Hakeem, Y. Sheikh, and M. Shah. $CASE^E$: A hierarchical event representation for the analysis of videos. In *Proceedings of the National Conference on Artificial Intelligence (AAAI Press)*, pages 263–268, 2004.
19. R. Hamid, S. Maddi, A. Johnson, A. Bobick, I. Essa, and C. Isbell. A novel sequence representation for unsupervised analysis of human activities. *Artificial Intelligence*, 173:1221–1244, 2009.

20. T. Hofmann. Probabilistic latent semantic analysis. In *Proceedings of the Conference on Uncertainty in Artificial Intelligence*, 1999.
21. S. Hongeng, F. Bremond, and R. Nevatia. Representation and optimal recognition of human activites. In *Proceedings of the IEEE Conference on Computer Vision and Pattern Recognition (CVPR)*, pages 818–825, 2000.
22. S. Hongeng and R. Nevatia. Multi-agent event recognition. In *Proceedings of the IEEE International Conference on Computer Vision (ICCV)*, 2:84–91, 2001.
23. S. Hongeng and R. Nevatia. Large-scale event detection using semi-hidden Markov models. In *Proceedings of the IEEE International Conference on Computer Vision (ICCV)*, volume 2, pages 1455–1462, 2003.
24. http://149.43.156.107/view/index.shtml.
25. http://161.28.134.223/view/index.shtml.
26. http://ftp.pets.rdg.ac.uk/PETS2001/.
27. http://www.axis.com/products/video/index.htm.
28. S. Intille and A. Bobick. Recognizing planned multiperson action. *Computer Vision and Image Understanding*, 81(3):414–445, 2001.
29. Y. Ivanov and A. Bobick. Recognition of visual activities and interactions by stochastic parsing. *IEEE Transactions on Pattern Analysis and Machine Intelligence*, 22(8):852–872, 2000.
30. H. Kang, Y. Matsushita, X. Tang, and X. Chen. Space-time video montage. In *Proceedings of the IEEE Conference on Computer Vision and Pattern Recognition (CVPR)*, 2:1331–1338, 2006.
31. V. Kettnaker. Time-dependent HMMs for visual intrusion detection. In *Proceedings of the IEEE CVPR Workshop on Event Mining: Detection and Recognition of Events in Video*, 2003.
32. D. Kuettel, M. D. Breitenstein, L. Van Gool, and V. Ferrari. Whats going on? Discovering spatio-temporal dependencies in dynamic scenes. In *Proceedings of the IEEE Conference on Computer Vision and Pattern Recognition (CVPR)*, 2010.
33. A. Madabhushi and J. Aggarwal. A Bayesian approach to human activity recognition. In *Proceedings of the International Workshop on Visual Surveillance*, 1999.
34. Y. Matsuo, K. Shirahama, and K. Uehara. Video data mining: Extracting cinematic rules from movie. In *Proceedings of the International Workshop on Multimedia Data Management (MDM-KDD)*, pages 18–27, 2003.
35. T. Mei, Y. Ma, H. Zhou, W. Ma, and H. Zhang. Sports video mining with mosaic. In *Proceedings of the International Multimedia Modelling Conference*, pages 107–114, 2005.
36. D. Minnen, I. Essa, and T. Starner. Expectation grammars: Leveraging high-level expectations for activity recognition. In *Proceedings of the IEEE Conference on Computer Vision and Pattern Recognition (CVPR)*, 2:626–632, 2003.
37. D. Moore and I. Essa. Recognizing multitasked activities from video using

stochastic context-free grammar. In *Proceedings of the National Conference on Artificial Intelligence, AAAI Press*, pages 770–776, 2002.

38. P. Natarajan and R. Nevatia. Coupled hidden semi Markov models for activity recognition. In *Proceedings of the IEEE Workshop on Motion and Video Computing (WMVC)*, page 10, 2007.
39. R. Nevatia, S. Hongeng, and F. Bremond. Video-based event recognition: Activity representation and probabilistic recognition methods. *Computer Vision and Image Understanding*, 96(2):129–162, 2004.
40. J. Niebles, H. Wang, and L. Fei-Fei. Unsupervised learning of human action categories using spatio-temporal words. In *Proceedings of the British Machine Vision Conference (BMVC)*, pages 299–318, 2006.
41. J. Oh and B. Bandi. Multimedia data mining framework for raw video sequence. In *Proceedings of the ACM Third International Workshop on Multimedia Data Management*, pages 1–10. Springer Verlag, 2002.
42. N. Oliver, B. Rosario, and A. Pentland. A Bayesian computer vision system for modeling human interactions. *IEEE Transactions on Pattern Analysis and Machine Intelligence*, 22(8):831–843, 2000.
43. S. Park and J. Aggarwal. A hierarchical Bayesian network for event recognition of human actions and interactions. *ACM Journal of Multimedia Systems*, 10:164–179, 2004.
44. Z. Pawlak. *Rough Sets: Theoretical Aspects of Reasoning About Data.* Kluwer Academic Publishing, 1991.
45. N. Petrovic, N. Jojic, and T. Huang. Adaptive video fast forward. *Multimedia Tools and Applications*, 26(3):327–344, 2005.
46. Y. Pritch, S. Ratovitch, A. Hendel, and S. Peleg. Clustered synopsis of surveillance video. In *Proceedings of the IEEE International Conference on Advanced Video and Signal Based Surveillance (AVSS)*, pages 195–200, 2009.
47. Y. Pritch, A. Rav-Acha, A. Gutman, and S. Peleg. Webcam synopsis: Peeking around the world. In *Proceedings of the IEEE International Conference on Computer Vision (ICCV)*, pages 1–8, 2007.
48. Y. Pritch, A. Rav-Acha, and S. Peleg. Non-chronological video synopsis and indexing. *IEEE Transactions on Pattern Analysis and Machine Intelligence*, 30(11):1971–1984, 2008.
49. P. Remagnino, T. Tan, and K. Baker. Multi-agent visual surveillance of dynamic scenes. *Image and Vision Computing*, 16(8):529–532, 1998.
50. A. Rosenfeld, D. Doermann, and A. Pentland (Eds.). *Video Mining.* Kluwer Academic Publishers, 2003.
51. N. Rota and M. Thonnat. Video sequence interpretation for visual surveillance. In *Proceedings of the IEEE International Workshop on Visual Surveillance*, pages 59–68, 2000.
52. M. S. Ryoo and J. K. Aggarwal. Recognition of composite human activities through context-free grammar based representation. In *Proceedings of the IEEE Conference on Computer Vision and Pattern Recognition (CVPR)*, 2:1709–1718, 2006.

53. S. Savarese, A. D. Pozo, J. C. Niebles, and L. Fei-Fei. Spatial temporal correlatons for unsupervised action classification. In *Proceedings of the IEEE Workshop on Motion and Video Computing*, 2008.
54. D. Simakv, Y. Caspi, E. Shechtman, and M. Irani. Summarizing visual data using bidirectional similarity. In *Proceedings of the IEEE Conference on Computer Vision and Pattern Recognition (CVPR)*, 2008.
55. M. A. Smith. Video skimming and characterization through the combination of image and language understanding. In *Proceedings of the IEEE Conference on Computer Vision and Pattern Recognition (CVPR)*, 1997.
56. T. Starner and A. Pentland. Visual recognition of American Sign Language using hidden Markov models. In *Proceedings of the International Workshop on Automatic Face and Gesture Recognition*, page 879, 1995.
57. Y.W. Teh, M.I. Jordon, M.J. Beal, and D.M. Blei. Hierarchical Dirichlet processes. *Journal of the American Statistical Association*, 101(476):1566–1581, 2006.
58. K. Tieu, G. Dalley, and W.E.L. Grimson. Inference of non-overlapping camera network topology by measuring statistical dependence. In *Proceedings of the IEEE Internatonal Conference on Computer Vision (ICCV)*, 2:1842–1849, 2005.
59. Pawan K. Turaga, Ashok Veeraraghavan, and Rama Chellappa. From videos to verbs: Mining videos for activities using a cascade of dynamical systems. In *Proceedings of the IEEE Conference on Computer Vision and Pattern Recognition (CVPR)*, pages 1–8, 2007.
60. S. Uchihashi, J. Foote, A. Girgensohn, and J. Boreczky. Video managa: Generating semantically meaningful video summaries. In *Proceedings of the ACM International Conference on Multimedia*, pages 383–392, 1999.
61. J. Varadaraja and J. M. Odobez. Topic models for scene analysis and abnormality detection. In *Proceedings of the Workshop on Visual Surveillance (VS)*, pages 1338–1345, 2009.
62. D. Wilson and A. Bobick. Non-linear PHMMs for the interpretation of parameterized gesture. In *Proceedings of the IEEE Conference on Computer Vision and Pattern Recognition (CVPR)*, 1998.
63. S. Wong, T. Kim, and R. Cipolla. Learning motion categories using both semantics and structural information. In *Proceedings of the IEEE Conference on Computer Vision and Pattern Recognition (CVPR)*, 2007.
64. L. Xie, S. Chang, A. Divakaran, and H. Sun. Structure analysis of soccer video with hidden Markov models. In *Proceedings of the IEEE International Conference on Acoustics, Speech and Signal Processing*, 2002.
65. L. Xie, S.-F. Chang, A. Divakaran, and H. Sun. *Unsupervised Mining of Statistical Temporal Structures in Video*. Kluwer Academic, 2003.
66. J. Yamato, J. Ohya, and K. Ishii. Recognizing human action in time-sequential images using hidden Markov model. In *Proceedings of the IEEE Conference on Computer Vision and Pattern Recognition (CVPR)*, pages 379–385, 1992.
67. J. Yu, Y. He, and S. Li. Content-based news video mining. In *Proceedings of*

the International Conference on Advanced Data Mining and Applications, pages 431–438, 2005.

68. L. A. Zadeh. Fuzzy sets. *Information and Control*, 8(3):338 –353, 1965.
69. J. Zhang and S. Gong. Action categorization by structural probabilistic Latent Semantic Analysis. *Computer Vision and Image Understanding*, 114(8):857–864, 2010.
70. X. Zhu, W. G. Aref, J. Fan, A. Catlin, and A. K. Elmagarmid. Medical video mining for efficient database indexing, management and access. In *Proceedings of the International Conference on Data Engineering*, page 569, 2003.
71. X. Zhu, J. Fan, W. G. Aref, and A. K. Elmagarmid. Classminer: Mining medical video content structure and events towards efficient access and scalable skimming. In *Proceedings of the ACM SIGMOD Workshop*, 2002.
72. X. Zhu, J. Fan, A. K. Elmagarmid, and W. G. Aref. Hierarchical video summarization for medical data. In *Proceedings of the IST/SPIE Storage and Retrieval for Media Databases*, volume 4676, pages 395–406, 2002.
73. X. Zhu and X. Wu. Mining video association for efficient database management. In *Proceedings of the International Joint Conference on Artificial Intelligence (IJCAI)*, pages 1422–1424, 2003.
74. X. Zhu and X. Wu. Sequential association mining for video summarization. In *Proceedings of the IEEE International Conference on Multimedia and Expo (ICME)*, pages 333–336, 2003.
75. X. Zhu, X. Wu, A. Elmagarmid, Z. Feng, and L. Wu. Video data mining: Semantic indexing and event detection from the association perspective. *IEEE Transactions on Knowledege and Data Engineering*, 17(5):665–677, 2005.

9

An Unsupervised Video Shot Boundary Detection Technique Using Fuzzy Entropy Estimation of Video Content

Biswanath Chakraborty
RCC Institute of Information Technology, Kolkata, India

Siddhartha Bhattacharyya
RCC Institute of Information Technology, Kolkata, India

Paramartha Dutta
Visva-Bharati University, Santiniketan, India

9.1 Introduction

Video classification and categorization have always been challenging propositions as far as the computer vision and image processing communities are concerned. Tracking and localization of moving objects within video sequences, leading to proper classification of the sequences by means of detecting video shot boundaries, are prerequisites in several applications. Some of these applications include video surveillance, target tracking, defense, robotic maneuvering, and the entertainment and advertising industries, to name a few [7, 22]. Boreczky and Rowe [7] presented a comprehensive comparison of several techniques of video shot boundary detection. In a later reporting in [22], the authors provided yet another excellent as well as compact study on the *pros* and *cons* of various methods of detection of change in video-shots in terms of their performances. Because video sequences entail a huge amount of data, estimation of the nature and degree of motion of the mobile objects within the motion scene is an uphill task in the image processing community. Moreover, video data is also continuous in nature. Hence, the commonly used estimation techniques in such a dynamic environment essentially comprise three approaches:

1. *Using object centric feature matching* in consecutive image frames [13], where Duncan and Chou exhibited the effectiveness of using optical flow for detection of video motion. Later on, the authors extended their work in [14].

2. *Using spatial and frequency component information* of the moving objects [29, 33]. Jacobson and Wechsler in [33] have shown the usefulness of spatio-temporal frequency measure for determining optical flow field in a video environment. On the other hand, in [29], Heeger proposed filters based on spatio-temporal features for the same purpose viz. determination of optical flow field.

3. *Using motion estimation using intensity flow/variations* in the consecutive image frames comprising moving objects [1, 14, 30]. In [1], Beauchemin and Barron took up the computational aspects of optical flow, whereas in [30], Horn and Schunck presented a general but comprehensive procedure for determination of optical flow. An interesting contribution in this respect is due to Pal et al. [46], in which they have focused on the effectiveness of fuzzy metrics used for the purpose of discrimination of frames in video motion.

It is needless to mention that but for effective and efficient demarcation of video shots emanating from different contexts, the above applications cannot perform as necessary. In other words, the problem of boundary detection is an unavoidable prerequisite when there are various video contexts present in the application.

In this chapter, we propose a novel method to detect the boundaries of video sequences by means of an estimation of the fuzzy entropy distributions of the sequences under consideration. The novelty of the method lies in the fact that no a priori knowledge is required for the detection and classification process. An application of the proposed method is demonstrated on several video sequences of different contexts. Results indicate encouraging avenues.

The chapter is organized as follows. In the next section, we consider a brief but compact survey of the literature. For the sake of completeness, the technical concepts prerequisite for the formalization of the present approach is provided in Section 9.3. In Section 9.4, the methodology is illustrated. The results are reported in the subsequent Section 9.5. Ultimately, in Section 9.6, we sum up the achievable and further research avenues in this direction.

9.2 A Brief Survey of Related Literature

This section is aimed at highlighting the path-breaking research initiatives in the field of soft computing-based image segmentation techniques applied to video shot detection. The section starts with the relevance and effectiveness of the soft computing paradigm in adjudging real-world understandings in data under consideration. It follows with a short and compact introduction of the fuzzy entropy measure (also an effective measure in quantifying uncertainty) relevant to this chapter. Finally, it points out the applications of soft computing techniques for the successful detection of video shot from video sequences of difference contexts.

Fuzzy entropy [36–40] has played a useful role in data segmentation. Some of the formidable reporting as per this approach include [44, 45, 52]. In [44], Pal tried to provide a comprehensive glimpse of the different metrics in quantitative terms for determining the performance of fuzzy-based image enhancement. Subsequently, in [45], Pal along with others used the twin ingredients in the form of index of fuzzyness and entropy to automate gray-level image thresholding. Later on, in a very recent development, Sen et al. utilized the strength of rough sets for ascertaining/handling ambiguities in an image content [52]. Moreover, in [41], Mokji and Bakar used the information extracted out of the co-occurrence matrix for tuning the threshold in gray images. As a follow-up of the concept of the co-occurrence matrix used by Mokji and Bakar, Das et al. used the fuzzy counterpart of the co-occurrence matrix for capturing the inherent uncertainties present in a gray image gamut [9]. In order to achieve this, they utilized the concept of a fuzzy histogram corresponding to a gray image drawing inspiration from a previous work by Jawahar and Ray [34]. We, on our part, considered the contribution due to Das et al. in [9] as the reference work for substantiating and justifying our contribution to the present scope. Another very recent reporting is due to Kucuktunk et al. [35], where the authors have used the extension of the fuzzy histogram to accommodate the various color components within the purview of their analysis for achieving segmentation. What is even more interesting is that their contribution is applicable to color video snaps. The first definition of fuzzy entropy on finite universal set was introduced by De Luca and Termini [39]. They used the basic union, intersection, and complement operators defined by Zadeh [57]. Fan and Ma [16] pointed out that fixing the fuzzy entropy at the fuzzy membership value 0.5 is fraught with some inherent drawbacks that can be overcome by extending the above definition of fuzzy entropy. In fact, they identified the shortcomings of generalized fuzzy entropy definition due to Zenzo et al. [58].

The detection of the boundary between two consecutive heterogeneous video streams has been tackled both in supervised as well as unsupervised manners, with reporting of the latter type being insignificant in comparison to the previous albeit [4,6,27,28,59]. There are different techniques reported to achieve this end, ranging over rough set-based approaches [12,27,47–49,51,54], hidden Markov model approaches [5], approaches based on latent semantic indexing [10,60], adaptive thresholding-based approaches [56], approach using multiple cues [42], generalized sequence trace-based approach [53], semantic object tracking-based approach [25], production model-based approach [26], feature-based hierarchical approach [55], to name a few. In addition to these, there is a recent work due to Fang et al. [17] capable of identifying video shot cuts both of abrupt as well as of a gradual nature.

9.3 Fuzzy Sets: Concepts and Terminologies

This section provides an overview of the basic concepts of fuzzy sets. It also discusses the properties of fuzzy sets and their representation forms. As a sequel to the concept, this section underscores some of the measures for quantitatively assessing the degree of fuzziness of fuzzy sets.

Fuzzy set theory was developed by L.A. Zadeh [57] in 1965 to explain the varied nature of ambiguity often encountered in real-life situations. Fuzzy set theory generalizes the notion of a set in that all the elements in the universe of discourse X belong to every fuzzy set therein, with varying degrees of membership. This membership or containment of an element in a fuzzy set A is decided by a characteristic membership function $\mu_A(x)$. The membership values of the elements lie in [0, 1]. The closer the membership value of an element to unity, the stronger the containment of the element within the fuzzy set. Similarly, a lower membership value implies a weaker containment of the element within the set. A fuzzy set A, characterized by a membership function $\mu_A(x_i)$ and comprising elements $x_i, i = 1, 2, 3, ...n$, is mathematically expressed as [50, 57]

$$A = \sum_i \frac{\mu_A(x_i)}{x_i}, i = 1, 2, 3, ..., n, \quad (9.1)$$

where $\sum_i$ represents a collection of elements.

9.3.1 Properties of a Fuzzy Set

The support $S_A \in [0, 1]$ of a fuzzy set A is defined as the set of all those elements whose membership values are greater than 0. Mathematically, it can be expressed as [50, 57]

$$S_A = \{\sum_{i=1}^{n} \frac{\mu_A(x_i)}{x_i} : x_i \in X \text{ and } \mu_A(x_i) > 0.\} \quad (9.2)$$

The resolution of a fuzzy set A is determined by the α-cut (or α-level set) of the fuzzy set. It is a crisp set A_α containing all the elements of the universal set U that have a membership in A greater than or equal to α, that is,

$$A_\alpha = \{x_i \in U |\ \mu_A(x_i) \geq \alpha\}, \alpha \in [0,1]. \tag{9.3}$$

If $A_\alpha = \{x \in U |\ \mu_A(x) > \alpha\}$, then A_α is referred to as a strong α-cut [50]. The set of all levels $\alpha \in [0,1]$ that represents distinct α-cuts of a given fuzzy set A is called a level set of A, that is,

$$\Lambda_A = \{\alpha |\ \mu_A(x) = \alpha, x \in U.\} \tag{9.4}$$

The maximum of all the membership values in a fuzzy set A is referred to as the height (hgt_A) of the fuzzy set [50]. If hgt_A is equal to 1, then the fuzzy set is referred to as a normal fuzzy set. If hgt_A is less than 1, then it is referred to as a subnormal fuzzy set. A normal fuzzy set is a superset of several nonempty subnormal fuzzy subsets [3]. A subnormal fuzzy subset (A_s) can be converted to its normalized equivalent by means of the normalization operator given by [3]

$$\mathrm{Norm}_{A_s(x)} = \frac{A_s(x)}{hgt_{A_s}}. \tag{9.5}$$

The corresponding denormalization operation is given by

$$DeNorm_{A_s} = hgt_{A_s} Norm_{A_s}. \tag{9.6}$$

In general, for a subnormal fuzzy subset with support, $S_{A_s} \in [L, U]$, the normalization and the denormalization operators are expressed as [3]

$$\mathrm{Norm}_{A_s(x)} = \frac{A_s(x) - L}{U - L} \tag{9.7}$$

and

$$\mathrm{Denorm}_{A_s(x)} = L + (U - L)\mathrm{Norm}_{A_s(x)}. \tag{9.8}$$

9.3.2 Measures of a Fuzzy Set

Several different measures for measuring the degree of fuzziness of a fuzzy set exist [24, 57]. These measures reflect the amount of ambiguity in a fuzzy set.

1. Index of fuzziness: The index of fuzziness $\nu(A)$ of a fuzzy set A having n elements is a distance metric between the set and its nearest ordinary set $\underline{A}$ defined as

$$\mu_{\underline{A}}(x) = \begin{cases} 0 & \text{if } \mu_A(x) \leq 0.5 \\ 1 & \text{if } \mu_A(x) > 0.5. \end{cases} \tag{9.9}$$

   The linear index of fuzziness $\nu_l(A)$ [24] of a fuzzy set A is the Hamming distance version of the index of fuzziness distance metric. It is given by

$$\nu_l(A) = \frac{2}{n}\sum_{i=1}^{n}[\min\{\mu_A(x_i), 1-\mu_A(x_i)\}]. \tag{9.10}$$

Similarly, the linear index of fuzziness for a subnormal fuzzy set A_s, $\nu_l(A_s)$ is defined as [3]

$$\nu_l(A_s) = \frac{2}{n}\sum_{i=1}^{n}[\min\{\mu_{A_s}(x_i) - L, U - \mu_{A_s}(x_i)\}]. \tag{9.11}$$

2. Fuzzy entropy: The entropy of a fuzzy set is also a measure of the fuzziness of the fuzzy set. Ebanks [15] suggested that a fuzzy entropy should satisfy five properties.

   The entropy E_A of a fuzzy set A, characterized by the membership function $\mu_A(x_i)$, is a measure of the degree of fuzziness in the fuzzy set. For a fuzzy set comprising n elements, it is represented based on Shannon's functional form as [37]

$$E_A = \frac{1}{nln2}\sum_{i=1}^{n} -\mu_A(x_i)ln(\mu_A(x_i)) - \{1-\mu_A(x_i)\}ln\{1-\mu_A(x_i)\}. \tag{9.12}$$

   The fuzzy entropy measure reflects the amount of ambiguity that corresponds to the randomness/disorder in an observation.

In 1993, Bhandari and Pal [2] extended the De Luca and Termini's formula by introducing the α-order fuzzy entropy, which used the α-order probability entropy form. A plethora of literature is available on different forms of evolved fuzzy entropy measures. A good survey of the fuzzy entropy for finite universal set can be found in [11, 43]. Interested readers may refer to [16] for details regarding the latest forms of entropy measures proposed in this direction.

9.4 Proposed Methodology

This section illustrates a detailed analysis of the proposed methodology of video shot boundary detection using fuzzy entropy estimation of video content. From the very objective of the chapter, it is evident that we are interested to identify whenever there is a change from one video context to another in the available sequence of image frames extracted from a mixture of a number of video files. If there are k video compositions with $n_1, n_2, \ldots n_k$ representing the respective number of image frames of these video types, then we are dealing with $m = \sum_{i=1}^{k} n_i$ image frames in the entire video spectrum. In this situation, the first n_1 frames are coming from the first video stream, the second n_2 images frames constitute the second video sequence, so on and so forth until the last video sequence comprises the last n_k image frames. Our task is to devise a mechanism that is capable of detecting a changeover in the image context from the last image frame of the previous video sequence to the first image frame of the next video sequence. There have been reports of some previous works in this context [27], along with comparisons among the competitive techniques [4,6,8,22,28,59] as well as figures of merit for their performance characterization [18,21–23]. However, most of these techniques are in need of external supervision for their implementation and unsupervised realization of the problem is inadequate, notwithstanding the reporting of a few techniques coming under this category [19,20]. Among other soft computing tools the rough set theoretic approach [12,47–49,51,54] has been found to be a formidable one [27].

In this chapter, our endeavor is to develop an unsupervised method of automatic detection of boundaries between heterogeneous video streams.

9.4.1 Our Methodology

We compute the entropy of each and every individual image frame in the composite video sequence. Let us represent these entropies by E_1, $E_2 \ldots E_m$. It has been observed that whenever there is a change from one video context to another, there is a sharp change in the corresponding fuzzy entropy measure. In other words, the fuzzy entropy measure appears to be quite sensitive to the change in video context. Based on this observation, we propose a fuzzy entropy spectrum computed in an online manner, which plays the role of maximum allowable limit, both in higher as well as lower senses, so that as and when any of the fuzzy entropy contents $\{E_k, k \geq 1\}$ move out of the spectrum (i.e., exceeds the limit induced), it is identified as an event of changeover from one video context to another. The beauty of this technique lies in the fact that while making such a decision, the fuzzy entropy spectra need not be provided in a precomputed manner. Rather, it is generated in an online fashion and is used concurrently whether or not there is any reporting of fuzzy entropy spike going beyond the generated spectrum. The advantage of the mechanism is twofold:

- Because the spectrum is generated online, the underlying computational overhead is reduced, making it congenial to video applications, and
- The method is absolutely devoid of any external intervention, implying that it works in an unsupervised manner absolutely. Moreover, there is strictly no logical upper limit of the number of image frames arriving/to be processed. Although we have represented the total number of image frames by m above, frames may keep arriving from possibly various video contexts. The capability of our decision process to respond in an online manner ensures that it will be able to detect whenever there is a video

context transition. This is irrespective of the number of frame arrivals and independent of their video origin.

It may be noted that even if the number of image frames is unknown beforehand, the method will work fine.

In order to generate the fuzzy entropy spectrum online, we make sure that the fuzzy entropy spectrum at instant r is determined by the average and standard deviation of fuzzy entropy measures up to instant $(r-1)$. For the sake of online realization of the same, we compute the average and standard deviation recursively as per Equations (9.13) and (9.14).

$$\mu_{n+1} = \frac{n}{n+1}\mu_n + \frac{1}{n+1}E_{n+1}, \tag{9.13}$$

$$\sigma_{n+1}^2 = \frac{n}{n+1}\sigma_n^2 + \frac{n}{(n+1)^2}(\mu_n - E_{n+1})^2 + \frac{n-1}{n+1}E_{n+1}^2, \tag{9.14}$$

where

$$\mu_n = \frac{\sum_{i=1}^{n} E_i}{n} \tag{9.15}$$

and

$$\sigma_n^2 = \frac{\sum_{i=1}^{n} E_i^2}{n} - \mu_n^2. \tag{9.16}$$

The recurrence relation for the average fuzzy entropy can be realized by the following treatment. From Equation (9.15), we can write

$$n\mu_n = \sum_{i=1}^{n} E_i. \tag{9.17}$$

Hence,

$$\sum_{i=1}^{n+1} E_i = (n+1)\mu_{n+1} = n\mu_{n+1} + \mu_{n+1}. \tag{9.18}$$

Again,

$$\sum_{i=1}^{n+1} E_i = \sum_{i=1}^{n} E_i + E_{n+1} = n\mu_n + E_{n+1} \tag{9.19}$$

Therefore, from Equation (9.19),

$$n\mu_n + E_{n+1} = (n+1)\mu_{n+1}. \tag{9.20}$$

So, we can write

$$\mu_{n+1} = \frac{n}{n+1}\mu_n + \frac{1}{n+1}E_{n+1}. \tag{9.21}$$

A similar relation exists for the standard deviation as well. Some computational efforts on Equations (9.15) and (9.16) will produce the recursive relation in the form of Equations (9.13) and (9.14). Now, if the entropy measure E_{n+1} lies within the fuzzy entropy spectrum corresponding to the interval $(\mu_n - \eta \frac{\sigma_n}{\sqrt{n}}, \mu_n + \eta \frac{\sigma_n}{\sqrt{n}})$, then it indicates no change in the video context and the present image frame belongs to the video shot under consideration. Otherwise, in case the entropy measure exceeds the spectrum, a possible change in video context is reflected. Here, η plays the role of a previously supplied regulation parameter. For the present, the authors have treated the value of the parameter as prefixed. However, ideally it should be adaptively determined from the underlying video application. However, the adaptive treatment is not within the purview of the present scope.

9.4.2 Previous Fuzzy Logic-Based Work due to Das et al.

We discuss the findings of Das et al. in [9] in the present section. This is because we, in the course of substantiating our present contribution, have earmarked the work in [9] as a recently published reference work. Naturally, a brief description of their work will make the present treatise complete and comprehensive. In their article, they converted the gray values retrieved from a gray image by a linear interpolation technique. If g_{min} and g_{max} represent the minimum and maximum intensities respectively of the input image, then the arbitrary gray value $g \in [g_{min}, g_{max}]$ is converted into its corresponding fuzzy membership value $\bar{g}$ as

$$\bar{g} = \frac{g - g_{min}}{g_{max} - g_{min}}. \tag{9.22}$$

Based on the above equation, the membership grade of $\bar{g}$ at a crisp real value x is given by

$$\mu_{\bar{g}x} = \max(0, 1 - \frac{| x - g |}{\alpha}), \tag{9.23}$$

where α is a positive real constant that is responsible for regulating the steepness of the membership characteristic. Based on the above definition and the one offered by Jawahar et al. in [34], a fuzzy histogram is defined as

$$H = \{H_s \mid s \in 0, \ldots L - 1\}, \tag{9.24}$$

L representing the number of gray levels under consideration. Now, H_s is the frequency of class s, $s \in 0, \ldots L - 1$ and is computed by

$$H_s = \sum_{i=0}^{M-1} \sum_{j=0}^{N-1} \mu_{\bar{g}I(i,j)}. \tag{9.25}$$

It is not necessary to say that H induces a fuzzy histogram profile corresponding to the input image I of size $M \times N$. Here, $I(i, j)$ represents the gray intensity value corresponding to pixel (i, j). In case of color extension in terms of the Red-Green-Blue (RGB) model, they have introduced the color histogram difference between two frames i and j as follows:

$$HD_{i,j} = \frac{1}{3C}[\sum_{n=0}^{L-1} \min(H^i_{r(n)}, H^j_{r(n)}) + \sum_{n=0}^{L-1} \min(H^i_{g(n)}, H^j_{g(n)}) + \sum_{n=0}^{L-1} \min(H^i_{b(n)}, H^j_{b(n)})], \tag{9.26}$$

where C is given by

$$C = \max\{\sum_{n=0}^{L-1} \min(H^i_{r(n)}, H^j_{r(n)}), \sum_{n=0}^{L-1} \min(H^i_{g(n)}, H^j_{g(n)}), \sum_{n=0}^{L-1} \min(H^i_{b(n)}, H^j_{b(n)})\}. \tag{9.27}$$

Here, $r(n)$, $g(n)$, and $b(n)$ are the bin number n corresponding to the red, green, and blue components of the color image, respectively. $H^i_{r(n)}$ is the frequency corresponding to the bin number n of the red component of frame number i. Similarly defined are $H^i_{g(n)}$ and $H^i_{b(n)}$. On the basis of this, the concept of a color fuzzy co-occurrence matrix (FCM) has been proposed, which is

$$FCM(i, j) = \frac{1}{3K}[\phi_R + \phi_G + \phi_B], \tag{9.28}$$

where

$$\phi_R = \sum_{m=0}^{L-1} \sum_{n=0}^{L-1} \min(f^i_{r(m,n)}, f^j_{r(m,n)}),$$

$$\phi_G = \sum_{m=0}^{L-1} \sum_{n=0}^{L-1} \min(f^i_{g(m,n)}, f^j_{g(m,n)}),$$

$$\phi_B = \sum_{m=0}^{L-1} \sum_{n=0}^{L-1} \min(f^i_{b(m,n)}, f^j_{b(m,n)}),$$

and K is defined as

$$FCM(i, j) = \max\{\phi_R, \phi_G, \phi_B\}. \tag{9.29}$$

Here, $f^i_{r(m,n)}$ is the frequency of occurrence of the gray value m followed by that of the gray value n for frame index i corresponding to red component of the color gamut. Similarly defined are the terms $f^i_{g(m,n)}$ and $f^i_{b(m,n)}$. What is really interesting is the use of interval type-2 fuzzy sets for construction of the corresponding fuzzy rule base. To achieve this, they proposed the following two metrics:

$$D_M(i) = \min\{HD_{i+j,i+1+j} \mid 0 \leq j \leq L-1\}, \tag{9.30}$$

$$D_{SM}(i) = \min\{\{FCM(i+j, i+1+j) \mid 0 \leq j \leq L-1\} - D_M(i)\}. \tag{9.31}$$

Accordingly, the Relative Change Index (*RCI*) has been defined as

$$RCI(i, i+1) = \frac{| D_M(i) - D_{SM}(i) |}{D_{SM}(i)}. \tag{9.32}$$

9.5 Results

We have conducted our experiment on an admixture of four different true color video sequences with each video sequence comprising five image frames as depicted sequentially in black and white in Figures 9.1 [32], 9.2, 9.3, and 9.4 [31]. The corresponding graphical representation of the fuzzy entropy

FIGURE 9.1 Sequence of image frames of first video shot.

FIGURE 9.2 Sequence of image frames of second video shot.

FIGURE 9.3 Sequence of image frames of third video shot.

FIGURE 9.4 Sequence of image frames of fourth video shot.

values across all image frames as well as the fuzzy entropy spectrum in the form of upper control limit (UCL) and lower control limit (LCL) are provided in Figure 9.5. It is observed from the graph that:

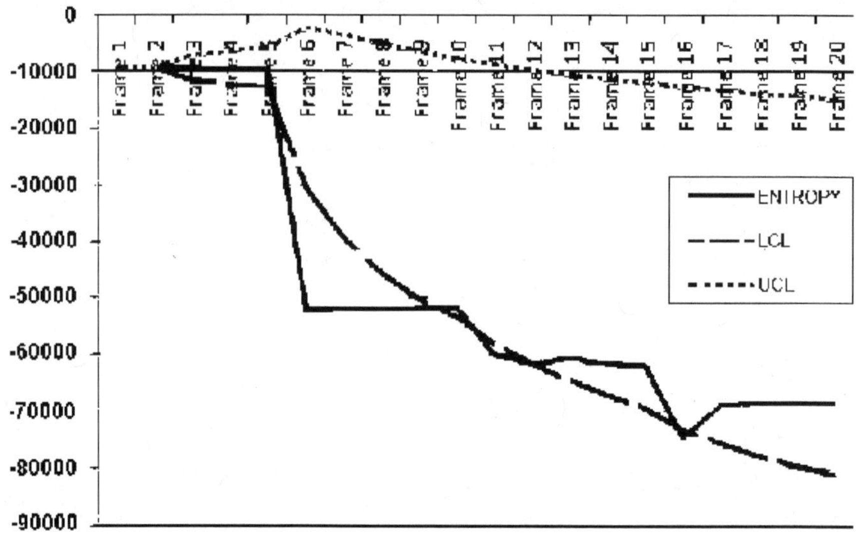

FIGURE 9.5 Entropy and fuzzy entropy spectrum.

- There is sharp change in the value of the fuzzy entropy after the fifth, tenth, fifteenth image frames in confirmation with the change in the video context around those frame positions.

- Every such changeover in the fuzzy entropy measure is being convincingly captured by the corresponding control limit of the fuzzy entropy spectrum.

We have taken the value of η as 0.82 for better visualization.

In order to justify the performance of our proposed technique, we considered the work due to Das et al. [9]. It has been found that their technique is capable of detecting shot boundaries/hard cut whenever there is a change in video context. Our technique has been found to be able to identify change in any context in a video sequence as desired, as evident in Figure 9.5.

It may be noted that we have compared the proposed technique with Das et al.'s work [9]. For this purpose, we have implemented their work and applied it on all the test video slots under consideration. Some of the representative results ($HD_{i,j}, FCM(i,j), RCI(i,i+1)$ between different frames i,j as given in Equations (9.26), (9.28), (9.32)) obtained after the application of their work on an admixture of different test video shots with the first three frames taken from the first video shot, are shown in Table 9.1 for the sake of self-sufficiency of the treatment. The **boldface** values in Table 9.1 indicate the detection of cuts by the method. Because the FCM values do not conform to the designed rule base for detection of cuts, the HD values can be used for the detection of cuts from an admixture of video sequences of video shots.

However, the real advantage of our proposed technique is computational efficiency in comparison to the other method referred to above. Whereas the

TABLE 9.1 Computed $HD_{i,j}, FCM(i,j), RCI(i,i+1)$ values for video sequences taken from first and third video shots using Das et al.'s method.

HD	0.93543	0.92947	0.96001	0.57476	0.68140	0.71399	0.67476	0.93568
RCI_{HD}	0.14819	0.14819	0.14819	**0**	0.00974	0.05495	**0**	0.13249
FCM	0.86058	0.84347	0.79285	0.46107	0.56204	0.59874	0.65987	0.70999
RCI_{FCM}	0.27649	0.27649	0.27649	0.27649	0.28996	0.28996	0.28996	0.28996

computational complexity of the method proposed by Das et al. is on the order of $O(L^2M^2N^2)$, L being the number of the histogram division and $M \times N$ being the image size, the computational complexity of our proposed technique is $O(MN)$. Naturally a gain in computational efficiency of such order proves to be of enormous importance in the context of a video environment. Moreover, our technique is capable of functioning in an online environment where computation is functionally independent of the inclusion of a newer image scene as data in nature.

9.6 Discussions and Conclusion

In the present chapter, we propose a very simple, yet very effective technique to identify the boundaries present in video sequence consisting of different video contexts. Each video context comprises a number of image frames. A fuzzy entropy measure of the image scenes is considered that has been shown to play a very effective role in determining the change in video context. The interesting part of the present work is that it is based on an online computation of the fuzzy entropy spectrum. This spectrum is utilized for the purpose of deciding as to whether a reported entropy measure of an image scene undergoes a drastic change or not. In case a change in the value of image entropy is reported surpassing the earmarked spectrum, it is inferred as an appearance of a new video context. The biggest advantage of this method is that its execution is devoid of any external intervention and hence it operates in a perfectly unsupervised manner.

References

1. S. S. Beauchemin and J. L. Barron. The computation of optical flow. *ACM Comput. Surv.*, 27(3):433–467, 1995.
2. D. Bhandari and N.R. Pal. Some new information measure of fuzzy sets. *Inform. Sci.*, 67:209–228, 1993.
3. S. Bhattacharyya, P. Dutta, and U. Maulik. Binary object extraction using bidirectional self-organizing neural network (BDSONN) architecture with fuzzy context sensitive thresholding. *Pattern Analysis and Applications*, pages 345–360, 2007.
4. J. Boreczky and L. Rowe. Comparison of video shot boundary detection tech-

niques. In *Proceedings of SPIE Conference on Storage and Retrieval for Video Databases IV*, San Jose, CA, 1995.

5. J. Boreczky and L. D. Wilcox. A Hidden Markov Model framework for video segmentation using audio and image features. In *International Conference on Acoustics, Speech, and Signal Processing*, pages 3741–3744, Seattle, WA, 1998.
6. J. S. Boreczky and L. A. Rowe. Comparison of video shot boundary detection techniques. *Journal of Electronic Imaging*, 5(2):122–128, 1994.
7. J.S. Boreczky and L.A. Rowe. Comparison of video shot boundary detection techniques. In *Proceedings of SPIE Conf. Storage & Retrieval for Image & Video Databases*, 2670:170–179, 1996.
8. A. Dailianas, R. B. Allen, and P. England. Comparison of automatic video segmentation algorithms. In *Proceedings of SPIE Photonics West*, Philadelphia, PA, 1995.
9. S. Das, S. Sural, and A. K. Majumdar. Detection of hard cuts and gradual transitions from video using fuzzy logic. *International Journal of Artificial Intelligence and Soft Computing*, 1(1):77–98, 2008.
10. S. Deerwester, S.T. Dumais, G.W. Furnas, T.K. Landauer, and R. Harshman. Indexing by latent semantic analysis. *Journal of the American Society for Information Science*, 41(6):391–407, 1990.
11. D. Dubas and H. Prade, Editors. N.R. Pal and J.C. Bezdek, *"Quantifying different facets of fuzzy uncertainty"*. Kluwer Academic Publishers, London, United Kingdom, 2000.
12. D. Dubois and H. Prade. Rough fuzzy sets and fuzzy rough sets. *International Journal of General Systems*, 17:191–209, 1990.
13. J. H. Duncan and T. Chou. Temporal edges: The detection of motion and the computation of optical flow. In *Proceedings of 2nd IEEE International Conference on Computer Vision (ICCV 88)*, 1988.
14. J. H. Duncan and T. Chou. On the detection of motion and the computation of optical flow. *IEEE Transactions on Pattern Analysis and Machine Intelligence*, 14(3):346–352, 1992.
15. B. R. Ebanks. On measures of fuzziness and their representations. *Journal of Math. Anal. Appl.*, 94:24–37, 1983.
16. J.-L. Fan and Y.-L. Ma. Some new fuzzy entropy formulas. *Fuzzy Sets and Systems*, 128:277–284, 2002.
17. H. Fang, J. Jiang, and Y. Feng. A fuzzy logic approach for detection of video shot boundaries. *Pattern Recognition*, 39(11):2092–2100, 2006.
18. R.M. Ford, C. Robson, D. Temple, and M. Gerlach. Metrics for shot boundary detection in digital video sequences. *Multimedia Syst.*, pages 37–46, 2000.
19. X. Gao and X. Tang. Automatic parsing of news video based on cluster analysis. In *Proceedings of 2000 Asia Pacific Conference on Multimedia Technology and Applications*, pages 17–19, Kaohsiung, Taiwai, China, 2000.
20. X. Gao and X. Tang. Unsupervised model-free news video segmentation. *IEEE Transactions on Circuits and Systems for Video Technology*,

12(9):765–776, 2002.

21. U. Gargi and R. Kasturi. An evaluation of color histogram based methods in video indexing. In *Proceedings of International Workshop on Image Databases and Multimedia Search*, pages 75–82, Amsterdam, The Netherlands, 1996.
22. U. Gargi, R. Kasturi, and S. H. Strayer. Performance characterization of video-shot-change detection methods. *IEEE Transactions on Circuits and Systems for Video Technology*, 10(1):1–13, 2000.
23. U. Gargi, S. Oswald, D. Kosiba, S. Devadiga, and R. Kasturi. Evaluation of video sequence indexing and hierarchical video indexing. In *Proceedings of SPIE Conference on Storage and Retrieval in Image and Video Databases*, pages 1522–1530, 1995.
24. A. Ghosh, N. R. Pal, and S. K. Pal. Self-organization for object extraction using a multilayer neural network and fuzziness measures. *IEEE Transactions on Fuzzy Systems*, (1):54–68, 1993.
25. B. Gunsel, A. M. Ferman, and A. M. Tekalp. Temporal video segmentation using unsupervised clustering and semantic object tracking. *Journal of Electronic Imaging*, 7(3):592–604, 1998.
26. A. Hampapur, R. Jain, and T. E. Weymouth. Production model based digital video segmentation. *Multimedia Tools and Applications*, 1(1):9–46, 1995.
27. B. Han, X. Gao, and H. Ji. A shot boundary detection method for news video based on rough-fuzzy sets. *International Journal of Information Technology*, 11(7):101–111, 2005.
28. B. Han, X.-B. Gao, and H.-B. Ji. An efficient algorithm of gradual transition for shot boundary segmentation. *Proceedings of SPIE Conference on MIPPR*, 5286:956–961, 2003.
29. D. J. Heeger. Optical flow using spatiotemporal filters. *International Journal of Computer Vision*, 1:279–302, 1988.
30. B. K. P. Horn and B. G. Schunck. Determining optical flow. *Artif. Intell.*, 17:185–204, 1981.
31. http://www.archive.org/details/FlyingtheP 61seriesAirplane.
32. http://www.fusionpicture.com.
33. L. Jacobson and H. Wechsler. Derivation of optical flow using a spatiotemporal frequency approach. *Comput. Vision Graph. Image Process.*, 38:29–65, 1987.
34. C. V. Jawahar and A. K. Ray. Fuzzy statistics of digital images. *IEEE Signal Processing Letters*, 2(8):225–227, 1996.
35. O. Kucuktunk, U. Gudukbay, and O. Ulusoy. Fuzzy color histogram-based video segmentation. *Computer Vision and Image Understanding*, 114(1):125–134, 2010.
36. A. De Luca and S. Termini. Algebric properties of fuzzy sets. *Journal of Mathematical Analysis and Applications*, 40:373–386., 1972.
37. A. De Luca and S. Termini. A definition of a nonprobabilistic entropy in the setting of fuzzy set theory. *Information and Control*, 20:301–312, 1972.

38. A. De Luca and S. Termini. Entropy of l-fuzzy sets. *Information and Control*, 24:55–73, 1974.
39. A. De Luca and S. Termini. Entropy measures in fuzzy set theory. *Systems and Control Encyclopedia*, pages 1467–1473, 1988.
40. A. De Luca and S. Termini. Vagueness in scientific theories. *Systems and Control Encyclopedia*, pages 4993–4996, 1988.
41. M. M. Mokji and S.A.R. Abu Bakar. Adaptive thresholding based on co-occurrence matrix edge information. *Journal of Computers*, 2(8):44–52, 2007.
42. M. R. Naphade, R. Mehrotra, A. M. Ferman, J. Warnick, T. S. Huang, and A. M. Tekalp. A high-performance shot boundary detection algorithm using multiple cues. In *Proceedings of IEEE International Conference on Image Processing*, 2:884–887, 1998.
43. N.R. Pal and J.C. Bezdek. Measuring fuzzy uncertainty. *IEEE Trans. Fuzzy Systems*, 2:107–118, 1994.
44. S. K. Pal. A note on the quantitative measure of image enhancement through fuzziness. *IEEE Trans. Pattern Anal. Machine Intell.*, 4(2):204–208, 1982.
45. S. K. Pal, R.A. King, and A.A. Hashim. Automatic gray level thresholding through index of fuzziness and entropy. *Pattern Recognition Letters*, (1):141–146, 1983.
46. S. K. Pal and A. B. Leigh. Motion frame analysis and scene abstraction: Discrimination ability of fuzziness measures. *Journal of Intelligent & Fuzzy Syatems*, (3):247–256, 1995.
47. Z. Pawlak. Rough set. *International Journal of Computer and Information Science*, 11(5):341–356, 1982.
48. Z. Pawlak. Vagueness and uncertainty: A rough set perspective. *ICS Research Reports, Warsaw University of Technology*, 19, 1994.
49. Z. Pawlak, J. Grzymala-Busse, R. Slowinski, and W. Ziarko. Rough sets. *Commun. ACM*, 38(11):89–95, 1995.
50. T. J. Ross. *Fuzzy Logic with Engineering Applications*. John Wiley, 2004.
51. M. Sarkar and B. Yegnanarayana. Rough-fuzzy membership functions. In *Proceedings of IEEE World Congress on Computational Intelligence and Fuzzy Systems*, 1:796–801, 1998.
52. D. Sen and S. K. Pal. Generalized rough sets, entropy and image ambiguity measures. *IEEE Trans. Syst, Man and Cyberns. Part B*, 39:117–128, 2009.
53. C. Tas, Kiran, and E. J. Delp. Video scene change detection using the generalized sequence trace. In *Proceedings of IEEE International Conference on Image Processing*, pages 2961–2964, 1998.
54. G.-Y. Wang, J. Zhao, J.-J. An, and Y. Wu. Theoretical study on attribute reduction of rough set theory: Comparison of algebra and information views. In *Proceedings of ICCI*, pages 148–155, 2004.
55. H. Yu, G. Bozdagi, and S. Harrington. Feature-based hierarchical video segmentation. In *International Conference on Image Processing*, pages

498–501, 1997.

56. Y. Yusoff, W. Christmas, and J. Kittler. Video shot cut detection using adaptive thresholding. In *The Eleventh British Machine Vision Conference*, pages 362–371, 2000.
57. L.A. Zadeh. Fuzzy sets. *Information and Control*, 8:338–353, 1965.
58. S.D. Zenzo, L. Cinque, and S. Levialdi. Image thresholding using fuzzy entropies. *IEEE Transactions on Systems, Man and Cybernetics.Part B*, 28:15–23, 1998.
59. H. J. Zhang, A Kankanhalli, and S. W. Smoliar. Automatic partitioning of full motion video. *Multimedia Systems*, 1(1):10–28, 1993.
60. R. Zhao and W.I. Grosky. A novel video shot detection technique using color anglogram and latent semantic indexing. In *ICDCS Workshops*, pages 550–555, 2003.

10

Multi-Robot and Multi-Camera Patrolling

Christopher King
University of Nevada, Reno, Nevada, USA

Maria Valera
Kingston University, London, United Kingdom

Raphael Grech
Kingston University, London, United Kingdom

Robert Mullen
Kingston University, London, United Kingdom

Paolo Remagnino
Kingston University, London, United Kingdom

Luca Iocchi
University of Rome "La Sapienza", Rome, Italy

Luca Marchetti
University of Rome "La Sapienza", Rome, Italy

Daniele Nardi
University of Rome "La Sapienza", Rome, Italy

Dorothy Monekosso
University of Ulster, Newtownabbey, United Kingdom

Mircea Nicolescu
University of Nevada, Reno, Nevada, USA

In this chapter we present a multi-camera platform to monitor the environment that is integrated to a multi-robot platform to enhance situation awareness. The multi-camera platform consists of two distinct stereo camera systems and use different vision approaches that will be described in detail. One of the stereo vision systems is applied to reason on object manipulation events, while the other system is used to detect an event such as a person leaving a bag in a corridor. The results from either of these two systems are encapsulated in a string message and sent via wireless network to the multi-robot system that, on alarm, will dispatch a robot to monitor the region of interest. Our ultimate goal is that of maximizing the quality of information gathered from a given area, thus implementing a heterogeneous mobile and reconfigurable multi-camera video-surveillance system.

10.1 Introduction

The problem of detecting and responding to threats through surveillance techniques is particularly well suited to a robotic solution comprised of a team of multiple robots. For large environments, the distributed nature of the multi-robot team provides robustness and increased performance of the surveillance system. Here we develop and test an integrated multi-robot system as a mobile, reconfigurable, multi-camera video-surveillance system.

The main stages of the pipeline in a video-surveillance system are moving object detection and recognition, tracking, and activity recognition. One of the most critical and challenging components of a semi-automated video surveillanceis the low-level detection and tracking phase. Data is frequently corrupted by the camera's sensor (e.g., CCD noise, poor resolution, motion blur, etc.), the environment (e.g., illumination irregularities, camera movement, shadows, reflections, etc.), and the objects of interest (e.g., transformation, deformation, occlusion, etc.). Even small detection errors can significantly alter the performance of routines further down the pipeline, and subsequent routines are usually unable to correct errors without using cumbersome, ad-hoc techniques. Compounding this challenge, low-level functions must process huge amounts of data, in real-time, over extended periods. To adapt to the challenges of building accurate detection and tracking systems, researchers are usually forced to simplify the problem. It is common to introduce certain assumptions or constraints that may include fixing the camera [30], constraining the background [29], constraining object movement, applying prior knowledge regarding object-appearance or location [28], assuming smooth object-motion, etc. Relaxing any of these constraints often requires the system to be highly specified for the given task. Active contours may be used to track non-rigid objects against homogeneous backgrounds [3], primitive geometric shapes for certain simple rigid objects [10], and articulated shapes for humans in high-resolution images [22]. There has been a push toward identifying a set of general features that can be used in a larger variety of conditions. Successful algorithms include the Maximally Stable Extremal Region (MSER), Harris-Affine, Hessian-Affine, and Salient Regions [21]. Despite their recent successes, each algorithm has its own weaknesses, and achieving flexibility still requires the combination of multiple techniques [26]. Because most of these approaches are either not real-time, or are barely real-time, running several in unison is usually not feasible on a standard processor.

Recently, to adapt to the challenges of building accurate detection and tracking systems, work has also been carried out using per-pixel depth information provided by stereo imagery devices to detect and track multiple objects [4, 9, 11, 16, 24, 32]. What is mainly thanks to improved performance in software computing for depth imagery [1, 2, 25] and also more affordable stereo imagery hardware [1, 2]. In [4, 9, 11] the detected and tracked features are directly applied to the depth information itself, while in [16, 24] the detection and tracking is done after the analysis of depth information is integrated with the color information.

In this chapter we mainly focus on two approaches to develop two different video surveillance systems. The first approach consists of applying a real-time, color-based, MSER detection and tracking algorithm. In the second method, a multi-object tracking system is presented based on a ground plane projection of real-time 3D data coming from stereo imagery, giving distinct separation of occluded and closely interacting objects. The rest of the chapter is structured as follows: Section 10.2 presents the architecture of the whole integrated system. In Section 10.3 the pipeline of processes of the first camera system is described and in Section 10.4 the second camera system is presented. In Section 10.5 the high-level process of the outputs from previous pipeline processes is provided. In Section 10.6, the results from the prototype system are provided to finish in Section 10.7 with the conclusions of this work.

10.2 System Architecture

We considered a highly heterogeneous system, where robots and cameras interoperate. These requirements make the problem significantly different from previous work. Figure 10.1 illustrates the architecture of the system. We also considered different events and different sensors and we will therefore consider different sensor models for each kind of event. We focused on the dynamic evolution of the monitoring problem, where at each time a subset of the agents will be in response mode, while the rest of them will be in patrolling mode. Therefore, the main objectives of the developed system are

1. Develop environment monitoring techniques through behavior analysis based on stereo cameras,
2. Develop distributed multi-robot coverage techniques for security and surveillance,

3. Validate our solution by constructing a technological demonstrator showing the capabilities of a multi-robot system to effectively self-deploy itself in the environment and monitor it.

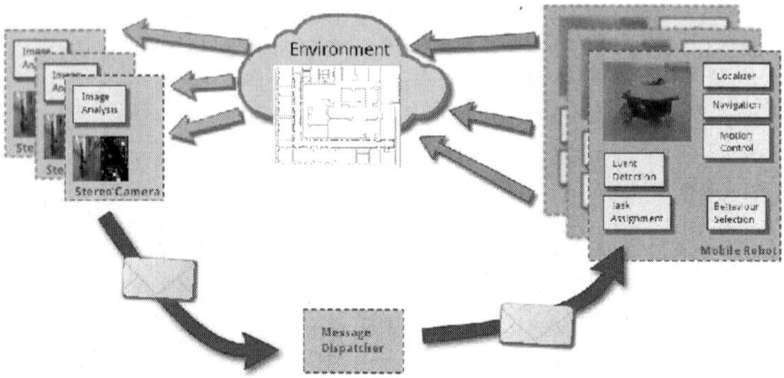

FIGURE 10.1 Block diagram of the proposed architecture.

10.2.1 Multi-Robot Monitoring Platform

As already mentioned, the problem of detecting and responding to threats through surveillance techniques is particularly well suited to a multi-robot platform solution comprised of a team of robots. Although this chapter does not focus on the description of these type of platforms, our approach has been concentrating on extending the work done in multi-robot patrolling, adding the capability for the robots to respond to events detected by visual and other sensors in a coordinated way.

Two issues are considered and solved through the project; developing a general algorithm for event-driven distributed monitoring based on our previous work has solved these problems. We already developed and successfully implemented new dynamic distributed task assignment algorithms for teams of mobile robots: applied to robotic soccer [17] and for foraging-like tasks [12]. More specifically, in [17] we proposed a greedy algorithm to effectively solve the multi-agent dynamic and distributed task assignment problem, which is very effective in situations where the different tasks to be achieved have different priorities. In [12] we also proposed a distributed algorithm for dynamic task assignment based on token passing that is applicable when tasks are not known a priori, but are discovered during the mission. The problem considered here requires both finding an optimal allocation of tasks among the robots and taking into account tasks that are discovered at run-time. Therefore it is necessary to integrate the two approaches. As a result, we do not only specialize these solutions to the multi-robot surveillance and monitoring task, but also study and develop extensions to these techniques in order to improve the optimality of the solutions and the adaptivity to an open team of agents, taking into account the physical constraints of the environment and of the task.

10.2.2 Multi-Camera Platform

The multi-camera platform consists of two stereo cameras. In one of the cameras, a novel, real-time, color-based, MSER detection and tracking algorithm is implemented. The algorithm synergistically combines MSER-evolution with image-segmentation to produce maximally stable segmentation. Our MSER algorithm clusters pixels into a hierarchy of detected regions using an efficient line-constrained evolution process. Resulting regions are used to seed a second clustering process to achieve image-segmentation.

The resulting region-set maintains desirable properties from each process and offers several unique advantages, including fast operation, dense coverage, descriptive features, temporal stability, and low-level tracking. Regions that are not automatically tracked during segmentation can be tracked at a higher-level using MSER and line-features. We supplement low-level tracking with an algorithm that matches features using a multi-phased, kd-search algorithm. Regions are modeled using transformation-invariant features that allow identification to be achieved using a constant-time hash-table. In the other stereo camera, a multi-object tracking system is implemented, based on a ground plane projection of real-time 3D data coming from stereo imagery, giving distinct separation of occluded and closely interacting objects. This approach consists of tracking, using Kalman filters [5], fixed templates that are created combining the height, and the statistical pixel occupancy of the objects in the scene. These objects are extracted from the background using a Gaussian Mixture Model combining luminance and chroma signals (YUV-color space) and depth information obtained from the stereo devices used in this work. The mixture model is adapted over time and is used to create a background model that is also upgraded using an adaptive learning rate parameter according to the scene activity level on a per-pixel basis. The results presented in Figures 10.9 and 10.10 illustrate the validity of both approaches.

10.3 Maximally Stable Segmentation and Tracking for Real-Time Automated Surveillance

The feature detection and tracking algorithm proposed in this section was specifically designed to satisfy the existing need for a system that can robustly track multiple deformable objects, in a variety of settings, in real-time (15 fps), on a modest processor (4 GHz). The algorithm can be used on both stationary and moving cameras, and provides seamless transitions between each. For increased flexibility, the algorithm tracks regions using additional features. These include color-blob features, which are typically more reliable for tracking unstructured or deformable objects through significant transformation. It also includes line-corner features, which offer better localization and are less affected by partial object-occlusion. Features are detected in a way that optimizes performance and feature stability (Section 10.3.1). Features are tracked using an optimized, multi-phased, kd-tree-based approach (Section 10.3.2). Discriminating between foreground and background regions is achieved using a unique background model consisting of high-level features (Section 10.3.3). Modeling and identification of object regions are achieved using a fast transformation-invariant modelling algorithm, and a constant-time hash-table-based search (Section 10.3.4).

10.3.1 Region Detection

The primary function of the region-detection phase is to massively reduce the amount of input data, while simultaneously preserving useful features. This is usually the most critical and error-prone step of processing. Even a modest 320×240 image contains 76,800 pixels, each of which can present 16,777,216 different values. To reduce unimportant data, detection algorithms typically search an input image for a set of patterns that are both stable and unique. Stability ensures that the same feature will be detected in future frames, while uniqueness ensures that a tracker can distinguish between the features. Mikolajczyk provides a comparison of the most promising feature-detection techniques [21]. Among those tested, the MSER detector was found to be superior in all scene types and for every type of transformation. Additionally, the MSER detector operated appreciably faster than the competing algorithms, processing 800×640 pixel images at sub-second frame rates using a 4.2-GHz processor.

The MSER algorithm was originally developed by Matas et al. [20] to identify stable areas of light-on-dark, or dark-on-light, in grayscale images. The algorithm is implemented by applying a series of binary thresholds to an image. As the threshold value iterates, areas of connected pixels grow and merge, until every pixel in the image has become a single region. During this process, the regions are monitored, and those that display a relatively stable size through a wide range of thresholds are recorded. This process produces a hierarchical tree of nested MSERs. Unlike other detection algorithms, the MSER identifies comparatively few regions of interest. This is beneficial in reducing computational costs of subsequent phases, but can be problematic when used for general object tracking because there is no guarantee that an object of interest will be represented by a MSER.

To increase the number of detections and improve coverage, Forssen [13] redesigned the algorithm to incorporate color information. Instead of grouping pixels based on a global threshold, Forssen incrementally clustered pixels using the local color gradient.Forssen's method is based on the extension to color but looking at successive time-steps of an agglomerative clustering of image pixels. Therefore, this process identifies regions of similar-colored pixels that are surrounded by dissimilar pixels. Although Forssen observed an increase in detections and an improvement in results, his algorithms had some limitations. First, the algorithm deteriorates quickly when confronted with noise or non-edge gradients (occurring on curved surfaces or lightly-textured objects). This deterioration occurs because, at the pixel level, these gradients are nearly indistinguishable from object boundaries. To limit this effect, Forssen applied multiple types of smoothing to his data. This improved stability of some regions, but at the expense of others. The second limitation resulted from Forssen's comparison of adjacent pixels to determine merge criteria. In most video feeds, the spatial correlation of color information is too high to offer reliable contrast, and MSER stability is greatly compromised. Forssen's response was to normalize edge weights in a way that ensured region growth occurred evenly across the maximum threshold iteration interval. Although this reduced missed detections, it greatly increased the extent that regions were detected multiple times at slightly different scales. Multiple detections require additional post-processing culling operations, and when combined with the natural inconsistencies of MSER detection, make reliable tracking between frames almost impossible.

Our approach takes advantage of the increased detection offered by Forssen's color-based approach, while greatly reducing the extent of compromise. Our algorithm offers the following improvements over Forssen's approach:

1. Region growth is constrained using detected lines. This improves segmentation results on objects with high-curvature gradients.

2. Our MSER evolution process merges three-pixel units, instead of two-pixel units. This reduces computation costs and allows the gradient to be measured with greater precision.

3. Our algorithm returns either a nested set of regions (traditional MSER-hierarchy formation), or a non-nested, non-overlapping set of regions (typical to image segmentation). Using non-nested regions significantly improves tracking speed and accuracy.

4. Regions in the flat MSER representation are completely filled in with pixels (every pixel in the image is assigned to exactly one region). This produces attractive segmentation and more accurate tracking.

5. Regions are constructed using both spatial and temporal information. This increases stability and speed of operation.

6. Region tracking is partially achieved at the lowest level of MSER formation. This reduces the number of regions that must be tracked in subsequent phases of the algorithm.

7. The Canny lines used in segmentation are available for other functions, such as tracking or structure analysis.

8. The MSER segmentation portion of our algorithm uses only one threshold, "MIN-SIZE" , which constrains minimum region size and MSER stability. This is an improvement over the traditional color-based MSER algorithm, which requires users to set separate thresholds for minimum size, MSER-stability, nested-region overlap, and others.

Our MSER algorithm is a multi-phase process involving Line Detection (Section 10.3.1), MSER-Tree Construction(Section 10.3.1), Region Expansion,(Section 10.3.1) and Region Feed-Forward (Section 10.3.1).

Line detection

The traditional color-based MSER algorithm is largely limited by its strict dependence on the color gradient. Theoretically, even if two regions have high gradient measurements spanning all but one pixel of their shared border, that one-pixel break will cause the regions to be detected as one. This characteristic is particularly limiting when the algorithm is applied to real-world videos because noise, movement-blur, shadows, reflections, etc. can all degrade the gradient. The Canny is much more effective in identifying a continuous border between objects because it considers a larger section of the gradient. If a low-gradient gap interrupts a high-gradient border, the gap is labeled as part of the border.

The Canny is also superior to the MSER in its ability to ignore gradients caused by curvature. For example, consider an image containing a non-textured background and a similarly colored, curved object (e.g., a hand). The MSER would form a region corresponding to the table, but before the object could form its own stable cluster, its pixels would be stripped away by the table region. In contrast, the Canny would likely produce its strongest response along the table-object border. The resulting outline would isolate pixels within the object and allow them to cluster independently of the table.

Our system processes each frame with the Canny algorithm. Canny edges are converted to line segments, and the pixels corresponding to each line segment are used to constrain MSER growth. Simply speaking, MSER evolution operates as usual, but is not permitted to cross any Canny lines. An example of detected lines is shown in Figure 10.3 (right). Detected lines are displayed in green.

MSER tree construction

Our MSER evolution algorithm uses the same basic principle as Forssen's approach [13]. For every current pixel p_c in the image, the color gradient is measured against adjacent pixels where p_{c-1} refers to the pixel on the left hand side of the current pixel p_c. Similarly, p_{c+1} is the adjacent pixel on the right. The outcome is then stored as horizontal (t_h) or vertical (t_v) texture elements using the following formula:

$$t_h = \sqrt{512 \times \sum_{c=\{r,g,b\}} \frac{(p_c - p_{c-1})^2}{(p_c + p_{c-1})} + \frac{(p_c - p_{c+1})^2}{(p_c + p_{c+1})}}.$$

Texture elements (ranging from 0 to 255) are sorted using a constant-time counting-sort algorithm. They are then processed in order, starting with 0-valued texture elements. For every processed element, the corresponding pixel is merged with its vertical or horizontal neighbors (depending on the direction of the element). If any of the three pixels belong to an existing region, the regions are merged. After all texture elements of a particular value (e.g. $0, 1, 2...255$) are processed, the rate-of-growth for all existing regions is measured for that iteration. As long as a region's growth consistently accelerates, or declines, it is left to evolve. If the rate of growth changes from decline (or stable) to acceleration (beyond a MIN-SIZE change), the state of the region before accelerated growth is stored as an MSER. The algorithm continues until all texture elements have been processed. At the end of the growth process, the set of all MSER regions will form a hierarchical tree. The tree-root contains the MSER node that comprises every pixel in the image, with incrementally smaller nested sub-regions occurring at every tree branch. The leaves of the tree contain the first-formed and smallest groups of pixels. To reduce memory and processing demand, the MIN-SIZE threshold is applied to these regions. Our implementation uses a MIN-SIZE of 24 pixels. Figure 10.2 shows three stages of the clustering process.

FIGURE 10.2 Pixel clustering during MSER formation. Clustered pixels are colored using the region's average color. Non-assigned pixels are shown in dark gray. Results represent clusters after iterations 2, 5, and 35 (left to right). (See color insert.)

Region expansion

The traditional MSER approach produces, as output, a hierarchical tree of nested nodes. Although this is desirable for certain applications (where over-detection is beneficial), we find that it does not provide any significant advantages and makes other tasks unnecessarily complicated. Traditional MSER approaches apply various ad-hoc devices to suppress the formation of nested regions, or to cull the regions once they occur. We choose to extract a segmented image from the MSER hierarchy instead of eliminating the problem by using a dual-pass MSER evolution process. Our algorithm can enforce that each pixel is contained in exactly one region instead of each image pixel belonging to zero or more different regions. Both MSER and segmentation representations provide certain unique advantages and our algorithm allows users to pick the representation that best suits their needs.

The first pass of our dual-pass algorithm was described in the previous section. This produces the traditional hierarchy of nested MSER regions, with tree leaves representing initial pixel clustering. These leaves are sparsely distributed within the image and are both nonoverlapping and non nested. Using a merging process similar to the one used in the first pass, we iteratively add pixels to the leaves until every pixel in the image is contained in exactly one leaf. During this process, we do not allow leaves to merge with one another. Once all pixels have been added, the hierarchy structure derived from the first pass is used to propagate pixel information up the tree, from the leaves to the root. At this point, every horizontal cross-section of the tree can produce a complete segmented image comprising all pixels. Although regions corresponding to non-leaf nodes may be useful, we choose to ignore them. Our image segmentation results are derived only from regions corresponding to the leaf nodes. Figure 10.2 shows segmentation from a tabletop scene. The center image displays the hierarchy of MSER regions, displayed as ellipses. The right image shows segmentation produced using the leaves of the MSER tree.

Region feed-forward

Most stable feature detection algorithms generate an entirely new set of features from every frame of the video sequence. Tracking algorithms are then required to match features between successive frames. Although this is a useful strategy for tracking small or fast-moving objects, it may be unnecessary when tracking large, textureless, objects that are slow moving or stationary. Without surface texture, pixels within the region's interior do not provide any useful information and re-computing their position every frame wastes resources. Resources would better be applied to pixels near the perimeter of a region, or to pixels that changed between frames. Because large textureless objects can make up significant portions of an image, we observe considerable performance increases using this approach.

In addition to speed advantages, our feed-forward algorithm improves spatial stability by integrating temporal information. Consider a slowly moving (or stationary) homogeneously colored object (like a person's wrinkled shirt) that contains enough surface texture to cause spurious MSER regions to form. The inherent instability of these regions makes them unsuitable for tracking or modeling, yet their removal is difficult without using ad-hoc strategies. Using our feed-forward approach, any region that cannot continually maintain its boundaries, will be assimilated into similarly colored adjacent regions. After several iterations of region competition, many unstable regions are eliminated automatically without any additional processing.

Our feed-forward algorithm is a relatively simple addition to our MSER algorithm. After every iteration of MSER generation, we identify pixels in the current frame that are nearly identical (RGB values within 1) to the pixel in the same location of the following frame. If the majority of pixels in any given MSER remain unchanged for the following video image, the matching pixels are pre-grouped into a region for the next iteration. This pixel cluster is then used to seed growth for the next iteration of MSER evolution.

It should be mentioned that an additional constraint must be added for this feed-forward strategy to work properly. To illustrate the problem, consider a stationary scene with an unchanging image. In this example, every pixel will be propagated, and there will be no pixels left for MSER evolution. Every region in the image will remain unchanged, and any errors in detection would be preserved indefinitely. A preferable strategy is to propagate pixels that contribute least to MSER evolution (low-gradient pixels), while allowing the MSER to evolve using more descriptive (high-gradient) pixels. To achieve this effect, we compute the average gradient value of pixels in each region and propagate pixels with gradient values below that average (as an optimization, we also propagate pixels with gradients below a predefined threshold). This technique approximately allows at least half the pixels in any non moving region to be propagated forward, while leaving the other half to reconstruct an updated stable region. Figure 10.3 (left) shows pixels designated for feed-forward. Dark gray pixels are propagated to the next frame. Light gray pixels are withheld.

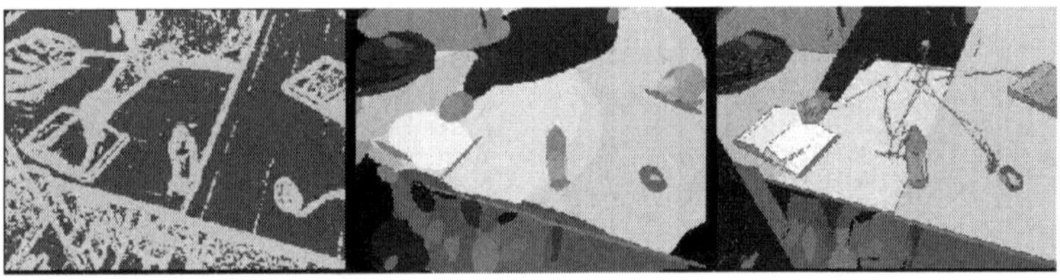

FIGURE 10.3 Left: An example of the feed-forward process. Dark gray pixels are preserved, light gray pixels are re-clustered. Center: MSERs are modeled and displayed using ellipses and average color values. Right: An example of MSER image segmentation. Regions are filled with their average color, detected lines are shown in gray, and the path of the tracked hand is represented as a dark gray line. (See color insert.)

10.3.2 Region Tracking

Region tracking can be defined simply as determining the optimal way detected regions in one frame match regions in subsequent frames. Despite the simple definition, tracking is a challenging problem due to

1. Object appearance changes: illumination, transformation, deformation, occlusion
2. Detection errors: false detections, multiple detections, missed detections

3. Detection inconsistencies: inaccurate estimation of position, size, or appearance

Yilmaz et al. [31] reviewed several algorithms, and listed the strengths and weaknesses of each. Yilmaz emphasized that each tracking algorithm inevitably fails under a certain set of conditions and that greater robustness can be obtained by combining strategies. Although this concept works well in theory, implementation can be difficult in real-time. Many of the available real-time region detection and tracking algorithms require a significant amount of computer resources to operate, often making the simultaneous operation of non related algorithms impractical. Additionally, fusing information obtained from several algorithms may create additional problems.

Our tracking algorithm was designed to specifically operate on the complementary set of features provided by our detection algorithm. As mentioned, our algorithm models regions using MSER features and line-corner features. Each feature type provides certain advantages and disadvantages, and our algorithm has been designed with the intent of exploiting the advantages of each. Our tracking algorithm applies four different phases. Each phase is best suited to handle a specific type of tracking problem, and if an object can be tracked in an early phase, later tracking phases are not applied to the object. By executing the fastest trackers first, we can further reduce resource requirements.

The four phases of our tracking algorithm include Feed-forward tracking(Section 10.3.2), MSER tracking(Section 10.3.2), Line tracking(Section 10.3.2), and Secondary MSER tracking(Section 10.3.2).

Feed-forward tracking

Traditional tracking algorithms match features between successive frames using descriptor similarity measures. This assumes that descriptors do not change significantly between frames. MSER regions can pose significant problems in this regard, as small changes in the image can cause large changes in the descriptors. For example, consider a video sequence taken of a person reaching completely across a table. Immediately before the person's arm bisects the table, a traditional MSER algorithm will detect the tabletop as a single region. Immediately afterward, the tabletop will appear as two smaller regions. Because a MSER tracker only receives information regarding the size and positions of the centroids, resolving the actual path of the region as it splits in two, would likely be a cumbersome process.

Using our pixel feed-forward algorithm, resolving the table-bisection scenario becomes a trivial matter. Since the majority of the table's pixels remain unchanged during the bisection, these pixels will maintain their exiting clustering. Even if the cluster of pixels is non-contiguous, MSER evolution will produce a single region. Tracking becomes a trivial matter of matching the pixel's donor region with the recipient region.

MSER tracking

Tracking MSERs has traditionally been an ill-posed problem. It is difficult to control the degree that similarly shaped regions are nested within one another and fluctuations make one-to-one region correspondences nearly impossible. As described in the Section 10.3.1, we eliminated the problem of nesting by reducing the hierarchy of MSERs to non hierarchical image segmentation. This representation theoretically makes one-to-one correspondences possible, and matches are identified using a greedy approach. The purpose of this phase of tracking is to match only those regions that have maintained consistent size and color between successive frames.

Each image region is represented by the following features:

1. Centroid (x,y) image coordinates
2. Height and width (second-order moment of pixel positions)
3. RGB color values

Matching is only attempted on regions that remained unmatched after the feed-forward tracking phase (Section 10.3.2). Matches are only assigned when regions have similarity measures beyond a predefined similarity threshold. Matching is conducted as follows; for every unmatched region in frame 't', a set of potential matches in frame '$t+1$' is identified using a kd-search tree. Regions matches that are not sufficiently similar in size, position, and color, are removed from consideration. All other region matches are sorted according to the feature similarity measures of size, position, and color. Potential matches are processed in order of their similarity measure (from most similar to least). If both regions are available to be matched, then a tracking link is provided to the pair. The algorithm proceeds until all potential matches have been considered.

Line-tracking

Because a primary component of the MSER descriptor is its vertical and horizontal size, tracking can be highly sensitive to occlusion and bifurcation. This makes MSER descriptors unsuitable for use as the sole feature used in tracking. Ideally, a second feature should be used, that does not require significant additional resources to detect. Because our algorithm already incorporates lines into region detection, the line features become an ideal candidate for complementing the MSER. Specifically, we use line segment end points, which typically occur on corners of image regions. Line corners are desirable because they are stable, they are unaffected if a different part of the region is occluded, and they provide good region localization.

In this tracking phase, line corners are matched based on their positions, the angles of the associated lines, and the colors of the associated regions. It should be mentioned that, even if a line separates (and is therefore associated with) two regions, that line will have different properties for each region. Specifically, the line angle will be 180 degrees rotated from one region to the other, and the left and right endpoints will be reversed.

Each line end is represented by the following features:

1. Position (x,y) image coordinates
2. Angle of the corresponding line
3. RGB color values of the corresponding region
4. Left/right handedness of endpoint (perspective of looking out from the center of the region)

Line corner matching is only attempted on regions that remained unmatched after the MSER tracking phase. Also, matches are only assigned for objects that have similarity measures beyond a predefined similarity threshold. Matching lines is conducted using the same strategy as described in Section 10.3.2.

Secondary MSER tracking

Tracking phases described in Sections 10.3.2 to 10.3.2 assume that features do not change significantly between frames. This may generally be the case, for example, noise, illumination changes, and occlusion may cause information to be degraded or lost in certain frames. To reduce the number of regions lost under these conditions, we conclude our tracking sequence by re-applying our MSER and line tracking algorithms using looser similarity constraints. This phase uses a greedy approach to match established regions (regions that were being tracked but were lost) to unassigned regions in more recent frames. Unlike the first three phases, which only consider matches between successive frames, the fourth phase matches regions within an n-frame window ("n" is usually fewer than 8). In this case, the established region's motion model is used to predict its expected location for comparison.

10.3.3 Foreground Detection

The sequence of steps in our approach to process a video feed is as follows:

1. Initially, construct a region-based background model
2. Cluster pixels in subsequent frames into regions
3. Track all regions
4. Identify regions in subsequent frames that differ from the background model
5. Update the background model using background regions

Because the background model in our approach comprises higher-level features, we can apply the algorithm in a greater variety of settings. For example, a motion model can be trained, allowing foreground detection to be performed on both stationary and panning surveillance systems. Additionally, because background features are continually tracked, the system is equipped to identify unexpected changes to the background. For example, if a background region moves in an unexpected way, our system can identify the change, compute the new trajectory, and update the background model accordingly.

Although there may be several ways to achieve foreground detection using our tracking algorithms, we feel it would be appropriate in these early stages of development to simply reproduce the traditional pipeline. To this affect, the first several frames in a video sequence are committed to building a region-based model of the background. Here, MSERs are identified and tracked until a reasonable estimation of robustness and motion can be obtained. Stable regions are stored to the background model using the same set of features listed in the tracking section. The remainder of the video is considered the operation phase. Here, similarity measurements are made between regions in the background model, and regions found in the current video frame. Regions considered sufficiently dissimilar to the background are tracked as foreground regions. Matching regions are tracked as background regions. Because we employ both tracking information and background-model comparison, our system can identify when background regions behave unexpectedly. We then have a choice to either update the background model, or to track the region as foreground.

10.3.4 Object Modeling

Once the foreground is segmented from the background, a color and shape-based model is generated from the set of foreground MSER features. This model is used to resolve collisions and occlusions, and to identify if a familiar object has re-entered the scene. A common modeling approach is to identify a set of locally invariant features. Lowe [18] proposed a technique where an image patch is sampled within a pre-specified distance around a detected region. The texture within the patch is binned into a 4×4 grid to form a Scale Invariant Feature Transform (SIFT). The resulting descriptor contains a 128-dimensional vector ($4 \times 4 \times 8$-bins). Despite its popularity, this technique is not effective when the object of interest undergoes significant transformation, or contains significant depth disparities. The 128-dimensional SIFT also requires significant computational resources for recording and matching.

Chum and Matas [8] describe a more efficient approach to modelling that reduces descriptor dimensionality to six features. The small dimensionality allows a constant-time hash table to be used for feature comparison. Chum's descriptors are based on MSER region pairs. Each MSER region is transformed into a locally invariant affine frame (LAF). The centroids are identified, as are two extremal points around the region's perimeter. The six-feature descriptor is formed using angles and distances between three point-pairs. Because there may be multiple transformations to the affine frame, each region may have multiple possible descriptors. A voting technique is implemented using the hash table to identify likely candidates. This is a constant time operation, making the technique orders of magnitude faster than patch-style algorithms. It is also less affected by depth discontinuities for foreground objects.

Our technique uses many of the principles presented by Chum, but our feature vectors were selected to provide improved robustness in scenes where deformable or unreliable contours are an issue. Chun was able to use descriptors with relatively low dimensionality because they provided a high degree of precision in estimating the transformation parameters of flat objects. We took the opposite approach by selecting a relatively large number of invariant features with low individual descriptive value. Even though individual features are likely to provide an inaccurate representation of our objects, the combined vote of many unrelated features should provide reasonably discriminatory abilities.

We propose an algorithm that represents objects using an array of features that can be classified into three-types: (1) MSER pairs, (2) MSE individuals, and (3) size position measure.

- Our MSER pair features are described using a four-dimensional feature vector. The first two dimensions (v_1, v_2) are computed by taking the ratio of color values (in RGB color space) between the two MSERs: $(red_1/grn_1) : (red_2/grn_2)$, and $(blu_1/grn_1) : (blu_2/grn_2)$. The third dimension (v_3) is the ratio between the area square roots: $\sqrt{area_1} : \sqrt{area_2}$. The fourth dimension (v_4) is the distance between ellipse centroids, divided by the sum of the ellipse diameters (cut along the axis formed by the line connecting the centroids). Descriptor values v for the MSER pair a and b are computed such that the ratio is between -1 and 1 for each vector dimension f:

$$v_f = \begin{cases} (1 - a_f/b_f) & \text{if } a_f < b_f \\ -(1 - b_f/a_f) & \text{otherwise} \end{cases}$$

- Our MSER individual features are described using a three-dimensional feature vector. The first two dimensions (v_1, v_2) are a measure of the region's color: red/grn, and blu/grn. The third dimension is a measure of curvature for the object's perimeter. Values range from one (parallelograms) to zero (regions not bound by lines).

- The final feature set is only used for computing vote tally. When models are generated, the relative size and position of the contained features are recorded. When features are tested against models in subsequent iterations, this information is used to approximate size and position for every object that receives a vote. To win the vote tally, an object must receive a sufficient number of votes, which agree on these approximations. The first feature dimension (v_1) is the ratio between the square root of the MSER's area, and the square root of the area containing all object-MSERs. The second dimension (v_2) represents the position of the MSER, in relation to the other MSERs in the object. This feature is only used when a consistent object orientation is expected. Because people in surveillance videos are not likely to display vertical symmetry, the value for the features is a function of its vertical position in the object (negative one at the bottom, positive one at the top).

10.4 Real-Time Multi-Object Tracking System Using Stereo Depth

In this section, a multi-object tracking system is presented based on a ground plane projection of real-time 3D data coming from a stereo imagery, giving distinct separation of occluded and closely interacting objects. Our approach consists of tracking, using Kalman filters [5], fixed templates that are created by combining the height and the statistical pixel occupancy of the objects in the scene. These objects are extracted from the background using a Gaussian Mixture Model combining luminance and chroma signals (YUV color space) and depth information obtained from the stereo devices used in this work. The mixture model is adapted over time and it is used to create a background model that is also upgraded using an adaptive learning rate parameter according to the scene activity level on a per-pixel basis. The results presented illustrate the validity of the approach.

The next section illustrates the segmentation algorithm used to achieve the foreground detection. Section 10.4.2 explains the idea behind the creation of plan views used to track objects. Section 10.4.3 presents the tracking procedure and data association.

10.4.1 Foreground Detection

The background subtraction model presented follows the excellent work by M. Harveille et al. [14, 15, 27]. It applies the well-known statistical method for clustering called the Gaussian Mixture Model, per-pixel, dynamically adapting the expected background using four channels: three color channels (YUV color space) and a depth channel. The input to the algorithm is a synchronized pair of color and depth images extracted from a single fixed stereo rig*. The depth is calculated by triangulation, knowing the intrinsic parameters and the disparity of the stereo rig, as shown in the first left part of the Equation 10.5. A 3D world point is projected at the same scan line to the left and right

*www.videre.com.

image of the stereo rig once it is calibrated. The displacement between each camera of theis projection point is called disparity. The dataset observation for a pixel i at time t is composed as follows: $X_{i,t} = [Y_{i,t} \quad U_{i,t} \quad V_{i,t} \quad D_{i,t}]$ and the observation history dataset for pixel i at the current observation is as follows: $[X_{i,1}...X_{i,t-1}]$. Therefore, the likelihood of $X_{i,t}$ taking into account the prior observations is defined as

$$P(X_{i,t}|X_{i,1}...X_{i,t-1}) = \sum_{j=1}^{K} \delta_{i,t-1,j}\varphi(X_{i,t};\theta_j(\mu_{i,t-1},\Sigma_{i,t-1})), \qquad (10.1)$$

where δ is the mixing weight of past observations:

$\sum_{j=1}^{K} \delta_{i,t-1,j} = 1$ and $\delta_j > 0$;

$\theta_j(\mu_{i,t-1},\Sigma_{i,t-1})$ is the Gaussian density function component.

The number of Gaussians used (i.e., K) initially was 5, although results obtained from posterior experiments showed that using 4 Gaussians were equally as good. Assuming the independence between measurements, the covariance matrix is constructed as a diagonal matrix whose diagonal components are the variance of each component in the dataset illustrated above. In order to reduce the computation time, the matching process between the current observation (per pixel) and the appropriate Gaussian is completed following an online K-means approximation, as it is done in [15]. The first step in the matching process is to sort all the Gaussians in a decreasing weight/variance order, which implies to give preference to the Gaussians that have been largely supported by previous consistency observations. Only the variance corresponding to luminance is used in the sorting as depth and chroma data may be unreliable sometimes. The second step in the matching process is to select the first Gaussian that is close enough to the new observation by comparing the square difference between the Gaussian's mean and the current observation with a fixed threshold value. If this difference is below the threshold, the Gaussian is selected. The value of the threshold, after several experiments, was set to 4. Then, if a match is found, the parameters of the selected Gaussian (i.e., the mean and its variance) are updated, taking into account the new observation. As stated before, the depth measurements are sometimes unreliable due to lighting variability or lack of texture in the scenes, which implies that the Gaussians used to represent the background can contain observations whose depth measurements can be valid or invalid. If many of these observations of a particular Gaussian are depth error measurements, the depth mean and variance of the Gaussian is considered unreliable and therefore its statistics cannot be used for the comparision with current observations. For that reason, only depth statistics of a Gaussian are taken into account if a fraction of its valid depth observations are above the fixed threshold (i.e., 0.2). The square difference regarding depth is calculated once the current depth observation and the depth statistics of the Gaussian are validated. If the difference is below the threshold, this indicates high probability that the pixel belongs to the background, so the fixed threshold is augmented by a factor 4, increasing the color matching tolerance. This addition allows dealing with cases, for example, where shadows appear, which match the background depth but not so well the background color. On the contrary, if the difference is above the threshold, this indicates a high probability that the pixel belongs to the foreground, and a foreground flag is set. Before proceeding to calculate the luminance difference, the luminance component of the current observation and the Gaussian's mean are checked to be above minimum luminance value, which would imply that the chroma data is reliable and therefore it can be used for the comparison. If a match is not found and the foreground flag has not been set, the last Gaussian in the sorting process is replaced by a new Gaussian with a mean equal to the new observation and low initial weight. The update equations for the selected Gaussian and for the weights for all the Gaussians are described as follows:

$$\begin{aligned}
\mu_{Y,i,t,k} &= (1-\alpha)\mu_{Y,i,t-1,k} + \alpha Y_{i,t,k} \\
\mu_{U,i,t,k} &= (1-\alpha)\mu_{U,i,t-1,k} + \alpha U_{i,t,k} \\
\mu_{V,i,t,k} &= (1-\alpha)\mu_{V,i,t-1,k} + \alpha V_{i,t,k} \\
\mu_{D,i,t,k} &= (1-\alpha)\mu_{D,i,t-1,k} + \alpha D_{i,t,k} \\
\sigma^2_{Y,i,t,k} &= (1-\alpha)\sigma^2_{Y,i,t-1,k} + \alpha(Y_{i,t} - \mu_{Y,i,t-1,k})^2 \\
\sigma^2_{C,i,t,k} &= (1-\alpha)\sigma^2_{C,i,t-1,k} + \alpha((U_{i,t} - \mu_{U,i,t-1,k})^2 + (V_{i,t} - \mu_{V,i,t-1,k})^2) \\
\sigma^2_{D,i,t,k} &= (1-\alpha)\sigma^2_{D,i,t-1,k} + \alpha(D_{i,t} - \mu_{D,i,t-1,k})^2
\end{aligned} \tag{10.2}$$

The weight update equation for all Gaussians is as follows:

$$\delta_{i,t,k} = (1 - \mathit{ActivityRecognition})\delta_{i,t-1,k} + \alpha M_{i,t,k}, \tag{10.3}$$

where $M_{i,t,k} = 1$ for the matched Gaussian and zero for the rest of Gaussians.

Finally, once the Gaussians are updated, every pixel in each processed frame is labeled as foreground if it was not matched to any Gaussian belonging to the background model. Morphological operations are applied to remove isolated regions to fill small foreground holes.

Adaptive learning rate parameter

In the equations illustrated above, α can be seen as a learning rate parameter as its value indicates how quickly the Gaussians will adapt to the current observation; if α has a big value, it implies that the Gaussians will get close to the new observations in faster incremental steps. In other words, static changes in the background are incorporated into the background model quickly. However, it also implies that foreground objects that have remained static for a certain time are quickly added to the background. A good compromise with α factor is found in [15] where its dynamic value is directly linked with the amount of activity level of the scene (as the authors called it). The activity level indicates the luminance changes between frames:

$$Ac_{i,t,k} = (1-\rho)Ac_{i,t-1,k} + \rho|Y_{i,t} - Y_{i,t-1}|. \tag{10.4}$$

Initially as follows in [16], the activity level (Ac) defined by Equation(10.4) is set to zero, and after it is computed as the difference in luminance between current frame and previous frame. Therefore, if the activity threshold is above the fixed threshold, which in this study was experimentally fixed to 5, the α factor used to update the Gaussians' statistics is reduced by an experimental factor of 5.

10.4.2 Plan View Creation

In this section the algorithm that renders 3D foreground cloud data as if the data was viewed from an overhead, orthographic camera is presented. The main reason to apply this transformation is for the computational performance increase, by reducing the amount of information when the tracking is done onto plan-view projection data rather than onto 3D data directly. The projection of the 3D data to a ground plane is chosen due to the assumption that people usually do not overlap in the direction normal to the ground plane. Therefore, this 3D projection allows us to separate and to track more easily than in the original camera view. Any reliable depth value can be back-projected to its corresponding 3D point knowing the camera calibration data and the perspective projection. Therefore, the first step in the creation of the plan views is to only back-project the foreground pixels detected in the previous algorithm, creating a 3D foreground cloud of visible points to the stereo camera. Then, the space of the 3D cloud points is quantized into a regular grid of vertically orientated bins. Looking at these vertical bins as the direction normal to the ground plane, some statistics regarding the 3D cloud points can be calculated within each bin. Therefore, a plan view image is constructed as a binary image where each pixel represents one vertical bin and the value of the pixel is some statistic of the 3D cloud point stored in the vertical bin. Two types of plan view images are created regarding the two interesting statistic of the 3D cloud points stored in the vertical bins: the occupancy,(i.e., the number of points accumulated in each vertical bin) and the height (i.e., the highest height of the 3D point cloud within each vertical bin). The first statistic indicates the amount of foreground projected on the ground plane and the second statistic indicates the shape of the 3D foreground could. In order to compensate for the smaller camera-view effect appearance of distant objects, the first statistic (i.e., the occupancy) is constructed as a weighted number of points accumulated in each bin. The factor used to calculate the occupancy map as suggested in [16, 23] is Z^2/f, where Z is the depth value and f is the focal length. The following equations describe the steps used to create the maps. Figure 10.4 and Figure 10.5 graphically describe the process of the projection onto the ground plane of the 3D foreground cloud points and the creation of the binary plan view image. Once the plan view binary images are created, a posterior refinement is applied to the images in order to remove much of the noise that appears in the occupancy and height maps (see the plan view occupancy on the right side of Figure 10.5).

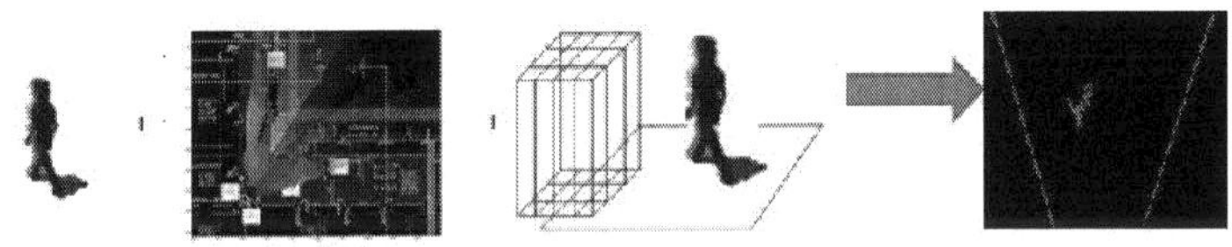

FIGURE 10.4 Illustrates the process of the creation of a plan view.

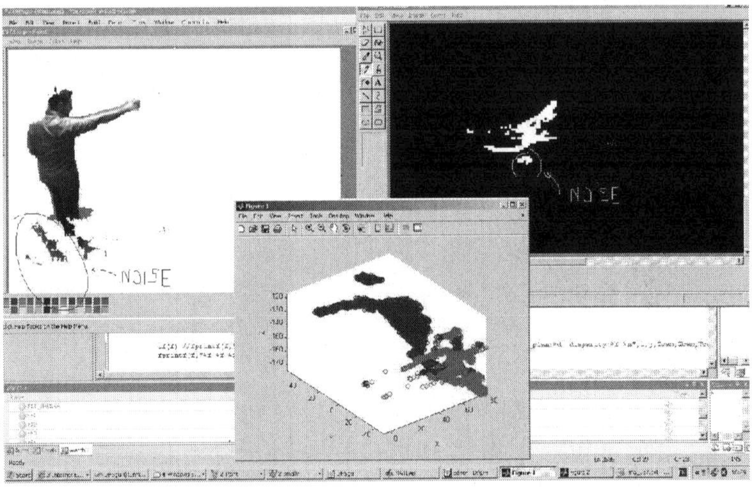

FIGURE 10.5 Illustrates the projection to the ground plane of the valid depth image points of the back-projected 3D foreground cloud of points.

Using the internal calibration parameters, any foreground pixel can be back-projected to a 3D cloud point:

$$Z_{cam} = \frac{bf_u}{disparity}, X_{cam} = \frac{Z_{cam}(u - u_o)}{f_u}, Y_{cam} = \frac{Z_{cam}(v - v_o)}{f_v}, \tag{10.5}$$

where,(u, v) is a pixel in the image plane, (u_o, v_o) is the image center of the projection, f_u and f_v are the horizontal and vertical focal lengths, b (baseline) is the distance between left and right stereo camera, and *disparity* is the difference between the pixel value seen from the left camera and the corresponding pixel value seen from the right camera.

We render the 3D cloud point obtained to an overhead camera view (X_w, Y_w, Z_w):

$$\begin{bmatrix} X_W & Y_W & Z_W \end{bmatrix}^T = -R_{cam} \begin{bmatrix} X_{cam} & Y_{cam} & Z_{cam} \end{bmatrix}^T - T_{cam}, \tag{10.6}$$

where Z_W is aligned with the direction normal to the ground plane, X_W and Y_W are the ground plane axis, and R_{cam} and T_{cam} are the rotation and translation matrices.

We discretize the vertical bins:

$$\begin{aligned} x_{plan} &= \lfloor (X_W - X_{min})/\lambda + 0.5 \rfloor \\ y_{plan} &= \lfloor (Y_W - Y_{min})/\lambda + 0.5 \rfloor , \end{aligned} \tag{10.7}$$

where λ is the resolution factor. In this case the value was set to 2cm/pixel.

10.4.3 Tracking Plan View Templates

This section describes the tracking algorithm applied, which is also based on the work presented in [16, 23]. The Gaussian and linear dynamic prediction filters used to track the occupancy and height statistics plan view maps are the well-known Kalman filters [5]. The state vector is conformed to the 2D position (center of mass) of the tracked object in the plan view, to the 2D velocity component of the object, and to the shape configuration of the object, which can be defined with the occupancy and height statistics, as described in the previous section. In this application, an object could be a robot, a person, or a bag. The input data to the filter consists of simple fixed templates of occupancy and height plan view maps. These templates (τ_H, τ_O) are small areas of the plan view binary images extracted at the estimated location of the object. To create these templates, it is assumed that the statistics of an object are quite invariant to ground plan location relative to the camera. Therefore, the size of the template (40 pixels) remains constant all the time for all the objects. Moreover, to avoid sliding the templates through time, after the tracking process has been applied, the template is centered back to the 2D position of the object i, rather than to the estimated position of object i.

Correspondence

The area to search is centered on the estimated 2D position of the object. The correspondence is resolved via match score, which is computed at all locations within the search zone. A lower match score value implies a better match. The following equation illustrates the computation of the match score:

$$\varphi(i, X) = \rho SAD(\tau_H, H_{masked}(X)) + \omega SAD(\tau_o, \theta_{sm}(X)) + \beta\sqrt{(x - x_{pred})^2 + (y - y_{pred}^2)} + \alpha \sum_{j<i} \theta_j(X, 40). \quad (10.8)$$

SAD refers to the *sum of absolute differences* between the height and occupancy templates and the occupancy and height plan view maps created from the current frame. The $(\rho, \omega, \beta, \alpha)$ weights are deduced in [16] based on physical principles. The first two weights are formulated in a way that the height and occupancy SADs have similar contributions. The third and fourth weights are formulated reflecting the area contribution of the foreground pixels to each vertical bin. It is deduced, from Equation (10.7), that the match score of person i at 2D position X is the sum of several contributions. It is the contribution of the weighted sum of the difference between the shape and visible region of tracked object i at location X from the shape and visible region of the object in the current frame. It is also the contribution of the distance between the current object's position and the estimated location of the object. The last contribution to the match score corresponds to the convergence distance of X to the rest of the predicted locations of the objects in the current frame. Once the best match score is chosen (i.e., the smallest value), it is compared to a threshold track, where it is related to the minimum amount of foreground pixel area needed to consider the tracked region a valid object. Only if the match score is below the threshold track will the object's Kalman state be updated with the new measurements. In Figure 10.6 some results are presented. A few key frames from a sequence (2,300 frames) are presented in this figure, where three different types of objects interact and an occlusion between two of the objects (a person and a robot) appears and the tracker is able to solve the occlusion. The images are captured using a videre stereo camera with VGA resolution.

10.5 Activity Recognition

One objective of a visual surveillance system is to identify when people leave, take, or exchange objects. To test the capabilities of our detection and tracking system, we are implementing a simple scenario involving the exchange of baggage. Specifically, the system will be designed to send a report if a person is observed leaving an object, taking another person's object, or removing something from the environment. Recognizing these types of actions can be done without sophisticated algorithms, so for this demonstration, we use simple rule sets based only on proximity and trajectories:

1. If an unknown object appears in the environment, models will be generated for that object and for the nearest person. If the associated person

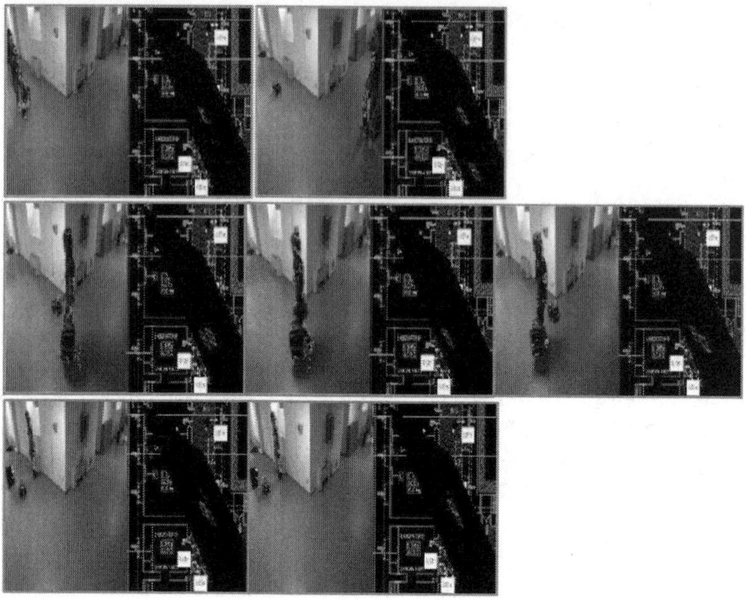

FIGURE 10.6 Seven frames of a sequence that shows tracking different types of objects (robot, person, and bag), including an occlusion. Each plan view map has been synchronized with its raw frame pair and back-projected to the real plan of the scene (right side of each image).

is observed moving away from the object, it will be considered "abandonment," and a report of the incident will be generated. If the same person is observed reacquiring the object, the report will be canceled.

2. If an object is associated with one person, and a second person is observed moving away with the object, it will be considered an "exchange" and a report will be generated containing models of the object and both involved people.

3. If a person is observed moving away with an object that was either present at the beginning of the sequence, or left by another, the incident will be considered a "theft" and a report will be generated containing the person and the object.

Figure 10.7 shows images taken during tabletop (left), and trash fire (right) scenarios. In both demonstrations, the system used color-based models to represent the object and a Hidden Markov Model was used to generalize object interactions. The system was able to accurately identify interactions over extended periods, in real-time.

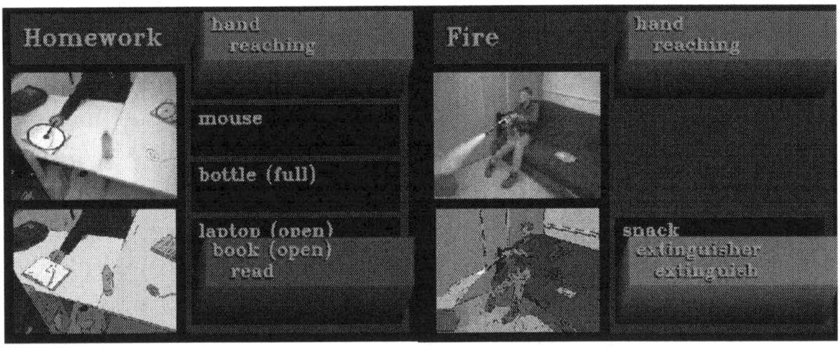

FIGURE 10.7 An example of activity recognition using our system. Each object is associated by a color bar at the right of the image. The apparent height of the bar corresponds to the computed probability that the person's hand is interacting with that object. In the scenario shown on the left, a person engaged in typical homework-type behaviors, including typing on a laptop, turning pages in a book, moving a mouse, and drinking from a bottle. In the scenario on the right, a person reached into a bag of chips multiple times, and extinguished a trash fire with a fire extinguisher. (See color insert.)

10.6 System Integration

In this chapter we present the integrated system on monitoring a large area with the following characteristics:

1. The system is composed of a number of agents, some of them having mobile capabilities (mobile robots) while others are fixed (video cameras).
2. The system is required to monitor and detect different kinds of predefined events at the same time.
3. Each agent has a set of sensors that is useful to detect some event. Sensors are of a different type within the entire system.
4. The system is required to operate in two modes:
   (a) Patrolling mode
   (b) Response mode

These requirements make the problem significantly different from previous work. First of all, we consider a highly heterogeneous system, where robots and cameras interoperate. Second, we consider different events and different sensors, and we will therefore consider different sensor models for each kind of event. Third, we will study the dynamic evolution of the monitoring problem, where at each time a subset of the agents will be in response mode, while the rest of them will be in patrolling mode.

10.6.1 Experimental Scenario

The scenarios used for the experimental validation were tested on the campus of the Department of Computer and System Science (DIS) of Sapienza University in Rome, Italy[§]. Two scenarios were used to test the capabilities of the multi-camera and robot-platform. In the first scenario (i.e., unattended baggage event), the system was designed to send a report if a person was observed leaving a bag. In the second scenario (i.e., object manipulation), the system should send a report if a person manipulated an unauthorized object from the environment. The selected scenario shown in Figure 10.8 was an indoor corridor (left) to simulate the unattended baggage event and a lab room (right) to simulate the object manipulation.

FIGURE 10.8 Experimental scenario at DIS.

Once a report is sent, a guard robot would be commissioned to go and take a high-resolution picture of the scenario. Recognizing these types of actions may be done without sophisticated algorithms, so for this demonstration, we use simple rule sets based only on proximity and trajectories:

Scenario 1:

- If a person is observed manipulating an object that was either present at the beginning of the sequence, or left by another person (i.e., unauthorized object), the incident will be considered an "alert," and a report will be generated and sent to the multi-robot system.

Scenario 2:

- If a bag appears in the environment, models will be generated for that bag and for the nearest person. If the associated person is observed moving away from the bag, it will be considered a "left bag," and a report of the incident will be generated.

[§]www.dis.uniroma1.it.

- If a bag is associated with one person, and a second person is observed moving away with the bag, it will be considered a "bag taken," and a report will be generated and sent to the multi-robot system.

10.6.2 Multi-Robot Environmental Monitoring

The implementation of the Multi-Robot Environmental Monitoring used in this project has been implemented on a robotic framework and tested both on 2 Erratic robots* and on many simulated robots in the Player/stage environment†. Figure 10.1 shows the block diagram of the overall system and the interactions among the developed modules. In particular, the team of robots monitors the environment while waiting for receiving event messages from the vision sub-system. In our system, we use a Bayesian filtering method to achieve sensor data fusion. In particular, we use a Particle filter for the sensor filters and event detection layer. In this way, the probability density functions (pdfs) describing the belief of the system about the events to be detected are described as sets of samples, providing a good compromise between flexibility in the representation and computational effort. The implementation of the basic robotic functionalities and of the services needed for multi-robot coordination are realized using the OpenRDK toolkit‡ [6]. The mobile robots used in the demonstrator have the following features:

- Navigation and Motion Control based on a two-level approach: a motion using a fine representation of the environment and a topological path-planner that operates on a less detailed map and reduces the search space; probabilistic roadmaps and rapid-exploring random trees are used to implement these two levels [7].
- Localization and Mapping based on a standard particle filter localization method and a well-known implementation GMapping¶ that has been succesfully experimented on our robots also in other applications [19].
- Task Assignment based on a distributed coordination paradigm using utility functions [17] already developed and successfully used in other projects.

Moreover, to test the validity of the approach, we replicate the scenarios in the Player/Stage simulator, defining a map of the real environment used for the experiments, and several software agents, with the same characteristics of the real robots. The combination of OpenRDK and Player/Stage is very suitable for development and experimentation in multi-robot applications, since

*www.videre.com.

†playerstage.sourceforge.net.

‡openrdk.sf.net.

¶openslam.org/gmapping.html.

they provide a powerful but yet flexible and easy-to-use robot programming environment.

10.6.3 Results

In this section we present the outputs of the recognized scenarios by both camera systems mentioned above. As stated, both systems are integrated to a multi-robot system to increase situation awareness. The integration of both platforms is done through a TCP client/server communication interface. Each static stereo camera is attached to a PC computer and they communicate between them and the robots via a private wireless network. Each static camera and its PC acts as a client and one of the PCs also acts as a server. The PC video-server is the only PC that communicates directly with the robots; the PC video-server becomes a client, however, when it communicates with the multi-robot system, and then the robots act as servers. Once the video surveillance has recognized an event, the PC client camera sends to the PC video-server the event name and the 3D coordinates. The video-server then constructs a string with this information (first transforming the 3D coordinates to a common platform coordinate system) and sends the message via the wireless network to the robots. Then, one of the robots is assigned to go and guard the area and take a high-resolution picture if the event detected is "bag taken."Figure 10.9 and Figure 10.10 show the results in Scenarios 1 and 2, respectively. Figure 10.9 illustrates a sequence of what may happen in Scenario 1. In the top-left image of the figure, a person with an object (bag) is walking through the corridor. On the top-right image of the figure, the video system detects that the person left the bag. Therefore a message is sent as "left bag."In the bottom-left image, another person walks very close to the bag. In the bottom-right image, the visual surveillance system detects that a person is taking a bag and a message "bag taken" is sent to the robots; and as can be seen, one of the robots is sent to inspect the risen event. Figure 10.10 illustrates a sequence of what may happen in Scenario 2. At the top left, a laptop is placed on the table and one of the robots can be seen patrolling. In the top right and bottom left images of the figure, there is a person who is allowed to manipulate different objects. In the bottom right the person is touching the only object, which is not allowed, therefore an alarm "alert" is raised.

10.7 Conclusion

The extent to which traditional surveillance system can cover large areas, is primarily limited by the number of video feeds a human operator can monitor. This limitation has generated demand for an automated surveillance system. This work is part of a funded project that aims to overcome the current limitations in static visual surveillance by incrementing situation aware-

FIGURE 10.9 This figure illustrates a sequence of what may happen in Scenario 1. Person A walks through the corridor with the bag and leaves it in the middle of the corridor. Person B approaches the bag and takes it, raising an alarm in the system causing the patrolling robot to go and inspect the area.

ness monitoring, making it flexible and dynamic. These enhanced, integrated multi-robot coordination and vision-based activity monitoring techniques advance the state-of-the-art in surveillance applications. Using a group of mobile robots combined with fixed surveillance camera systems has several significant advantages over solutions that only use fixed surveillance camera systems. For example, the robots in the team have the power to collaborate on the monitoring task and are able to preempt a potential threat. Moreover, the multi-robot platform can communicate with a human operator and receive commands about the goals and potential changes in the mission, allowing for a dynamic, adaptive solution. In this chapter, we introduced a maximally stable segmentation algorithm that efficiently divides image sequences into spatially and temporally stable regions. By tracking these regions, our system can more quickly discriminate between local and global changes in the image, and can use that information to intelligently update environment and object models. We have successfully tested our system in a number of real-world activity-recognition scenarios, and are currently working to apply it to a multi-camera surveillance system. Also, a real-time multi-object tracking system for a stereo camera has been presented. Furthermore, the computer vision

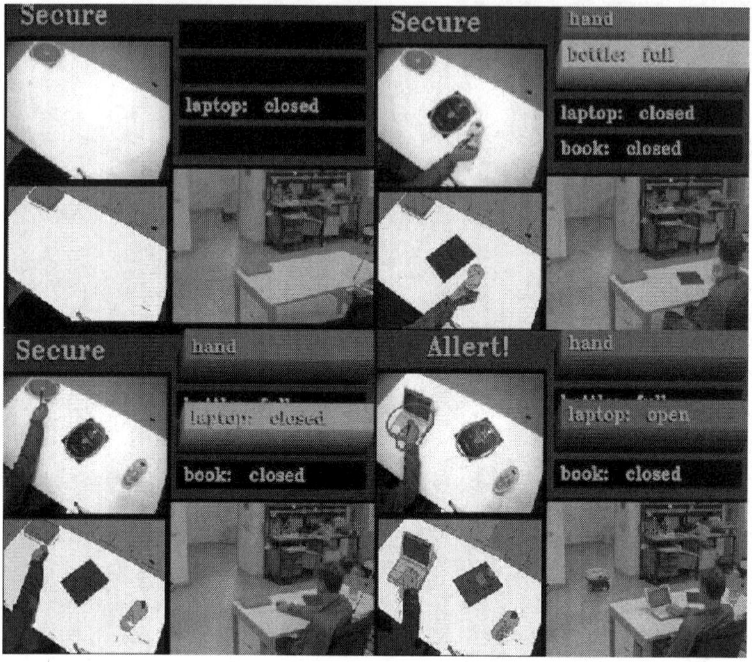

FIGURE 10.10 This figure illustrates a sequence of what may happen in Scenario 2. Person B places a book and a bottle on the table and manipulates them under the surveillance of the system; until Person B decides to touch an unauthorized object (i.e., laptop) raising an alarm in the system causing the patrolling robot to go and inspect the area.

algorithms presented in Sections 10.3 and 10.4 can eventually be portable to a mobile robot platform. Therefore our solution could be used in environments that previously have not been equipped with a camera-based monitoring system: the robot team could be deployed quickly to obtain information about an unknown environment, allowing the robots to position themselves within the environment in order to best acquire the necessary information.

References

1. Point grey stereo cameras. `http://www.ptgrey.com/`.
2. Videre stereo cameras. `http://www.videredesign.com/`.
3. A. Baumberg and D.C. Hogg. Learning deformable models for tracking the human body. In M. Shah and R. Jain, Editors, *Motion-Based Recognition*, pages 39–60. Kluwer Academic Publishers, 1996.
4. D. Beymer and K. Konolige. Real-time tracking of multiple people using continous detection. In *Proc. International Conference on Computer Vision (ICCV'99), Frame-Rate Workshop*, 1999.
5. R.G. Brown and P.Y.C. Hwang. *Introduction to Random Signals and Applied*

Kalman Filtering. 2nd ed,John Wiley & Sons, 1997.

6. D. Calisi, A. Censi, L. Iocchi, and D. Nardi. OpenRDK: A modular framework for robotic software development. In *Proc. of Int. Conf. on Intelligent Robots and Systems (IROS)*, pages 1872–1877, 2008.
7. D. Calisi, A. Farinelli, L. Iocchi, and D. Nardi. Autonomous navigation and exploration in a rescue environment. In *Proceedings of the 2nd European Conference on Mobile Robotics (ECMR)*, pages 110–115, 2005.
8. O. Chum and J. Matas. Geometric hashing with local affine frames. *CVPR*, pages 879–884, June 2006.
9. R. Cipolla and M. Yamamoto. Stereoscopic tracking of bodies in motion. *Image Vision Comput.*, 8(1):85–90, 1990.
10. D. Comaniciu, V. Ramesh, and P. Meer. Kernel-based object tracking. *IEEE Transactions on Pattern Analysis and Machine Intelligence*, 25(5):564–575, 2003.
11. T. Darrell, D. Demirdjian, N. Checka, and P. Felzenszwalb. Plan-view trajectory estimation with dense stereo background models. In *Eighth IEEE International Conference on Computer Vision (ICCV 2001)*, 2:628–635, 1990.
12. A. Farinelli, L. Iocchi, D. Nardi, and V. A. Ziparo. Assignment of dynamically perceived tasks by token passing in multi-robot systems. *Proceedings of the IEEE*, 94(7):1271–1288, 2006.
13. P.-E. Forssén. Maximally stable colour regions for recognition and matching. In *IEEE Conference on Computer Vision and Pattern Recognition*, MN, USA, June 2007. IEEE Computer Society, IEEE.
14. M. Harville. Stereo person tracking with adaptive plan-view statistical templates. *Image and Vision Computing*, 22:127–142, 2002.
15. M. Harville, G. Gordon, and J. Woodfill. Foreground segmentation using adaptive mixture models in color and depth. In *IEEE Workshop on Detection and Recognition of Events in Video*, 0:3, 2001.
16. M. Harville and D. Li. Fast, integrated person tracking and activity recognition with plan-view templates from a single stereo camera. In *IEEE Computer Society Conference on Computer Vision and Pattern Recognition*, 2:398–405, 2004.
17. L. Iocchi, D. Nardi, M. Piaggio, and A. Sgorbissa. Distributed coordination in heterogeneous multi-robot systems. *Autonomous Robots*, 15(2):155–168, 2003.
18. D. Lowe. Distinctive image features from scale-invariant keypoints. *IJCV*, 60(2):91–110, 2004.
19. L. Marchetti, G. Grisetti, and L. Iocchi. A comparative analysis of particle filter based localization methods. In *Proc. of RoboCup Symposium*, 2006.
20. J. Matas, O. Chum, M. Urban, and T. Pajdla. Robust wide baseline stereo from maximally stable extremal regions. In *Proc. of British Machine Vision Conference*, 1:384–393, 2002.
21. K. Mikolajczyk, T. Tuytelaars, C. Schmid, A. Zisserman, T. Kadir, and L. Van Gool. A comparison of affine region detectors. *International Journal of*

Computer Vision, 65(1-2):43–72, November 2005.

22. H. Z. Ning, L. Wang, W. M. Hu, and T. N. Tan. Articulated model based people tracking using motion models. In *Proc. Int. Conf. Multi-Model Interfaces*, pages 115–120, 2002.
23. R. Mu noz Salinas, E. Aguirre, and M. García-Silvente. People detection and tracking using stereo vision and color. *Image Vision Comput.*, 25(6):995–1007, 2007.
24. R. Mu noz Salinas, R. Medina-Carnicer, F.J. Madrid-Cuevas, and A. Carmona-Poyato. People detection and tracking with multiple stereo cameras using particle filters. *J. Vis. Comun. Image Represent.*, 20(5):339–350, 2009.
25. P. Pritchett and A. Zisserman. Wide baseline stereo matching. In *ICCV '98: Proceedings of the Sixth International Conference on Computer Vision*, pages 754–760, Washington, DC, USA, 1998. IEEE Computer Society.
26. J. Sivic and A. Zisserman. Video google: A text retrieval approach to object matching in videos. *ICCV*, 2, 2003.
27. C. Stauffer and W.E.L. Grimson. Adaptive background mixture models for real-time tracking. *Computer Vision and Pattern Recognition, IEEE Computer Society Conference on*, 2:246–252, August 1999.
28. T. N. Tan, G. D. Sullivan, and K. D. Baker. Model-based localization and recognition of road vehicles. *International Journal Computer Vision*, 29(1):22–25, 1998.
29. T. Tian and C. Tomasi. Comparison of approaches to egomotion computation. *Computer Vision and Pattern Recognition*, pages 315–320, 1996.
30. K. Toyama, J. Krumm, B. Brumitt, and B. Meyers. Wallflower: Principles and practice of background maintenance. In *Intl. Conf. on Computer Vision*, pages 255–261, 1999.
31. A. Yilmaz, O. Javed, and M. Shah. Object tracking: A survey. *ACM Computer Survey*, 38:13, 2006.
32. T. Zhao, M. Aggarwal, R. Kumar, and H. Sawhney. Real-time wide area multi-camera stereo tracking. In *CVPR '05: Proceedings of the 2005 IEEE Computer Society Conference on Computer Vision and Pattern Recognition*, 1:976–983, Washington, DC, USA, 2005. IEEE Computer Society.

11

A Network of Audio and Video Sensors for Monitoring Large Environments

Claudio Piciarelli
University of Udine, Udine, Italy

Sergio Canazza
University of Padova, Padova, Italy

Christian Micheloni
University of Udine, Udine, Italy

Gian Luca Foresti
University of Udine, Udine, Italy

11.1 Introduction

During past years, the need for security-oriented surveillance systems has grown larger and larger. Nowadays many public environments, such as airports, train stations, etc., are monitored by some sort of video-surveillance system in order to detect or prevent security issues. The involved technology ranges from the use of plain closed-circuit cameras (CCTV) to sophisticated computer-based video processing systems. The CCTV approach has been the only feasible choice in the past, and it is still widely used; however, its limits are becoming more and more evident: the increase in the number of sensors (modern surveillance systems can use hundreds of cameras) is often not matched by an adequate number of human operators, whose attention is spread over many different tasks and quickly decreases over time. Modern computer-based systems try to face these problems using automatic video analysis and understanding techniques, in order to highlight only the potential security issues and thus requiring the attention of a human operator only in a limited number of cases. The research in this field has been very active and has produced many techniques for video analysis and interpretation, yet many works are limited to the use of static cameras. Only recently has the research community started focusing on more sophisticated sensors like Pan-Tilt-Zoom (PTZ) cameras, and very few have considered the advantages of using heterogeneous sensors, such as cooperation of audio and video sensors. In particular, audio microphones have several advantages, like wide omnidirectional coverage and low price, and thus could be particularly useful in covering large environments. Audio data is, of course, less discriminative than cameras for a human operator required to classify the detected activities; this is why the simultaneous, coordinated use of both audio and video sensors could lead to a real advancement in the field of surveillance systems. This work proposes a novel audio- and video-based surveillance system where audio sensors are used to cover large portions of the environment, and the detected audio sources are further analyzed by means of PTZ cameras.

11.1.1 Audio Sensors

Acoustic Source Localization (ASL) allows us to extract information about the space location of one or more sources using microphone arrays and signal processing techniques. The ASL techniques are applicable in various contexts as, for example, the tracking of the speaker during a conference [31], the reduction of noise coming from concurrent sources [9], or the acoustical analysis of a mechanical device [34]. Recently, the growing demand for automatic surveillance systems (networks of video cameras integrated with other types of sensors) has sparked interest in systems capable of monitoring the presence and movement of sound sources in public places [33]. It is therefore necessary to adapt and optimize the already studied ASL techniques to meet the constraints imposed by the new application scenario: (1) the far-field condition (it is often necessary to locate sources at a distance of tens of meters), in which the acoustic pressure wave can be approximated to a plane wave; (2) the need to monitor sources that are moving on a two-dimensional space (the plane of a square, a street or a monitored park); (3) the need to place sensors on a plane different from that monitored, in order to avoid damage by pedestrians or vehicles; and (4) the need to have a reduced number of arrays, not to invade the public spaces in an excessive way. In fact, whereas in the near-field case a linear array of at least three microphones should be sufficient to locate the sources position in a two-dimensional space, in the far-field case the estimation of the source position is extremely difficult, if not almost impossible, using a single array; from the Time Difference Of Arrival (TDOA) among the microphones we can estimate the Direction Of Arrival (DOA) of the sound, but not its distance. Therefore, the two-dimensional position of the source can be estimated using at least two linear arrays, by means of the triangulation of the DOA estimations (see Figure 11.3(b)). As our aim is to test a network configuration with a minimum number of arrays (see the point (4) above), we will refer to a capture system composed of only two arrays. This system works efficiently in the case of a single source; but if there is more than one source, we have the problem of what are the correct angles to link with each of the others (see Figure 11.3(b)).

In the literature, several works address similar problems using an approach based on the tracking of the sources: in [14] [27] [35] by means of a particle filter, also known as sequential Monte Carlo method, and in [11] [32] by means of the Kalman filter theory. Methods based on movement tracking can fail in some specific situations: (1) during the initialization phase of the filter, (2) in the presence of sources with unpredictable trajectory (e.g., in the case of rapid changes of the velocity vector), and (3) when two sources have intersecting trajectories.

This chapter explores a new approach, which can be applied in a manner complementary to the filter-based methods, enhancing the performance of the system when the movement tracking is difficult. Our approach is based on source separation and the comparison of similarity among sounds.

Assume that n sound sources have been identified for each array, coming from n different DOAs. Because the localization of any source in the two-dimensional plane requires the triangulation of information measured by the two arrays, we must associate each DOA estimated by the first array with one corresponding to the same source estimated by the second array. In total we have n^2 possible combinations, but we do not know a priori which are the right n pairs. To this end, the sources estimated by each array are separated by means of beamforming techniques, maximizing the SNR of the sources and minimizing the interferences and sounds coming from other directions. Then, we proceed to compare the signal coming from all the n directions identified by the first array with those identified by the second one. Finally, among the n^2 possible pairs, the n pairs whose signals have a greater similarity are selected.

To check the similarity of acoustic sources, we use a method based on the spectral distance measure, which allows us to identify sounds by comparing the spectral power in such a way that their difference spectrum power minimizes an error criterion. We describe the spectral distance function in Section 11.3.3.

11.1.2 Video Sensors

Concerning video sensors, countless works have been proposed on surveillance systems based on static cameras; however, the research on dynamic, active networks of PTZ cameras is still limited (for an example of some recent work in this field, see [1]). Some of the proposed works focus on optimizing the camera coverage of the monitored area according to specific criteria. Angella et al. [3] propose a method to maximize the area coverage using a 3D model of the observed zone, but their work only aims at finding a good initial camera displacement, which cannot be dynamically modified according to the observed data. Mittal and Davis [17, 18] also consider the presence of dynamic occluding objects in order to evaluate the visibility of the scene. Piciarelli et al. [25] propose a method to automatically and dynamically reconfigure the camera orientations and zoom levels using an Expectation-Maximization-based approach.

However, up to now, sensor reconfiguration techniques have been mainly proposed to improve tracking performances. Jain et al. [7] propose a master-slave architecture involving both static and dynamic cameras. Soto et al. [30] developed a system for multi-target tracking with a network of self-reconfiguring PTZ cameras without the need of a central processing unit by using a distributed version of the Kalman filter. Karuppiah et al. [10] propose two new metrics that, based on the dynamics of the scene, allow us to choose the pair of cameras that maximize the detection probability of a moving object. In [21], Park et al. discuss a distributed look-up table-based approach to determine the cameras' viewing frustums that allow us to select the best cameras for tracking purposes. Qureshi and Terzopoulos [26] propose a proactive control of multiple PTZ cameras through a solution that plans assignment and handoff. In particular, the authors consider the problem of controlling multiple cameras as a *multibody* planning problem in which a central planner controls the actions of multiple physical agents. In the context of person tracking, their approach computes the relevance of a PTZ camera to an observation task by considering five factors: (1) camera-pedestrian distance, (2) frontal viewing direction, (3) PTZ limits, (4) observational range, and (5) handoff success probability. The planning is then achieved by employing a greedy best-first search to find the optimal sequence of states. Other approaches to sensor reconfiguration for people tracking have been developed by employing game theory. Arslan et al. [4] demonstrate that the problem of finding an optimal configuration can be expressed in terms of a game whose solution is expressed by the Nash equilibrium. From this formulation, different approaches [13, 29] solve the camera assignment problem by maximizing a global utility function. Different mechanisms to compute the utilities can be provided as in [13, 29]; then a bargaining process is executed on the predictions of person utilities at each step. The cameras with the highest probabilities are used to track the target, thus providing a solution to the handoff problem in a video network. On the other hand, when a PTZ camera is reconfigured to track an object or it is switched on/off to save power, the topology of the network is modified. As a consequence, a new configuration is required to provide optimal coverage of the monitored environment. Song et al. [29] adopt a uniform distribution of the targets and the coverage resolution utility to negotiate the new network reconfiguration.

The chapter is structured as follows: After presenting the system architecture in Section 11.2, the audio subsystem is described in Section 11.3. The adopted algorithm for Time-Delay Estimation is described in Section 11.3.1, while in Section 11.3.3 we illustrate how the two-dimensional position of a source can be evaluated starting from the TDOAs estimated by two arrays. Section 11.4 describes how the system can achieve an optimal video coverage of the monitored environment using PTZ cameras. In Section 11.5 we show how the video and audio subsystem can interact, by focusing a single camera on an audio source while the remaining video sensors automatically compute a new optimal coverage configuration. Finally, Section 11.6 illustrates the prototype of the real-time system and some preliminary experimental results obtained in a real-world scenario.

11.2 System Architecture

The logical architecture of the proposed system can be roughly subdivided into two main modules, the audio and the video.

11.2.1 Audio Subsystem Architecture

To solve the problem of multi-source localization in a two-dimensional space, we propose the logical architecture shown in Figure 11.1. Basically, it consists

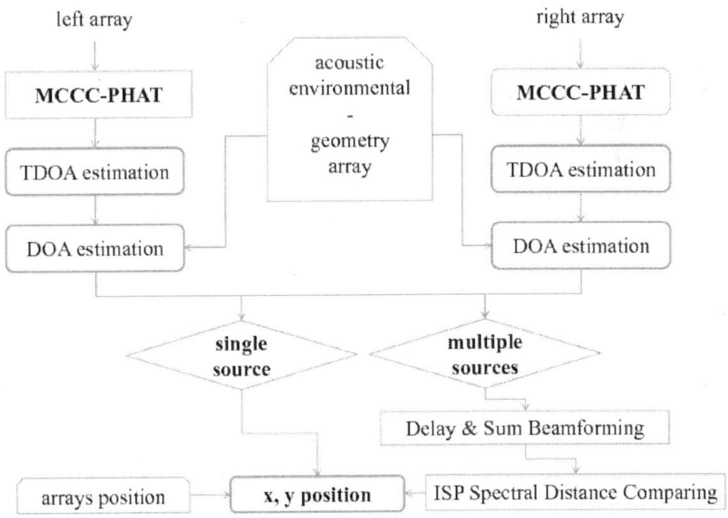

FIGURE 11.1 The block diagram of the processor, showing the data flow of all the tasks of the experimental system implementation.

of two parallel processing lines, corresponding to the left and right arrays. The first processing step is the TDOA estimation, based on the measurement of the time difference between the signals received by different microphones.

We use the Multichannel Cross-Correlation Coefficient (MCCC) method [5] to calculate the TDOA, because this method allows us to take advantage of the redundant information provided by multiple sensors. In addition, to improve the resolution of the peaks for the TDOA estimations and minimize the influence of noise and interference, we apply a Phase Transform (PHAT) filter [12] before calculating MCCC. Processing the TDOA information with the knowledge of the array geometry and the acoustic environment (far-field and free-field), we can calculate the DOA of the sound source. Finally, the two-dimensional coordinates of the source can be estimated by combining the DOAs at the left and right arrays (see Figure 11.3(b)). If more than one source is identified, a beamformer and a spectral distance comparison provide a guide to solve the problem of associating the DOAs of the left array with those of the right array. In the case of stationary noisy environments, correlated among the array microphones, the ASL can be improved by means of a de-noise task [28].

11.2.2 Video Subsystem Architecture

Figure 11.2 shows the logical architecture of the video subsystem and how it integrates with the audio subsystem. The system is composed of several

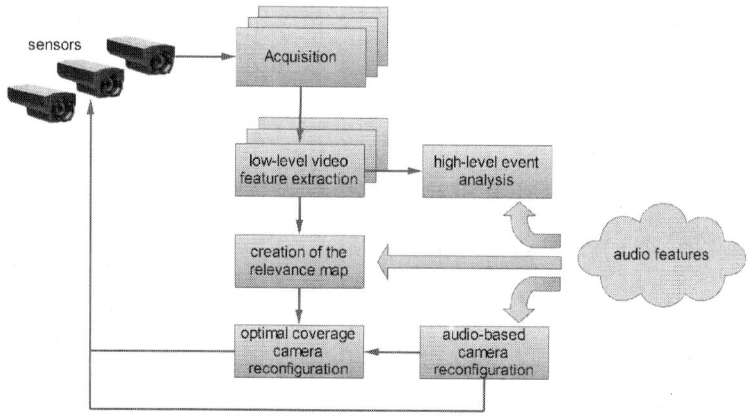

FIGURE 11.2 Logical architecture of the video subsystem.

PTZ cameras with fixed position and reconfigurable pan, tilt, and zoom parameters. Video sequences are acquired from each camera and processed in order to extract relevant video features. The feature extraction phase typically involves the use of change detection, object classification, and tracking techniques in oder to collect basic information on the moving objects in the scene. Because image transmission can be bandwidth consuming, this step is ideally computed locally, for example using embedded systems for on-camera processing. The video features, together with audio features, are sent to a high-level event analysis and recognition engine. The aim of the event engine is to combine audio and video features, together with spatial, temporal, and logical constraints, in order to detect complex events that could be relevant for surveillance purposes and signal them to a human operator. The exact nature of the events is dependent on the specific application, for example, people leaving unattended baggage in an airport hall, people moving on the railways in a train station, etc. This chapter does not describe the event engine itself; readers interested in the topic can refer to previous works of the authors [23, 24].

Video sensors can be reconfigured by means of two approaches. The audio-based reconfiguration relies on audio sources localization computed by the audio subsystem. In this case the video subsystem is given a specific location in the monitored environment, and one or more cameras are oriented in order to observe that specific zone. This way it is possible to gather visual information about audio sources that can be exploited by the event analysis module. The optimal coverage reconfiguration approach instead aims to reconfigure the sensors to maximize the coverage of the monitored area. The optimal coverage is computed on the basis of *relevance maps*, that is, maps denoting which zones are most actively used by moving objects and thus possibly relevant for surveillance purposes. The relevance maps are built on the basis of both audio and video localization features. The two reconfiguration tasks can operate simultaneously: If one or more sensors are focused on an audio source, the remaining ones are automatically reconfigured in order to optimally cover the rest of the environment.

11.3 Audio Localization

11.3.1 Time-Delay Estimation

The MCCC algorithm is a spatial correlation-based method. According to [6], we consider a linear array of N ($N \geq 2$) microphones. The discrete-time signal k received by the n-th microphone in a free-field environment with M multiple sources can be modeled as

$$y_n[k] = \sum_{m=1}^{M} \alpha_{nm} s_m[k - t_m - F_n(\tau_m)] + v_n[k], \tag{11.1}$$

where α_{nm} is the attenuation of the sound propagation (inversely proportional to the distance from source m to microphone n), $s_m[k]$ are the unknown source signals, t_m is the propagation time from the unknown source m to the reference sensor, $F_n(\tau_m)$ is the TDOA of the m-th signal between the n-th microphone and the reference, and $v[k]$ is an additive noise signal at the n-th sensor, which is assumed to be uncorrelated with not only all the source signals but also with the noise observed at the other sensors. The function $F_n(\tau_m)$ depends on the microphone array geometry. In our case, for a linear and equispaced arrays, that is, a Uniform Linear Array (ULA), and for a single source we have

$$\mathcal{F}_n(\tau) = (N-1)\tau, \quad n = 2, \ldots, N. \tag{11.2}$$

If we consider a single source and neglect the noise terms, we have

$$y_n[k + \mathcal{F}_n(\tau)] = \alpha_n s[k - t]. \tag{11.3}$$

Therefore, $y_1[k]$ is aligned with $y_n[k + \mathcal{F}_n(\tau_n)]$, and the new signal vector can be written

$$\mathbf{y}[k,p] = \left[y_1[k] \quad y_2[k+\tau] \ldots y_n[k + (N-1)\tau]\right]^T, \tag{11.4}$$

where p is a dummy variable for the hypothesized TDOA τ. The spatial correlation matrix of N microphones array is

$$\mathbf{R}[p] = \begin{bmatrix} \sigma_{y_1}^2 & r_{y_1 y_2}[p] & \cdots & r_{y_1 y_N}[p] \\ r_{y_2 y_1}[p] & \sigma_{y_2}^2 & \cdots & r_{y_2 y_N}[p] \\ \vdots & \vdots & \ddots & \vdots \\ r_{y_N y_1}[p] & r_{y_N y_2}[p] & \ddots & \sigma_{y_N}^2 \end{bmatrix}, \tag{11.5}$$

where $\sigma_{y_i}^2 = E[y_i]^2$ is the variance of signal y_i and $r_{y_i y_j}[p]$ is the cross-correlation between y_i and y_j. The spatial correlation matrix can be factored as

$$\mathbf{R}[p] = \mathbf{\Sigma} \widetilde{\mathbf{R}}[p] \mathbf{\Sigma}, \tag{11.6}$$

where

$$\mathbf{\Sigma} = \begin{bmatrix} \sigma_{y_1} & 0 & \cdots & 0) \\ 0 & \sigma_{y_2} & \cdots & 0 \\ \vdots & \vdots & \ddots & \vdots \\ 0 & 0 & \ddots & \sigma_{y_N} \end{bmatrix}, \tag{11.7}$$

and

$$\widetilde{\mathbf{R}}[p] = \begin{bmatrix} 1 & \rho_{y_1 y_2}[p] & \cdots & \rho_{y_1 y_N}[p] \\ \rho_{y_2 y_1}[p] & 1 & \cdots & \rho_{y_2 y_N}[p] \\ \vdots & \vdots & \ddots & \vdots \\ \rho_{y_N y_1}[p] & \rho_{y_N y_2}[p] & \cdots & 1 \end{bmatrix} \tag{11.8}$$

is a symmetric matrix, and

$$\rho_{y_i y_j}[p] = \frac{r_{y_i y_j}[p]}{\sigma_{y_i} \sigma_{y_j}} \tag{11.9}$$

is the Pearson Correlation Coefficient (PCC) between the i-th and j-th aligned microphone signals. The MCCC algorithm can be used to estimate the TDOA between the first two microphone signals as

$$\widehat{\tau}^{\text{MCCC}} = \arg(\text{local})\min_{p} \det[\widetilde{\mathbf{R}}[p]]. \tag{11.10}$$

11.3.2 MCCC-PHAT Method

In our system we use a filtered cross-correlation function, the Generalized Cross-Correlation (GCC) [12], the most common technique employed for TDOA estimation of microphone pairs. The GCC in the frequency domain is

$$r_{y_1 i y_j}^{\mathrm{GCC}}[p] = \sum_{f=0}^{L-1} \Psi[f] S_{y_i y_j}[f] e^{\frac{j2\pi pf}{L}}, \tag{11.11}$$

where L is the number of samples of the observation time, $\Psi[f]$ is the frequency domain weighting function, and the cross-spectrum of the two signals is defined as

$$S_{y_i y_j}[f] = E\{Y_i[f] Y_j^*[f]\}, \tag{11.12}$$

where $Y_i[f]$ and $Y_j[f]$ are the Discrete Fourier Transform (DFT) of the signals and * denotes the complex conjugate. GCC is used for minimizing the influence of uncorrelated noise and interferences, and maximizing the peak in correspondence of the time delay. In order to improve their robustness to additive noise, see [20]. The PHAT weighting function normalizes the amplitude of the spectral density of the two signals and uses only the phase information to compute the GCC:

$$\Psi_{\mathrm{PHAT}}[f] = \frac{1}{|S_{y_i y_j}[f]|}. \tag{11.13}$$

It is widely acknowledged that GCC is able to provide consistent performance when the characteristics of the source signal change over time. Thus its performance is dramatically reduced in the case of pseudo-periodic sounds.

Now we can write the new spatial correlation matrix for MCCC-PHAT:

$$\mathbf{R}^{\mathrm{PHAT}}[p] = \begin{bmatrix} 1 & r_{y_1 y_2}^{\mathrm{GCC}}[p] & \cdots & r_{y_1 y_N}^{\mathrm{GCC}}[p] \\ r_{y_2 y_1}^{\mathrm{GCC}}[p] & 1 & \cdots & r_{y_2 y_N}^{\mathrm{GCC}}[p] \\ \vdots & \vdots & \ddots & \vdots \\ r_{y_N y_1}^{\mathrm{GCC}}[p] & r_{y_N y_2}^{\mathrm{GCC}}[p] & \ddots & 1 \end{bmatrix}. \tag{11.14}$$

The TDOA estimation between the first two microphones in the general case of multiple sources is

$$\hat{\tau}^{\mathrm{MCCC-PHAT}} = \arg(\mathrm{local})\min_{p} \det[\mathbf{R}^{\mathrm{PHAT}}[p]]. \tag{11.15}$$

11.3.3 Sources Localization

Single source localization

The location of one sound source depends on the estimation of its DOA at the microphone arrays. In a far-field condition, the DOA value θ can be calculated as

$$\theta = \arcsin\left(\frac{\tau c}{d}\right), \tag{11.16}$$

where c is the speed of sound and d the distance between microphones. The assumed DOA range is: $[-90°, +90°]$, where zero is in front of the array and the microphone reference is the first from the left.

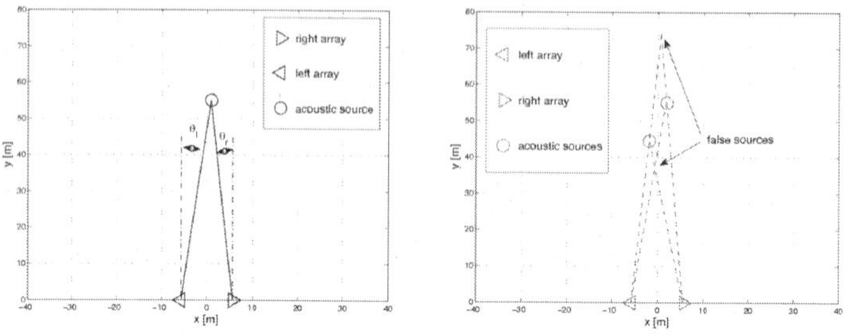

(a) (b)

FIGURE 11.3 (a) Single source localization; x, y axes reference. (b) The problem of multiple source localization.

As shown in Figure 11.3(a), the calculation of the two-dimensional position of the source is a simple trigonometric problem. However, we must consider that the two arrays are not coincident with the plane of interest, but they are placed at a certain height. Referring to Figure 11.3(a), we have to consider that the possible points identified by DOA are located on a cone surface. Each microphone array presents a cone. The intersection of the two cones is represented by a circumference. The intersection point between the circumference and the plane of interest is the estimation of the source distance from arrays.

Hence, we consider d_a the distance of the arrays, h the height of arrays above the plane of interest, θ_r and θ_l the DOA estimated on the right and left arrays:

$$x = \frac{d_a}{2}\left(\frac{\tan\theta_l + \tan\theta_r}{\tan\theta_l - \tan\theta_r}\right) \tag{11.17}$$

$$y = \sqrt{\left(\frac{d_a}{\tan\theta_l - \tan\theta_r}\right)^2 - h^2} \tag{11.18}$$

Innovative approach for multiple source localization

In case of multiple sources, the wavefront of each source will arrive at each of the two arrays with a different angle. We therefore have the problem of how to correctly associate the DOA angles relative to the same source, otherwise we may incur the location of false sources (Figure 11.3(b)), that is, a wrong estimate of the sound sources position.

To solve this problem, we propose the use of a beamforming technique to separate sound sources, and the comparison among the power spectrum of the different sources to determine the more consistent pairs of angles. Suppose that left and right arrays detect the presence of M sources, and then M DOAs are calculated for each of the two arrays. Steering the beamformer to these

M directions, M signals can be obtained for each of the arrays. The power spectrum of each signal calculated from one array is compared with that of the M signals related to the other array. Then, among the M^2 possibilities, the M couples with less distance between their power spectrum are chosen.

The beamforming [8] can be seen as a combination of the delayed signals from each microphone in such a way that an expected pattern of radiation is preferentially observed. The process can be subdivided in two sub-tasks: synchronization and weight-and-sum. The synchronization task consists of delaying (or advancing) each sensor output of an adequate interval of time, so that the signal components coming from a desired direction are synchronized. The information required in this step is the TDOA estimation. The weight-and-sum task consists of weighting the aligned signals and then adding the results together to form a single output. The output signal of beamformer allows us to enhance a desired signal from its detection corrupted by noise or competing sources. Delay & Sum Beamforming (DSB) is the classical technique for realizing directional array systems. In general, the DSB output y is formed as

$$y[k] = \frac{1}{N} \sum_{n=1}^{N} y_n[k + \Gamma_n(\tau)] \tag{11.19}$$

and the power spectrum output, Incident Signal Power (ISP), of ULA on the desired direction is

$$P[\theta, f] = \left| \sum_{n=1}^{N} Y_n[f] e^{\frac{-j2\pi f(n-1)d \sin\theta}{c}} \right|^2, \tag{11.20}$$

where N is the number of microphone signals, $Y_n[f]$ is the DFT of the signal, d is the distance between the microphones, c is the speed of sound, and θ is the DOA of the interesting sound.

The comparison between the M^2 couples of ISPs obtained by beamforming requires the definition of a Spectral Distance Function (SDF), to be used as error criteria. Among several possibilities (for comparison see [36]), we considered five of the most used functions (presented below) in order to verify how our system performance varies as a function of SDF.

The first selected function is a simple Spectral Difference (SD),

$$SD(\theta_l, \theta_r) = \frac{1}{L} \sum_{f=0}^{L-1} (P[\theta_l, f] - P[\theta_r, f]), \tag{11.21}$$

where θ_l and θ_r are the DOA measured by the left and right arrays.

A classic spectral estimation method is Linear Prediction (LP) [15], where we insert the "minus one" to standardize the minimum to zero as the other SDF

$$LP(\theta_l, \theta_r) = \frac{1}{L} \sum_{f=0}^{L-1} \left(\frac{P[\theta_l, f]}{P[\theta_r, f]} - 1 \right). \tag{11.22}$$

The others functions are the Itakura-Saito (IS) distance measure [16],

$$IS(\theta_l, \theta_r) = \frac{1}{L} \sum_{f=0}^{L-1} \Big(\frac{P[\theta_l, f]}{P[\theta_r, f]} - \log \frac{P[\theta_l, f]}{P[\theta_r, f]} - 1 \Big); \tag{11.23}$$

the Root Mean Square (RMS) log [22],

$$RMS(\theta_l, \theta_r) = \frac{1}{L} \sum_{f=0}^{L-1} \Big(\log \frac{P[\theta_l, f]}{P[\theta_r, f]} \Big)^2; \tag{11.24}$$

and the COSH measure [2],

$$\begin{aligned} COSH(\theta_l, \theta_r) = \frac{1}{L} \sum_{f=0}^{L-1} \Big(& \frac{P[\theta_l, f]}{P[\theta_r, f]} - \log \frac{P[\theta_l, f]}{P[\theta_r, f]} \\ & + \frac{P[\theta_r, f]}{P[\theta_l, f]} - \log \frac{P[\theta_r, f]}{P[\theta_l, f]} - 2 \Big). \end{aligned} \tag{11.25}$$

Assuming there are M sources, a list of M DOAs from each of the two arrays will be estimated; in particular, $\bar{\theta}_l = [\theta_{l_1}, ..., \theta_{l_M}]$ and $\bar{\theta}_r = [\theta_{r_1}, ..., \theta_{r_M}]$ are the vectors containing the angles determined from the left and right arrays, respectively. To locate the sources it is therefore necessary to find, among the $M!$ possibilities, the correct combination of M pairs between elements of $\overline{\theta}_l$ and $\overline{\theta}_r$. Let i be one of these combinations and $i(m)$ the m-th pair of the combination, we define the Spectral Distance Estimation (SDE) as the mean value of the SDFs calculated on the M pairs of the combination:

$$SDE(i) = \frac{1}{M} \sum_{m=1}^{M} |SDF(i(m))|, \tag{11.26}$$

and the vector Spectral Difference Mean Estimation (SDME) as

$$\mathbf{SDME} = \big[SDE(1) \quad SDE(2) \dots SDE(M!) \big]. \tag{11.27}$$

Finally, the index of the minimum value of the vector **SDME** identifies the target combination:

$$\widehat{i} = \underset{index}{\text{argmin}} \mathbf{SDME}. \tag{11.28}$$

11.4 Video Reconfiguration for Optimal Coverage

The coverage of a wide area with static video sensors is a problematic task. Full coverage could require an impractical number of sensors to be installed, while partial coverage would leave some zones totally unsecured. The use of PTZ cameras could mitigate the problem, reducing the number of sensors to be installed while maintaining a wide coverage, even though not all the zones can be monitored at the same time. However, it is not clear how the PTZ cameras should be configured (in terms of pan, tilt, and zoom parameters) in order to guarantee good coverage. The proposed method relies on previous work by Piciarelli et al. [25], and it is based on the concept of *relevance maps*. A relevance map is a map of the monitored environment highlighting the zones that are more relevant for surveillance purposes, and thus requiring higher priority in terms of camera coverage. The way relevance maps are defined is heavily dependent on the practical application context of the system. In our system we hypothesized that relevant objects always emit a sound signature (e.g., consider a system whose aim is to detect vehicles approaching a building). In this case the relevance map is a matrix like the one shown in Figure 11.4, where each cell counts how many audio sources have been detected in that zone by the audio localization subsystem described in Section 11.3 over a large amount of time.

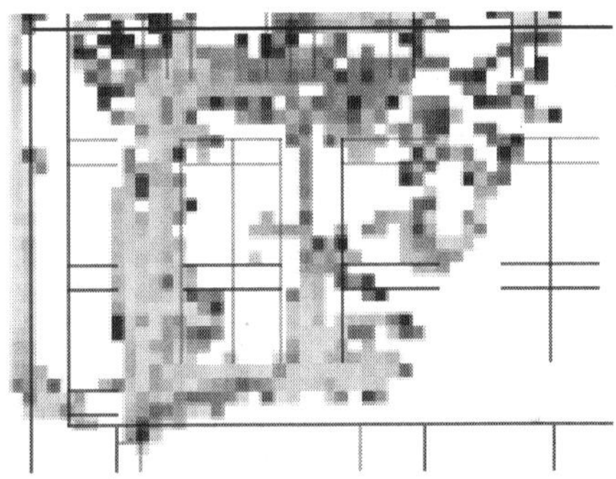

FIGURE 11.4 A relevance map defined by audio sensors, superimposed on a map of the observed environment of a parking lot. (See color insert.)

The basic idea is to consider the cone of view of each video sensor and its intersection with the ground plane, which is a conic section (and specifically an ellipse if the camera is not looking above the horizon) as shown in figure 11.5. Solving for optimal camera coverage thus means to find a set of ellipses that encloses the most relevant zones. This could be achieved by a popular data-fitting technique, the Gaussian-mixture-based Expectation-Maximization (EM) algorithm [19], where ellipses can be interpreted as support regions for a given quantile for each Gaussian model. However, not all ellipses found by EM can be interpreted as valid solutions for the reconfiguration problem: figure 11.5 intuitively shows that the major axis of the ellipse must be oriented toward the camera. Rather than solving a complex constrained EM problem, this can be achieved using a simple geometrical consideration. Figure 11.5 again gives an intuitive interpretation of the proposed solution: Rather than working with ellipses on the ground plane, we can consider the surface of a sphere enclosing each camera and with radius equal to the camera height. As can be seen, the intersection of the cone of view with the sphere has a circular shape, and any circle corresponds to a valid PTZ configuration of the sensor. Working in the new reference system of the sphere surface thus removes the constraints on ellipse orientation, and standard EM with mixture of isotropic Gaussians can be applied.

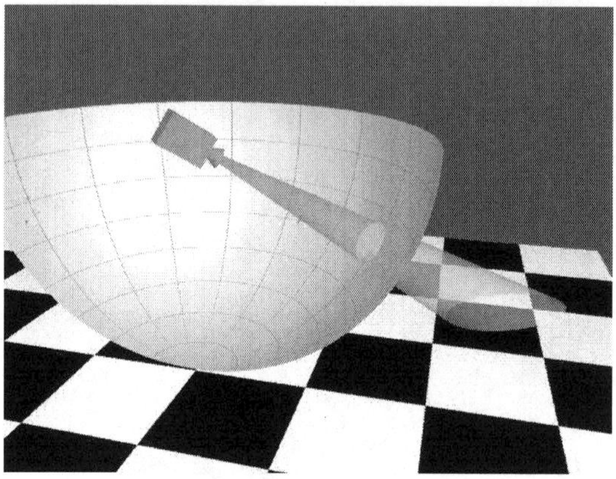

FIGURE 11.5 Intersection of the camera cone-of-view with the ground plane and a sphere centered in the camera.

More formally, a coordinate change $f : \mathbb{R}^2 \rightarrow \mathbb{R}^2$ is needed to switch from the standard 2D coordinates on the ground plane to its projection on the surface of the sphere. Let $\mathbf{x} = (x_1, x_2)$ be a point in the original ground plane coordinate system, and $\hat{\mathbf{x}}$ its projection in the new system: that is $\hat{\mathbf{x}} = f(\mathbf{x})$. We define $\psi(\mathbf{x})$ and $\theta(\mathbf{x})$ respectively as the pan and tilt angles of point $\mathbf{x}$ in the spherical coordinate system, defined as

$$\begin{cases} \psi(\mathbf{x}) = \arctan\left(\frac{x_2 - X_2}{x_1 - X_1}\right) \\ \theta(\mathbf{x}) = \arctan\left(\frac{\sqrt{(x_1 - X_1)^2 + (x_2 - X_2)^2}}{X_3}\right) \end{cases}, \tag{11.29}$$

where (X_1, X_2, X_3) are the 3D world coordinates of the camera. The final coordinate change f is then defined as

$$\hat{\mathbf{x}} = f(\mathbf{x}) = [\theta(\mathbf{x})\cos(\psi(\mathbf{x})), \theta(\mathbf{x})\sin(\psi(\mathbf{x}))]. \tag{11.30}$$

In the new coordinate system, ellipses become circles, as shown in Figure 11.6.

Observe that the coordinate change depends on the camera position, thus K different functions $f_k(\mathbf{x}), 1 \leq k \leq K$ must be defined. Using the coordinate change described above, the coordinates $\mathbf{x}_n$ of the n-th cell of the relevance map as observed by the k-th camera are defined as

$$\hat{\mathbf{x}}_{nk} = f_k(\mathbf{x}_n). \tag{11.31}$$

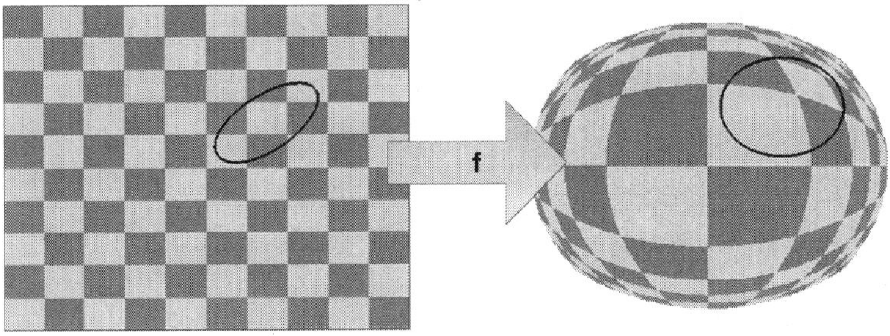

FIGURE 11.6 By using a proper change of coordinate system, a constrained ellipse-fitting problem is transformed into an unconstrained circle-fitting one.

At this point, the standard EM equations can be applied. The expectation step computes the conditional probability $p(k|n)$ that a point n is generated by the k-th Gaussian model, and it is defined as

$$p(k|n) = \frac{w_n G(\hat{\mathbf{x}}_{nk}, \mu_k, \sigma_k) c_k}{\sum_{z=1}^{K} w_n G(\hat{\mathbf{x}}_{nz}, \mu_z, \sigma_z) c_z}. \tag{11.32}$$

The maximization step computes the mean μ, the isotropic variance σ, and the mixing probability c for each Gaussian model:

$$\begin{aligned} \mu_k &= \frac{\sum_{n=1}^{N} w_n \; p(k|n) \; \hat{\mathbf{x}}_{nk}}{\sum_{n=1}^{N} w_n \; p(k|n)}, \\ \sigma_k^2 &= \frac{\sum_{n=1}^{N} w_n \; p(k|n) \; \|\hat{\mathbf{x}}_{nk} - \mu_k\|^2}{2 \sum_{n=1}^{N} w_n \; p(k|n)}, \\ c_k &= \frac{\sum_{n=1}^{N} w_n \; p(k|n)}{\sum_{n=1}^{N} w_n}. \end{aligned} \tag{11.33}$$

Here, w_n is the relevance of each map cell, as defined in the relevance map. Equations (11.32) and (11.33) are iterated until convergence to a stable solution. Once a solution is found, the pan and tilt angles for each camera can be computed by inverting the polar transformation,

$$\begin{cases} \psi_k = \arctan\left(\frac{\mu_k(2)}{\mu_k(1)}\right) \\ \theta_k = \sqrt{\mu_k(1)^2 + \mu_k(2)^2} \end{cases}, \tag{11.34}$$

while the width angle of the cone of view, expressed in radians, is defined as $\sigma_k / X_{3,k}$ ($X_{3,k}$ is the height of camera k, and thus the sphere radius).

11.5 Focusing Video Sensors on Audio Targets

Using the video coverage technique described in Section 11.4, an optimal coverage of the monitored area is achieved. However, "optimal coverage" does not necessarily mean that the entire area is covered; actually, this might not even be possible if the number of video sensors is not sufficient. As mentioned before, audio arrays are more suitable for full area coverage, especially because of their omnidirectionality properties and lower cost with respect to video sensors. We thus propose an integrated system where cameras are focused on audio target locations whenever a new audio source is being detected.

The only prerequisite here is to know the position of video sensors in the audio reference system. Under this assumption, audio sources can be searched for by means of the techniques described in Section 11.3. Once a source is detected, its position is passed to the video subsystem. Here, the camera k is chosen to be focused on the audio source, where $\hat{k}$ is defined as

$$\hat{k} = \underset{k}{\operatorname{argmin}} \left(\sqrt{(X^c_{k,1} - x^a_1)^2 + (X_{k,2} - x^a_2)^2} \right), \tag{11.35}$$

where X^c_k and x^a respectively are the position of the k-th camera and the audio source in the audio reference system.

Once the closest camera to the audio source has been selected, the proper pan and tilt angles required to focus on the target can be found by means of simple geometric considerations. As shown in Figure 11.7, the pan ($\psi_{\hat{k}}$) and tilt ($\theta_{\hat{k}}$) values for camera $\hat{k}$ can be easily defined as

$$\psi_{\hat{k}} = \arctan\left(\frac{X^c_{\hat{k},2} - x^a_2}{X^c_{\hat{k},1} - x^a_1}\right), \tag{11.36}$$

$$\theta_{\hat{k}} = \arctan\left(\frac{X^c_{\hat{k},3}}{\sqrt{(X^c_{\hat{k},1} - x^a_1)^2 + (X^c_{\hat{k},2} - x^a_2)^2}}\right). \tag{11.37}$$

The zoom parameter cannot be computed a priori, because the audio subsystem cannot estimate the size of the detected object. A proper zoom level can possibly be estimated by means of video processing techniques once the camera has been focused on the target.

Once a camera has been focused on an audio target, it can be considered temporarily unavailable for the area coverage task, and the optimal area coverage technique proposed in Section 11.4 can be applied. This way, the system will reconfigure the network of video sensors in order to observe the most relevant zones with a lower number of sensors (the same technique could also be applied in the case of a sensor failure).

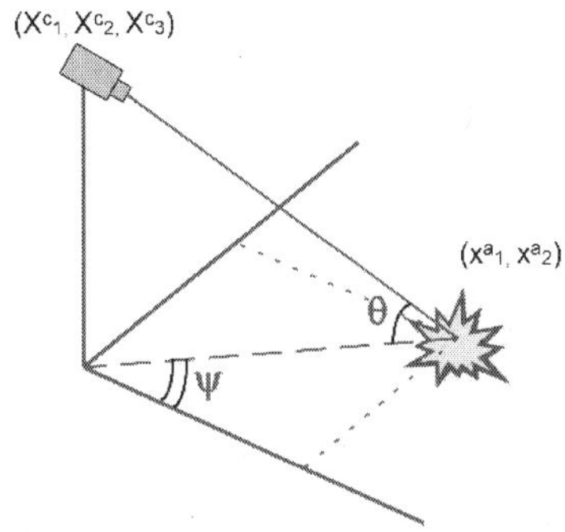

FIGURE 11.7 Focusing a camera on an audio source.

11.6 Experimental Results

This section discusses the experimental results and performance of the proposed system. The algorithms for the multi-source localization of the audio subsystem are tested in a real scenario (see Section 11.6.1); the experimental results show the efficiency of our framework. The experiments about the video subsystem are first carried out using synthetic data in order to have ground truth data for performance measurements. Experiments show that the reconfiguration algorithm can achieve good coverage results. The integration with the audio subsystem has been tested on the real-world scenario of a parking lot. The audio/video integration is tested both in the case of audio-based relevance maps for sensor reconfiguration and in the case of cameras focusing on audio sources, with subsequent reconfiguration of the remaining sensors.

11.6.1 Audio Subsystem

To test in a real scenario the algorithms for the multi-source localization, we made a prototype that has been installed on the roof of the building that houses the computer science department in Udine (see Figure 11.8). The prototype includes two linear arrays, each one composed of four microphones. The arrays are located with a distance of 11.4 m between them and a height of 12.1 m above the plane. The sample rate of digital system is 48 kHz and the microphone distance is 25 cm. We use for our experiment the sound sources of human voice, scream, shotgun, car, bus, horn. Though the localization is theoretically possible with just two microphones, more sensors allow ASL to make a quite robust time delay estimation, using the redundant information coming from the six TDOAs calculated for each array; in general, $N(N-1)/2$, where N is the number of microphones. Moreover, using only four microphones in each array, the computational load is not too high, both for the MCCC-PHAT algorithm and DSB, and the prototype also works in real-time with

FIGURE 11.8 The prototype installed on the roof of the university building. Inside the circles, the two arrays. The microphones are protected from the weather by a waterproof box.

an entry-level personal computer. The improvement due to the redundant information coming from the four microphones can be noticed by comparing Figures 11.9 and 11.10. In particular, the GCC-PHAT calculated for each

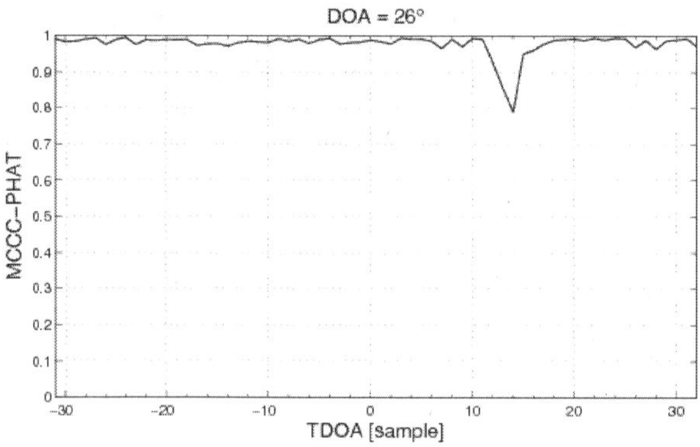

FIGURE 11.9 MCCC-PHAT algorithm performance with a single source.

pair of microphones M1M2 (GCC-PHAT of microphone 1 and 2) seems to present two sources, with two peaks, whereas M1M3 has two peaks very close together. This ambiguity disappears when analyzing the minimum peak of the MCCC-PHAT (see Figure 11.9), which considers all six combinations between the four microphones. Finally, Figure 11.10 shows the comparison among the MCCC-PHAT calculated with a different number of microphones, from which

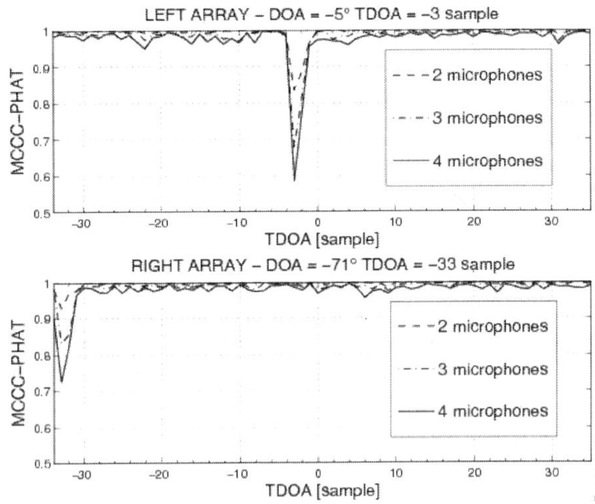

FIGURE 11.10 Comparing the microphone number of arrays and the MCCC-PHAT algorithm performance with a single source.

we can see that the location of the source is more evident as the number of microphones increases.

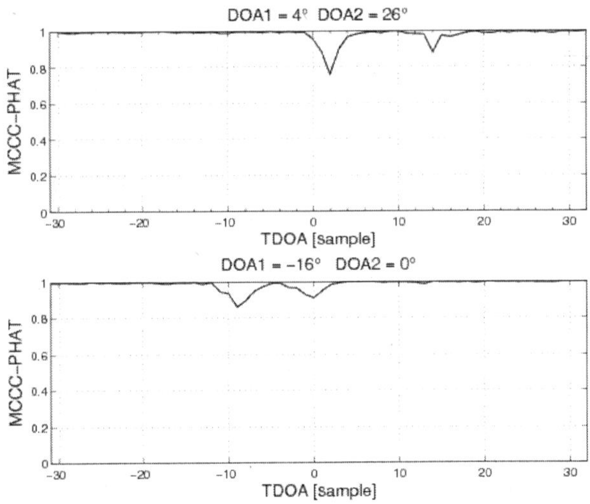

FIGURE 11.11 MCCC-PHAT algorithm performance with two sources.

In case of multiple sources, Figure 11.11 shows the presence of two peaks for each array related to the DOAs of two sources. The real sources are positioned with respect to the xy coordinates $(-3, 30)$ and $(5, 20)$ in meters. By estimating the position of local minima we can calculate the TDOAs. In this application scenario, the real sources are found at matching angles $\theta_{r1}\theta_{l1}$ and $\theta_{r2}\theta_{l2}$, otherwise we fall into the error of false localizing sources.

Applying the SDF, we get the vector **SDME** reported in Table 11.1.

TABLE 11.1 SDME of multiple sources referring to Figure 11.11 (M=2).

SDF	$\theta_{l1}\theta_{r1} - \theta_{l2}\theta_{r2}$	$\theta_{l1}\theta_{r2} - \theta_{l2}\theta_{r1}$
SD	278	1019
LP	0.4	5.4
IS	0.6	5
RMS	7	189
COSH	7	10

Each row represents the SDME calculated using a different distance function. Each column represents one of the different $M!$ angle combinations.

All the spectral distance functions minimize the vector **SDME** with the correct angle combination. The limitation of this approach relates to cases of acoustic sources with a similar spectral content, where it is very difficult to link the angles between arrays correctly. Figure 11.12 shows the case of three sources, two of which have similar spectral contents, a car, located at $(-5, 47)$, and a bus, located at $(0, 6)$; the third is a human voice, positioned at $(3, 35)$. In

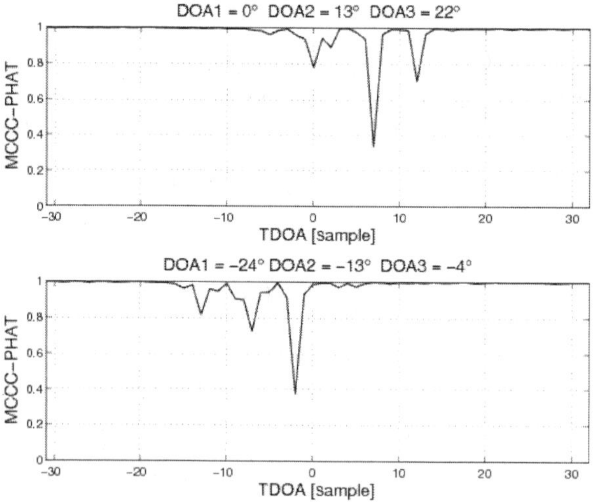

FIGURE 11.12 MCCC-PHAT algorithm performance with three sources: car sound $(DOA1_l, DOA2_r)$, human voice $(DOA2_l, DOA3_r)$, and bus sound $(DOA3_l, DOA1_r)$.

this case, applying the task of identification, we obtain the ISP for each DOA

(Figure 11.13) and the vector **SDME** of the six possible combinations. The

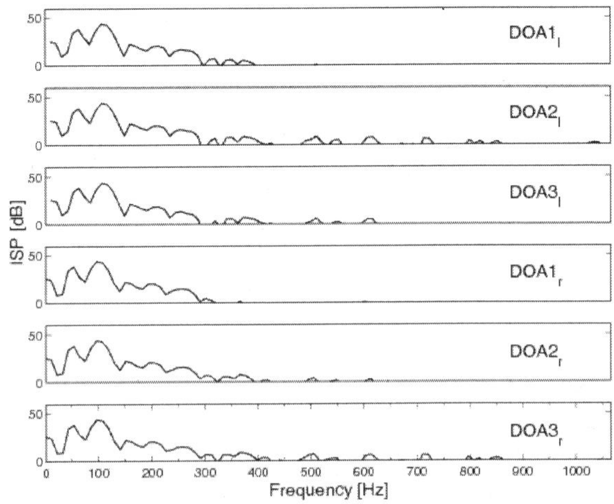

FIGURE 11.13 Comparing the ISP of different DOA estimations of Figure 11.12.

estimated minimum value of the vector **SDME** actually identifies the correct angle combination, $(DOA1_l - DOA2_r; DOA2_l - DOA3_r; DOA3_l - DOA1_r)$; looking at Figure 11.13, the similarity of human voice ISP is well marked but, on the contrary, it is difficult to check the matching pairs of the other two, so that we could make the mistake of locating the sources in a wrong position. We can see, even in Figure 11.13, that having index = 2, corresponding to the combination $(DOA1_l - DOA1_r; DOA2_l - DOA3_r; DOA3_l - DOA3_r)$, there is a second peak due to the similarity of the two sounds, and angle $(DOA2_l - DOA3_r)$ refers to the human voice. In this example, the RMS log is the only spectral function that gives an incorrect result.

Once the correct DOA pairs are identified, the xy coordinates can be estimated. Localization resolution depends, regarding the DOA estimation, on the accurate calculation of cross-correlation, which is linked to, see Equation (11.16), the sampling rate and the distance between microphones. Of course, this also influences the minimum resolvable distance between two sources. It is important to highlight that the distance of the microphones determines the minimum frequency beyond which spatial aliasing can occur. It means that a sound that does not contain spectral components below the minimum frequency cannot be uniquely localized. Instead, the resolution whereby the xy coordinates are calculated is also related to the distance between the arrays.

Regarding the computational complexity, the algorithm requires us to calculate the $M!$ elements of the vector Spectral Difference Mean Estimation, where M is the number of sound sources. In reality, not all the M^2 pairs of angles are geometrically consistent, only those that meet the condition $\theta_l > \theta_r$. Thus, the vector will have $M!$ elements only in the worst case. Nevertheless, the order of complexity of the algorithm makes it suitable only in contexts where the number of sources to be localized is limited. Alternatively, if the number of sources to localize is high, the algorithm can be integrated with a traditional tracking system, based on filtering. In this case, the localization algorithm based on the comparison of the ISPs would come into action only for those sources where the tracking system is unable to respond with a sufficient likelihood of success.

11.6.2 Video Subsystem

Concerning the video sensor network reconfiguration, in order to gather enough data for meaningful results, we first tested the system with synthetic data. A trajectory generator** has been used to create random trajectory sets; each set has a varying number of clusters of trajectories (from 2 to 6) and each cluster is composed of a random number of trajectories, between 50 and 150. Fifty relevance maps have been built by discretizing the space of each trajectory set with a 64×48 grid and counting how many trajectories pass within each cell. The counter is used as a relevance criterion, thus giving more importance to the areas crossed by many moving objects. The number and position of the cameras is randomly generated, from a minimum of two to a maximum of six cameras, always placed along the map borders and at a fixed height. Figure 11.15 shows an example of the optimal reconfiguration found after twelve iterations for a system with three sensors. Map colors represent the relevance (blue: low relevance, red: high relevance, white: no relevance). For each camera, the observed elliptic region and the direction of observation are shown.

In order to measure the quality of the data coverage, a score s_n for each data point $\mathbf{x}_n$ can be defined as

$$s_n = \sum_{k=1}^{K} w_n \; G(\hat{\mathbf{x}}_{nk}, \mu_k, \sigma_k). \tag{11.38}$$

This way, the score reflects the fact that the Gaussians should be centered on the most relevant points (with high weights w_n); moreover, the score is higher if high-relevance points are covered by more than one Gaussian. A global configuration score can then be defined as

$$\frac{1}{N} \sum_{n=1}^{N} s_n. \tag{11.39}$$

This score allows us to monitor the iterative EM process, thus making it possible to check if the algorithm really converges to a better solution with

**`http://avires.dimi.uniud.it/papers/trclust/`.

respect to the initial configuration. Figure 11.14 shows the score values at each iteration for the experiment shown in Figure 11.15. As it can be seen, the score has increased up to convergence, meaning that the algorithm has really found a better solution at each iteration step. This behavior has been observed in all the performed tests.

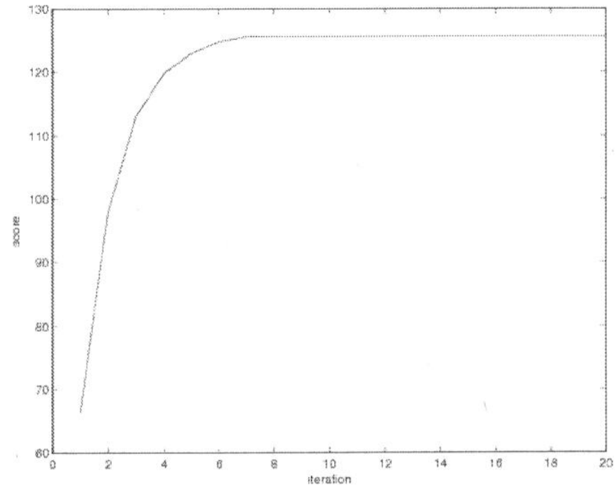

FIGURE 11.14 The overall score at different iteration steps for the experiment shown in Figure 11.15.

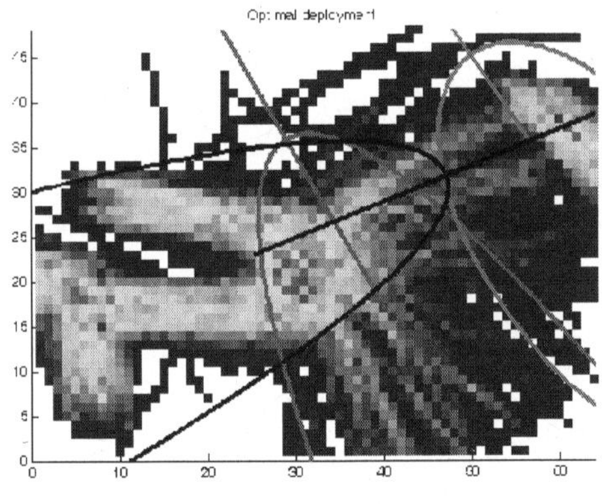

FIGURE 11.15 Optimal configuration for a camera network composed of three video sensors. (See color insert.)

The score metric shows how the iterative process converges to a solution by monothonically improving the performance at each iteration; however, it is not suitable for performance comparisons between different datasets. Because of this, we defined another performance measurement metric, the coverage c, defined as

$$c = \frac{\sum_{n \in E} w_n}{\sum_{n=1}^{N} w_n}, \tag{11.40}$$

where E is the set of map points falling within the observed area. In other words, c measures the obtained coverage, weighted according to the relevance values. We have run fifty experiments using the above-mentioned datasets and evaluated the coverage. As it can be seen in Table 11.2, the proposed method gives good results in terms of coverage, with a minimum coverage of 0.9333 and a maximum coverage of 1, with a mean of 0.9757 and a standard deviation of 0.0157. Observe that the score values are not easily comparable between different tests, as they depend also on the number of Gaussians (cameras), which is randomly chosen for each test. The scores here are reported only for comparison with other techniques applied on the same datasets[††].

TABLE 11.2 Test results over 50 random datasets.

test	score	coverage	test	score	coverage
1	66.7272	0.9851	26	149.7472	0.9773
2	19.0007	1.0000	27	34.8725	0.9920
3	192.4728	0.9799	28	24.9652	0.9695
4	19.9578	0.9958	29	122.9014	0.9670
5	49.7638	0.9840	30	126.9778	0.9632
6	67.3638	0.9800	31	75.1907	0.9833
7	143.3025	0.9983	32	184.6799	0.9584
8	72.9799	0.9973	33	151.1612	0.9905
9	126.6761	0.9868	34	34.8747	0.9512
10	233.0987	0.9654	35	33.6040	0.9870
11	137.9270	0.9830	36	58.9277	0.9925
12	78.4417	0.9798	37	256.8763	0.9630
13	41.7085	0.9746	38	58.1713	0.9778
14	71.5236	0.9757	39	144.4666	0.9726
15	41.4097	0.9758	40	39.0397	0.9333
16	120.6578	0.9678	41	72.6748	0.9588
17	26.5087	0.9986	42	99.7195	0.9596
18	35.2319	0.9792	43	50.1147	0.9839
19	26.6857	0.9909	44	109.4096	0.9883
20	128.2982	0.9436	45	33.1720	0.9639
21	189.8183	0.9760	46	116.3147	0.9597
22	17.9085	0.9399	47	163.1278	0.9887
23	188.4652	0.9624	48	69.1296	0.9778
24	155.6098	0.9652	49	67.8829	0.9992
25	191.8158	0.9610	50	194.3093	0.9814

For each dataset (available online) the score and coverage metrics are given.

The system has also been tested on the real-world scenario of the parking lot shown in Figure 11.8. Here, two color PTZ cameras are mounted on the roof of a building facing the parking lot at height of 12.1 meters, and the observed area is roughly 80×60 meters. The two cameras can span full 360° in pan and 0° to 120° in tilt direction, and thus can achieve any feasible solution computed by the system. The minimum and maximum zoom levels imposed a lower and an upper bound to the computed values of σ^2. The relevance map shown in Figure 11.16 has been built using the audio information acquired by the system described in Section 11.6.1 over 4 hours. Figure 11.16 also shows

[††]Publicly available at `http://avires.dimi.uniud.it/papers/EEH/dataset.zip`.

the final configuration computed by the proposed system, achieving a final coverage of 0.9537.

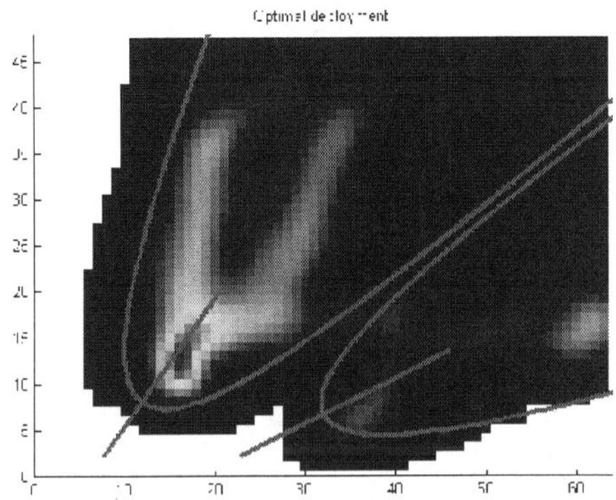

FIGURE 11.16 A real-world camera reconfiguration example. The relevance map is built according to audio data acquired during a 4-hour monitoring session. The ellipses show the areas observed by two PTZ cameras after reconfiguration. (See color insert.)

Finally, regarding audio-video integration, we performed real-world tests involving the parking lot shown in Figure 11.17. The environment is the same as the previous real-world experiments, with a map of roughly 80×60 meters and all the sensors mounted on top of a building in the lower-left part of the map. The sensors are composed of two microphone arrays, each one with four microphones, and three color PTZ cameras. The test was performed with a target emitting different sounds (clapping hands, shouting) and measuring the distance between its real position and the intersection point of the camera's optical axis and the ground plane. Thirty different tests have been run, the audio source has been localized with the techniques described in Section 11.3, and both cameras have been oriented toward the source using the audio-based reconfiguration technique discussed in Section 11.5. The audio source was placed in different points of the parking lot, with an average microphones-target distance of 45 meters. For each test, the Euclidean distance between the real object position and the aimed point was measured. Under these conditions, the measured average error was 1.53 meters.

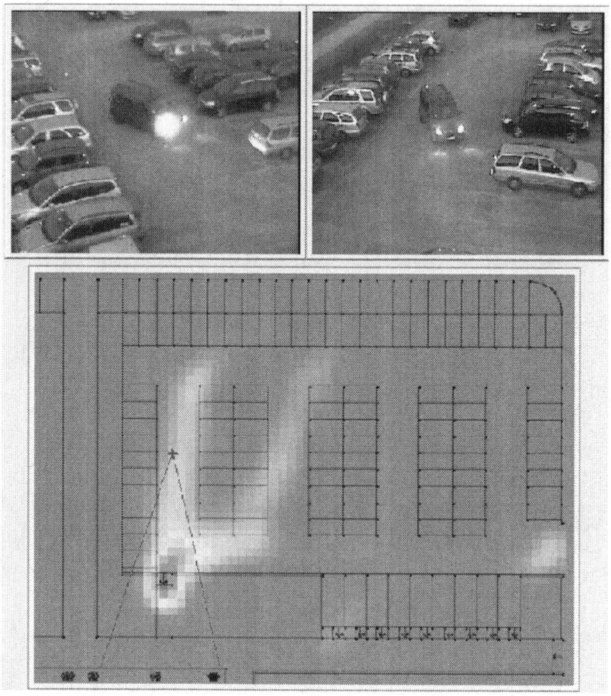

FIGURE 11.17 Two PTZ cameras are automatically focused on an audio source detected by the audio subsystem. (See color insert.)

References

1. Special issue on video analytics for surveillance: Theory and practice. *IEEE Signal Processing Magazine*, 27(5), 2010.
2. Jr. A. H. Gray and J. D. Markel. Distance measures for speech processing. *IEEE Transactions on Acoustics, Speech, and Signal Processing*, ASSP-28 (4):380–391, October 1976.
3. F. Angella, L. Reithler, and F. Gallesio. Optimal depolymont of cameras for video surveillance systems. In *IEEE International Conference on Advanced Video and Signal based Surveillance*, London, UK, 2007.
4. G. Arslan, J. Marden, and J. Shamma. Autonomous vehicle-target assignment: A game-theoretical formulation. *ASME Journal of Dynamic Systems, Measurement and Control*, 129(5):584–596, 2007.
5. J. Chen, J. Benesty, and Y. Huang. Robust time delay estimation exploiting redundancy among multiple microphones. *IEEE Transactions on Speech and Audio Processing*, 11:549–557, November 2003.
6. J. Chen, J. Benesty, and Y. Huang. Time delay estimation using spatial correlation techniques. In *Proc. International Workshop on Acoustic Echo and Noise Control*, pages 207–210, Kyoto, Japan, September 17–23, 2003.
7. A. Jain, D. Kopell, K. Kakligian, and Y. F. Wang. Using stationary-dynamic camera assemblies for wide-area video surveillance and selective attention. In *IEEE Computer Society Conference on Computer Vision and Pattern Recognition*, 2006.
8. H. Johnson and D.E. Dudgeon. *Array Signal Processing: Concepts and Techniques*. Simon & Schuster, 1993.
9. Y. Kaneda and J. Ohga. Adaptive microphone-array system for noise reduction. *IEEE Transactions on Acoustics, Speech and Signal Processing*, 34(6):1391–1400, 1986.
10. D. Karuppiah, R. Grupen, A. Hanson, and E. Riseman. Smart resource reconfiguration by exploiting dynamics in perceptual tasks. In *IEEE International Conference on Intelligent Robots and Systems*, 2005.
11. U. Klee, T. Gehrig, and J. McDonough. Kalman filters for time delay of arrival-based source localization. *EURASIP Journal on Applied Signal Processing*, 2006:1–15, 2006.
12. C. Knapp and G. Carter. The generalized correlation method for estimation of time delay. *IEEE Transactions on Acoustics, Speech, and Signal Processing*, 24(4):320–327, May 1976.
13. Y. Li and B. Bhanu. Utility-based dynamic camera assignment and handoff in a video network. In *IEEE/ACM International Conference on Distributed Smart Cameras*, pages 1–9, Stanford, CA, 7–11 Sep. 2008.
14. W.-K. Ma, B.-N. Vo, S. Singh, and A. Baddeley. Tracking an unknown time-varying number of speakers using tdoa measurements a random finite set approach. *IEEE Transactions on Acoustics, Speech, and Signal Processing*, 54(9):3291–3304, September 2006.

15. J. Makhoul. Linear prediction: A tutorial review. *Proc. IEEE*, 63(4):561–580, April 1975.
16. R. J. McAulay. Maximum likelihood spectral estimation and its application to narrow-band speech coding. *IEEE Transactions on Acoustics, Speech, and Signal Processing*, ASSP-32(2):243–251, April 1984.
17. A. Mittal and L. S. Davis. Visibility analysis and sensor planning in dynamic environments. In *European Conference on Computer Vision*, Prague, CZ, May 2004.
18. A. Mittal and L. S. Davis. A general method for sensor planning in multi-sensor systems: extension to random occlusion. *International Journal of Computing Vision*, 76:31–52, 2008.
19. T.K. Moon. The expectation-maximization algorithm. *IEEE Signal Processing Magazine*, 13(6):47–60, 1996.
20. M. Omologo and P. Svaizer. Acoustic event localization using a crosspower-spectrum phase based technique. In *Proc. IEEE ICASSP*, 2:273–276, 1994.
21. J. Park, P. C. Bhat, and A. C. Kak. A look-up table based approach for solving the camera selection problem in large camera networks. In *Workshop on Distributed Smart Cameras*, Boulder, CO, USA, Oct. 31, 2006.
22. L.L. Pfeifer. Inverse filter for speaker identification. Technical report, RADCTR-74-214, Speech Communications Research Lab Inc Santa Barbara, CA, 1974.
23. C. Piciarelli, S. Kumar, C. Micheloni, and G.L. Foresti. Event recognition with ptz cameras. In *3rd International Conference on Imaging for Crime Detection and Prevention*, London, UK, 2009.
24. C. Piciarelli, C. Micheloni, and G.L. Foresti. Trajectory-based anomalous event detection. *IEEE Trans. on Circuits and Systems for Video Technology*, 18(11):1544–1554, 2008.
25. C. Piciarelli, C. Micheloni, and G.L. Foresti. Occlusion-aware multiple camera reconfiguration. In *4th ACM/IEEE International Conference on Distributed Smart Cameras*, pages 88–94, Atlanta, GA, USA, 2010.
26. F.Z. Qureshi and D. Terzopoulos. Planning ahead for ptz camera ssignment and handoff. In *International Conference on Distributed Smart Cameras*, pages 1–8, Como, Italy, Aug-Sep, 2009.
27. D. Riva, D. Saiu, A. Sarti, M. Tagliasacchi, S. Tubaro, and F. Antonacci. Tracking multiple acoustic sources using particle filtering. In *Proc. European Signal Processing Conference*, Florence, Italy, September 4–8, 2006.
28. D. Salvati and S. Canazza. Improvement of acoustic localization using a short time spectral attenuation with a novel suppression rule. In *Proc. International Conference on Digital Audio Effect*, pages 150–156, Como, Italy, September 1–4, 2009.
29. Bi Song, C. Soto, A. K. Roy-Chowdhury, and J.A. Farrell. Decentralized camera network control using game theory. In *IEEE/ACM International Conference on Distributed Smart Cameras*, pages 1–8, Stanford, CA, 7–11 Sep. 2008.

30. C. Soto, B. Song, and A.K. Roy-Chowdhury. Distributed multi-target tracking in a self-configuring camera network. In *IEEE Conference on Computer Vision and Pattern Recognition*, Miami, FL, USA, 2009.
31. N. Strobel and R. Rabenstein. Robust speaker localization using a microphone array. In *Proceedings of the X European Signal Processing Conference*, III:1409–1412, 2000.
32. D.E. Sturim, M.S. Brandstein, and H.F. Silverman. Tracking multiple talkers using microphone-array measurements. In *Proc. International Conference on Acoustics, Speech, and Signal Processing*, pages 371–374, Munich, Germany, April 21–24, 1997.
33. G. Valenzise, L. Gerosa, M. Tagliasacchi, F. Antonacci, and A. Sarti. Scream and gunshot detection and localization for audio-surveillance systems. In *AVSS '07: Proceedings of the 2007 IEEE Conference on Advanced Video and Signal Based Surveillance*, pages 21–26, Washington, DC, USA, 2007. IEEE Computer Society.
34. S.R. Venkatesh, D.R. Polak, and S. Narayanan. Beamforming algorithm for distributed source localization and its application to jet noise. *AIAA Journal*, 41(7):1238–1246, 2003.
35. D.B. Ward, E.A. Lehmann, and R.C. Williamson. Particle filtering algorithms for tracking an acoustic source in a reverberant environment. *IEEE Transactions on Acoustics, Speech, and Signal Processing*, 11:826–836, 2003.
36. B. Wei and J.D. Gibson. Comparison of distance measures in discrete spectral modeling. In *Proc. IEEE Digital Signal Processing Workshop*, Hunt, TX, Oct. 15–18, 2000.

Index